MW00838110

A Primer on Pontryagin's Principle in Optimal Control

Second Edition

I. Michael Ross

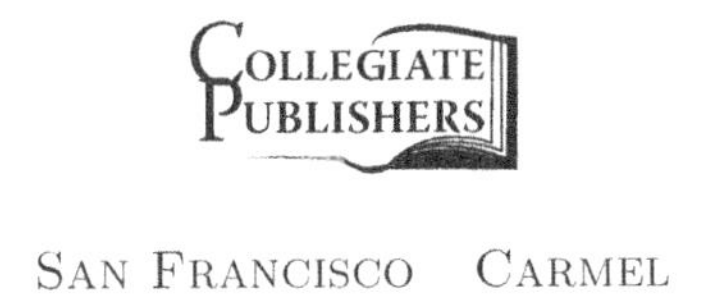

SAN FRANCISCO CARMEL

Cover Design: B. Hadden
Cover Images: I. M. Ross
Copy Editor: E. Solon

First Edition: 2009
Second Edition: 2015

10 9 8 7 6 5 4 3 2 1

Printed in the United States of America.

ISBN-13: 978-0-9843571-1-6
ISBN-10: 0-9843571-1-4

To Linda, Ben and Joe

I optimize,
therefore I think.

Preface

Caution: Do not read this book cover-to-cover!
If you do, you might end up a little smarter.

I wrote the first edition, not knowing it was going to be the first of two editions. It was simply meant to be a "gap filler" between elementary texts and advanced texts. I was pleasantly surprised at how well the first edition was received with calls to write a second one to fill in a few more gaps. This second edition is largely a response to this call.

Each chapter in this book is simply an expanded version of the first edition. As a result, the book got substantially fatter.

As I noted in the first edition, *proving Pontryagin's Principle is not easy.* But the good news is that you do not need to know how to prove Pontryagin's Principle to apply it, and apply it effectively. This is why I did not even attempt a proof in the first edition. I have remained faithful to this contrarian view in offering no proof.

The absence of a proof has allowed me to explain, "what the heck are those λ's," in starkly different terms than almost any other book. I have put in print what I've been teaching my own students for over a decade. I hope my intuitive presentation of those "darn covectors" will get you so comfortable with these concepts that you'll start seeing them everywhere! You may not know it, but you actually use covectors in your everyday life!!

Applying Pontryagin's Principle is very easy! Unfortunately, too many authors make it sound hard. It is not! What is hard is not the application of Pontryagin's Principle, but trying to solve nonlinear differential equations . . . by hand, and then pinning the difficulty on optimal control theory!! The nonlinear differential equations are a result of applying Pontryagin's Principle, not the principle itself. Once this distinction is clear, the application of Pontryagin's Principle is, quite literally, a step-by-step process that is *only as hard as the problem data.* See Chapter 3 for several worked-out examples, and Chapter 4 for additional problems and solutions.

There is also a widespread misconception that Pontryagin's Principle does not generate closed-loop solutions. What is ironic about this misconception is that Pontryagin's Principle can generate both open- and closed-loop solutions. To add insult to injury, Pontryagin et al showed[75] how to generate closed-loop solutions in more or less the same approach I've used in Chapter 3. I have adopted and extended this idea[92] through the concept of real-time optimal control (RTOC).

For all these reasons, I hope this second edition will help students, educators and researchers acquire a different perspective on the subject matter than what is traditionally found in the thousands of books on optimal control.

The first edition had a few errors. I have fixed all those I found; however, I'm quite positive I have, inadvertently, created new ones. I appreciate all the wonderful comments sent by so many of you. I certainly do not have plans to write a third edition, but, I welcome feedback and other constructive comments.

I. Michael Ross
2015

A Note About DIDO©

Warning: Do not use DIDO;
if you do, you might solve problems.

You can understand optimal control theory quickly and effectively by using the software package, DIDO.* It does not take too long to get acquainted with DIDO because it uses a style of coding that is nearly identical to writing the problem on a piece of paper. I created this style of coding all the way back in 2001; since then, I have maintained nearly the same intuitive "objects" in keeping things simple.

Since its inception in 2001, DIDO has been used to solve a large number of problems around the world: from the control of the very small (quantum particles) to a historic flight implementation on the very large (International Space Station). I've been fortunate to have been a part of many of these celebrated applications. The DIDO discussions in this book provide snippets of how Pontryagin's Principle is used in industrial applications.

An important learning outcome from DIDO is that you can see Pontryagin's Principle in action without going through the arduous process of solving the equations! This implies that the need for Pontryagin's Principle is greater than ever before because it supports and strengthens a deeper analysis of a given problem, particularly through the notion of the *problem of problems.* See Fig. 1.50 on page 84.

Much of a student's frustration in learning optimal control comes from the dual problems of inadequate hands-on experience in applying Pontryagin's Prin-

*A free copy of DIDO can be downloaded from the Elissar Global Website: http://www.ElissarGlobal.com.

ciple and an absence of an easy-to-use tool that illustrates the concepts. This book addresses the former problem while DIDO addresses the latter.

To achieve the intended learning objectives, it is not entirely necessary to use DIDO. Any other software, particularly the ***DIDO clones*** (i.e., those that are structured like DIDO) will also suffice. If you use such tools, you can still achieve a majority of the intended learning objectives because the pseudocode I have used in this book to illustrate the concepts can be quickly mapped to an actual code on a DIDO clone, as well. Note, however, that if you use a DIDO clone, it may not exhibit the same efficiency or output as DIDO. In this case, I recommend you use the free version of DIDO. Many academic problems discussed in this book can be solved using this free version.

DIDO Alert: Do not confuse an easy-to-use tool as being bullet-proof or "idiot-proof." In fact, it is very easy to write a bad DIDO code precisely *because* it is so easy to use. For example, if you are unable to reproduce one or more of the DIDO plots discussed in this book, it is a good indication that:

1. You may be writing a bad or poor DIDO code.
2. You may be using improper scaling or balancing techniques.
3. You may be using an older version of DIDO.
4. You are not using MATLAB properly.
5. You may be limiting your experience to only the free version.
6. You may be using a DIDO clone.
7. You do not know much about optimal control.
8. All of the above.

To The Student

Welcome ... to a theory of everything!
Use it wisely, grasshopper.

It is quite possible that optimal control theory is the operating system of the universe. Problems in artificial intelligence, economics, engineering, management, medicine, physics, sociology and much more can all be formulated as just another "app" over the "operating system" of optimal control theory. Given this phenomenal scope and power of optimal control theory, it should not be a surprise that the subject might be difficult to understand at first; however, the payoff of patience is quite enormous.

A minimum-time path to understanding optimal control theory is to pick your favorite problem from your field of expertise (i.e., pick an "app") and insert it wherever Brachistochrone is mentioned in this book; then, follow along. As you follow along, if you work out at least a few of the *Study Problems*, it will quickly clear any difficulties you may have along the way, or reinforce the material learned. Many students, especially beginners, must totally avoid those problems and text marked with the dangerous bend of Knuth: ☡.

Applying Pontryagin's Principle does not mean solving the problem. A deeper understanding of "the" problem as well as Pontrygain's Principle itself is gained by analyzing the *problem of problems*. Problem solving is indeed hard, at least to the person who is claiming it, but you might be surprised how easy it can get once you use the right concepts and tools.

Finally, be forewarned: Once you understand these concepts, you might see optimal control problems everywhere. Your friends might accuse you of being a know-it-all. Use this power wisely.

To The Instructor

RIP COV.

Ask your students if they can prove $(-1) \times (-1) = +1$. If they can, then they are smarter than the average bear!

A long time ago, I used to teach a course on optimal control the standard way: Prove first, explain afterwards. By the time I got to the end of "theorems and proofs," there was very little time to demonstrate how wonderful and powerful the subject was. All I could do was solve an academic problem here and a "trick" problem there. The students now had to trust me to take a follow-on course to see the amazing power of whatever I was talking about. Question is, why should they trust me given the pain I put them through in the first place? And how was I going to show them the amazing power of optimal control theory? Particularly in a non-trivial setting?

I solved the second problem by creating DIDO©; however, the first one remained. One year, on a whim, I abandoned tradition and embarked on a new philosophy: Teach students how to use mathematics intuitively. To heck with the proof, unless, of course, they wanted to see it. Wonderful things happened: They learned! And learned how to learn!! They asked me great questions and helped me become a better educator. Some even taught me new ideas because they saw things without the clutter of preconceptions and the overhead of mathematical minutia. Many even wanted to take an advanced course! When this happens for you, you may then proceed to open the torture chamber and inflict the calculus of variations. Meanwhile ... RIP COV.

Contents

Notation

If 1 + 1 = 0,
What is + = ?

E	endpoint or event cost function (a.k.a. "Mayer" cost function)
$\boldsymbol{e}$	endpoint or event constraint function
$\overline{E}$	endpoint Lagrangian; i.e., for the pair $\{E, \boldsymbol{e}\}$
F	running cost function (integrand of the Lagrange cost)
f	generic scalar function
$\boldsymbol{f}$	function for the right-hand side of a differential equation
g (*or* $\boldsymbol{g}$)	generic scalar (or vector) function
H	control Hamiltonian (Pontryagin's Hamiltonian)
$\overline{H}$	Lagrangian of the Hamiltonian; i.e., for the pair $\{H, \boldsymbol{h}\}$
$\mathcal{H}$	lower Hamiltonian (minimized Hamiltonian)
$\boldsymbol{h}$	path constraint function
J	scalar cost functional in optimal control
$N_{(\cdot)}$	generic positive integer; with appropriate subscript, $N_{(\cdot)} \in \mathbb{N}$ denotes the dimension of the relevant vector or vector function; for example, N_x is the dimension of the state vector $\boldsymbol{x}$; thus, $\boldsymbol{x} \in \mathbb{R}^{N_x}$
$\boldsymbol{q}$	generic vector; quaternion
t	independent variable; time
$\boldsymbol{u}$	control variable
$\boldsymbol{u}(\cdot)$	control trajectory/function; typically, a piecewise continuous function $t \mapsto \boldsymbol{u} \in \mathbb{R}^{N_u}$
$\boldsymbol{v}$	generic vector function; velocity variable

x	generic unknown variable
$\boldsymbol{x}$	state vector; generic unknown vector
$\boldsymbol{x}(\cdot)$	state trajectory/function; typically, an absolutely continuous function $t \mapsto \boldsymbol{x} \in \mathbb{R}^{N_x}$
x, y, z	Cartesian axes; generic variables

Greek

Θ	zero function, either scalar- or vector-valued
$\boldsymbol{\lambda}$	generic covector; or, adjoint covector; or, Lagrange multiplier in static optimization; or, costate variable in dynamic optimization
$\boldsymbol{\lambda}(\cdot)$	costate or adjoint covector function; typically, an absolutely continuous function $t \mapsto \boldsymbol{\lambda} \in \mathbb{R}^{N_x}$
$\boldsymbol{\mu}$	path covector associated with path constraints
$\boldsymbol{\nu}$	endpoint covector associated with endpoint constraints

Sets/Spaces

L^p	Lebesgue space of functions
$\mathbb{N}$	set of natural numbers $\{1, 2 \ldots\}$
$\mathbb{R}$	space of real numbers
$\mathbb{R}_+$	nonnegative (i.e., ≥ 0) real numbers
$\mathbb{U}$	$\subseteq \mathbb{R}^{N_u}$, constraint set for $\boldsymbol{u}$; often called the control space
$\mathcal{U}$	function space for the control function $\boldsymbol{u}(\cdot)$; typically, $\mathcal{U} = L^\infty$
$W^{m,p}$	Sobolev space of functions
$\mathbb{X}$	$\subseteq \mathbb{R}^{N_x}$, constraint set for $\boldsymbol{x}$; if $\boldsymbol{x}$ is the state variable, $\mathbb{X}$ is often called the state space
$\mathcal{X}$	function space for the state function $\boldsymbol{x}(\cdot)$; typically, $\mathcal{X} = W^{1,\infty}$

Subscripts

0	variable initial value; for example, x_0 is a variable initial condition
f	variable final value; for example, x_f is a variable final condition
∞	max norm

Superscripts

$*$	the optimal point; or, a reference point
0	given initial value; for example, x^0 is a given number for x_0
f	given final value; for example, x^f is a given number for x_f

Abbreviations

BVP	boundary value problem
CU	common unit; cost unit
HMC	Hamiltonian minimization condition
KKT	Karush-Kuhn-Tucker
NLP	nonlinear programming (problem)
PS	pseudospectral (Note: Pseudospectral is not hyphenated!)
RTOC	real-time optimal control
V&V	verification and validation

Other

$\arg\min$	argument of the minimum
$@t$	evaluation of several variables at t
$\underset{\boldsymbol{u}}{\mathsf{Min}}$	Minimization is to be performed with respect to $\boldsymbol{u}$ only
∂_x	$:= \dfrac{\partial}{\partial x}$, unless otherwise stated
$:=$	is equal to, by definition
$\equiv$	is identically equal to
$\lVert\cdot\rVert$	norm, with subscript denoting the computational method
$\lvert\cdot\rvert$	absolute value
$\dagger$	complementary; e.g., $\boldsymbol{\mu} \dagger \boldsymbol{h} \Rightarrow \boldsymbol{\mu}$ and $\boldsymbol{h}$ satisfy complementarity conditions

Optimal Control Made Difficult

Go ahead, skip this section.
You have the right to be confused.

In his satirical book, *Mathematics Made Difficult*, Linderholm writes[63],

> Mathematicians always strive to confuse their audiences; where there is no confusion there is no prestige.

A significant part of a beginning student's difficulty with optimal control theory can be traced to poor, inconsistent or confusing notation. In this book we follow a standard notational style that allows most students to quickly and correctly "guess" the meaning of a new symbol.

Notational Philosophy

1) We use uppercase letters (e.g., E, H) for cost or cost-like functions and lowercase letters (e.g., $\boldsymbol{e}, \boldsymbol{h}$) for the corresponding constraints.

2) We use overbars (e.g., $\overline{E}, \overline{H}$) to denote the Lagrangians of the problems defined by the relevant cost-constraint pair (e.g., $\overline{E}$ for the E-$\boldsymbol{e}$ pair, $\overline{H}$ for the H-$\boldsymbol{h}$ pair).

3) We use $N_{(\cdot)}$ with the appropriate subscript to denote the dimension of the relevant vector or vector function. For example, N_x is the dimension of the state vector $\boldsymbol{x}$; thus, $\boldsymbol{x} \in \mathbb{R}^{N_x}$. Now guess what N_u is.

Know the Differences Between $\boldsymbol{x}, \boldsymbol{x}(\cdot), \boldsymbol{x}(t), \boldsymbol{x}(t_0), \boldsymbol{x}_0$ and $\boldsymbol{x}^0$

Optimal control theory deals with functions of functions called functionals. As a result, there is a critical need to clearly distinguish between:

- variables,
- functions,
- values of functions,
- equations, and
- constraints

In addition, the mathematics of optimal control theory requires the use of values of functions as variables. Got that? This is one of the primary sources of confusion among beginning students. One way to avoid this confusion is to use different symbols for functions, their values, and variables; however, this would require a learner to keep track of a very large number of quantities. We adopt the philosophy that reusing symbols is conducive to learning. It requires less memory. In this spirit, we adopt the following convention:

1. The symbol $\boldsymbol{x}$ denotes a point in N_x-dimensional real space, $\mathbb{R}^{N_x}$.
2. The symbol $\boldsymbol{x}(\cdot)$ denotes a function that lives in some function space, $\mathcal{X}$.
3. The symbol $\boldsymbol{x}(t)$ denotes the value of the function $\boldsymbol{x}(\cdot)$, evaluated at the point t. In many introductory texts, the notation $\boldsymbol{x}(t)$ is used to denote the function $\boldsymbol{x}(\cdot)$, rather than its value at t. It is possible to understand ordinary calculus without such a distinction; however, in optimal control theory, conflating a function with its value can lead to substantial confusion which can be easily avoided.
4. From the preceding point, it follows that $\boldsymbol{x}(t_0)$ means the value of the function $\boldsymbol{x}(\cdot)$ evaluated at the point t_0. In optimal control, we need to use $\boldsymbol{x}(t_0) \in \mathbb{R}^{N_x}$ as a variable, as well (e.g., variable initial condition); hence, at various instances, it becomes convenient to abbreviate $\boldsymbol{x}(t_0)$ to $\boldsymbol{x}_0$.
5. We use the notation $\boldsymbol{x}^0$ to denote the numerical value of $\boldsymbol{x}_0$. Hence, $\boldsymbol{x}_0 = \boldsymbol{x}^0$ is a shorthand notation for $\boldsymbol{x}(t_0) = \boldsymbol{x}^0$, which means that $\boldsymbol{x}(t_0)$ is given by some number $\boldsymbol{x}^0$.

If $\mathcal{X}$ is the function space where $\boldsymbol{x}(\cdot)$ lives, such as the space of differentiable functions, we write $\boldsymbol{x}(\cdot) \in \mathcal{X}$, not $\boldsymbol{x}(t) \in \mathcal{X}$; however, $\boldsymbol{x}(t) \in \mathbb{R}^{N_x}$ because $\boldsymbol{x}(t)$ is the value of $\boldsymbol{x}(\cdot)$ at time t.

Note that we do not use the notation $f(\cdot)$ to mean a function if the symbol f already means a function.

The Map Notation

Among other things, the map notation or the arrow notation provides additional clarity in separating the notion of a function from its value. The symbol,

$$f : A \to B \qquad \text{or} \qquad f : A \mapsto B$$

means f ***is the name of the rule*** (i.e., function ... like the name of a computer file) that takes elements from the set A (called the domain ... like allowable inputs to a function file) and sends it (via the symbol $\to$ or $\mapsto$) to the set B (called the codomain ... like the class of possible outputs from a function file). Frequently, it is convenient to define a function without giving it a name. In this case, we use a barred arrow ($\mapsto$) symbol. For instance, the notation,

$$x \mapsto y$$

refers to a function (with no name) that sends the element x (from the domain of the unnamed function) to the element y (in the codomain of the nameless function). For example, if we want to refer to a function $f(x) = \sqrt{x}$, $x \in \mathbb{R}_+$, but do not wish to use the name f for the function (because, for instance, we might have already used f to mean something else), then it is convenient to say the function $x \mapsto \sqrt{x}$.

Thus, the notation $t \mapsto \boldsymbol{x}$ refers to the state trajectory which we have also abbreviated as $\boldsymbol{x}(\cdot)$ when it is more convenient to use $\boldsymbol{x}(\cdot)$ instead of $t \mapsto \boldsymbol{x}$.

Stop With All This Set Notation ... Arrgh!

Yes, Virginia, it is possible to "teach" and "learn" optimal control theory without set notation, but that's the equivalent of hammering a nail with a screwdriver: Yes it's possible, but pretty soon you'll need a better tool. A small amount of early investment in using set notation will pay off immensely in the immediate future.

At multiple places throughout this book I have emphasized clarity in notation with some element of mathematical precision. It is possible to violate these "rules" and still do well; however, when confusion rears its ugly head, it may be wise to get back to the basics.

The Dangerous Bend of Knuth

Text marked with the dangerous bend, ☡, means that a beginning student is well advised to skip that section at first reading. Sections and problems marked with the dangerous bend sign are not critical to understanding the concepts that follow it. The symbol is due to Donald Knuth, the inventor of TEX, the typesetting language used in the creation of this book. Needless to say, a double dangerous bend, ☡☡, is even more dangerous than the single dangerous bend.

Chapter 1

Problem Formulation

If problem–solving is hard,
formulating the right problem is harder.

Formulating an optimal control problem is the most important step an analyst takes. Too often, too many people formulate, solve and perpetuate the wrong problem simply because they believe the correct problem is not "solvable." As a means to introduce some basic concepts in optimal control, we take several perspectives on a famous "unsolved" problem posed by Bernoulli.

1.1 The Brachistochrone Paradigm

In 1696, in the June issue of the journal, *Acta Eruditorum*, Johann Bernoulli, challenged* mathematicians to solve what he called, at Leibniz' suggestion, the Brachistochrone Problem. In Greek, *brachistos* = shortest, and *chronos* = time. The problem may be physically described as follows (see Fig. 1.1).

*"I, Johann Bernoulli, address the most brilliant mathematicians in the world. Nothing is more attractive to intelligent people than an honest, challenging problem, whose possible solution will bestow fame and remain as a lasting monument. Following the example set by Pascal, Fermat, etc., I hope to gain the gratitude of the whole scientific community by placing before the finest mathematicians of our time a problem which will test their methods and the strength of their intellect. If someone communicates to me the solution of the proposed problem, I shall publicly declare him worthy of praise."[102] It is doubtful that such an introduction to a technical paper would be acceptable to a present-day journal.

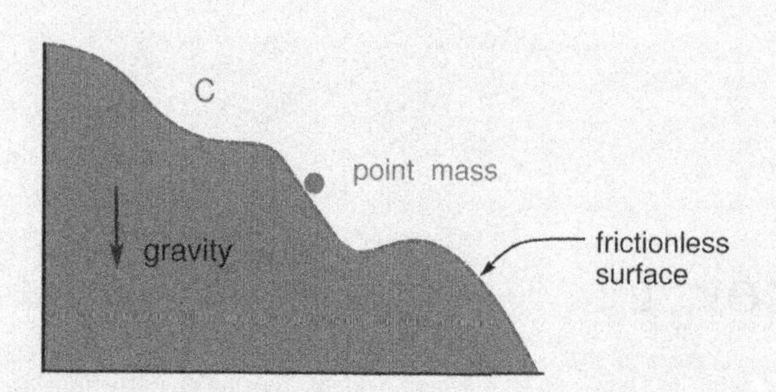

Figure 1.1: An illustration for the Brachistochrone problem. The symbol, C, is an abstract mathematical representation of the frictionless rigid surface.

A point mass is allowed to slide over a frictionless, rigid, surface (in two-dimensions) under the action of gravity alone. What is the shape of the surface that makes the mass move from one point (at the top) to another point (at the bottom) in minimum time?

The unknown is not a point or a number; it is an entire function, namely a "curve," C, that defines the frictionless rigid surface. This curve is our ***decision variable***: We are free to *choose* or "control" it at will. For instance, we may chip some other curves as illustrated in Fig. 1.2.

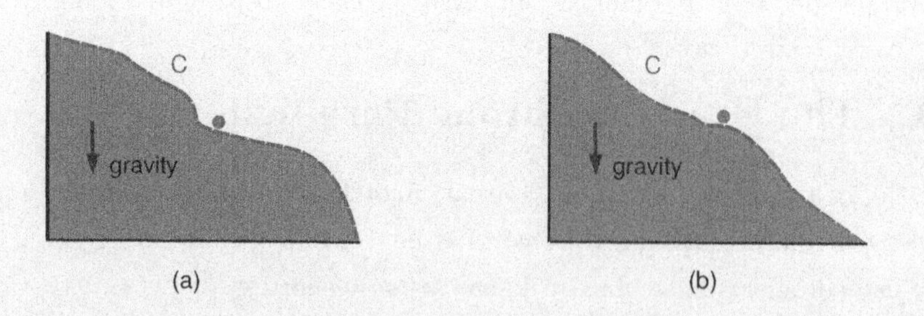

Figure 1.2: Two additional choices for the curve, C, for the Brachistochrone problem.

To every curve, C, we can associate a number, namely, the time of travel, t_f. That is, given any curve, C, we can, in principle, construct a "formula," J, which we can use to find the travel time, t_f,

$$J(C) = t_f \tag{1.1}$$

A conceptual block diagram for constructing the formula is shown in Fig. 1.3.

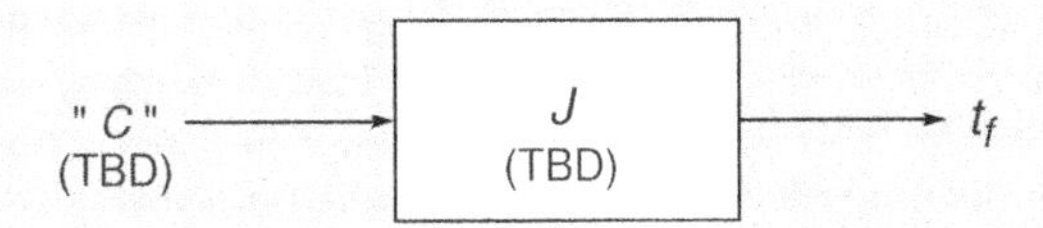

Figure 1.3: Conceptual block diagram for constructing the formula: $J(C) = t_f$. The input is "C"; the name of the function "file" is J; and the output is t_f. The process by which J takes the input to compute t_f (such as lines of code) is "to be determined" (TBD), and so is the input C itself. These TBDs are determined later in Eqs. (1.7) and (1.8) on page 11.

We can condense the entire block diagram by using the notation,

$$C \xrightarrow{J} t_f$$

which essentially removes the box and places J on top of the arrow. A more common condensation places J to the left of C, followed by a colon:

$$J : C \mapsto t_f \tag{1.2}$$

Thus, everything to the left of the colon (:) is the name of the *function file*, the symbols in between the colon and the ***barred arrow*** ($\mapsto$) denote the *inputs* and the symbols to the right of the arrow are the *outputs*.

The problem is to find a particular curve, $C = C^*$, which gives us the smallest value of t_f. How do we go about filling in the details for the problem; that is, a precise mathematical description?

In the conceptual formula given by Eq. (1.1), "C" is an abstract representation of the "curve" to be designed. How do we mathematically articulate C and the formula J?

First, note that C is not a point but a function (or, loosely speaking, a curve). Hence, J is a *function of a function*. Recall that a function is a point-to-point map: That is, a function is a "rule" that takes a point as an input and outputs a point. The concept can be imagined as a plot on a graph paper. The formula J, on the other hand, is "***implottable***": It takes an entire function ("C") as an input an outputs a point (t_f). The input cannot be identified as a point on the abscissa of a graph paper. To distinguish such functions of functions from just functions, we use the word ***functional***.[†] The analysis of functionals is called functional analysis and optimal control theory is a branch of functional analysis.

Although optimal control theory was not known to the scientists of the seventeenth century, six papers by Jakob Bernoulli (Johann's brother), Newton,[‡] Leibniz, L'Hôpital and Tschirnhaus appeared in the May 1697 issue of *Acta Eruditorum* containing the solution[95]. A solution to this problem using Pontryagin's principle is discussed in §3.1, page 171. This problem attracted the attention of these giants because they recognized its importance: *the solution, not being a straight line* (already proved by Johann, who also knew the solution at the time of his challenge) must mean something more ... something significant. This problem paved the way for the invention of the calculus of variations, a predecessor to optimal control theory. In turn, the calculus of variations spawned many other branches of mathematics (like differential geometry), just as optimal control theory continues to inspire new mathematics (like nonsmooth analysis[21]) and new applications (like quantum control[62]).

1.1.1 Development of a Problem Formulation

To properly formulate an optimal control problem, a key step in the process is the development of the ***system dynamics*** in a *standard form* called the ***state space form***. This process is best illustrated starting with the schematic shown in Fig. 1.4. A straightforward application of Newton's second law of motion ("$F = ma$") along the velocity direction yields,

---

[†] The word *functional* appeared in the mathematical literature in the early part of the 20th century, well after Bernoulli's era. Prior to this usage, functionals were called *functions of lines*[9].

[‡] This was published anonymously, but Johann is said to have remarked, "you know the lion from its claws[95]." Bernoulli's challenge did not please Newton as he wrote afterwards, "I do not have to be pestered and teased by foreigners about mathematical things." After a particularly tired day, Newton solved the problem, in twelve hours, between four in the afternoon and four the next morning[102].

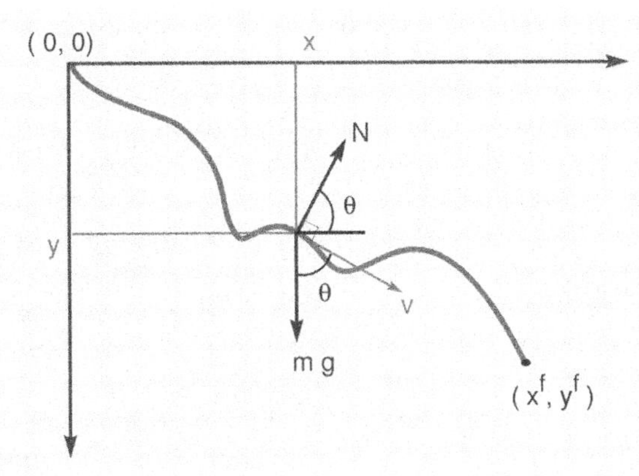

Figure 1.4: A schematic for a mathematical formulation of the Brachistochrone problem.

$$\overbrace{mg\cos\theta}^{\text{“}F\text{”}} = \overbrace{m\dot{v}}^{\text{“}ma\text{”}}$$

where m is the mass of the particle, g is the constant gravitational acceleration, θ is the angle made by the normal force, N, with respect to the horizontal, and v is the speed. The kinematics for this problem follow from the definition of the variables illustrated in Fig. 1.4,

$$\begin{aligned} \dot{x} &= v\sin\theta \\ \dot{y} &= v\cos\theta \end{aligned}$$

By appending the Newtonian dynamical equations to the kinematics, we have the following model for the ***system dynamics***,

$$\begin{aligned} \dot{x} &= v\sin\theta \\ \dot{y} &= v\cos\theta \\ \dot{v} &= g\cos\theta \end{aligned} \tag{1.3}$$

It is apparent that if we choose a vector, $\boldsymbol{x} \in \mathbb{R}^3$, defined by

$$\boldsymbol{x} := \begin{bmatrix} x \\ y \\ v \end{bmatrix}$$

then, its time derivative

$$\dot{\boldsymbol{x}} := \begin{bmatrix} \dot{x} \\ \dot{y} \\ \dot{v} \end{bmatrix}$$

is a vector form of the left-hand side of Eq. (1.3). The variables x, y and v are called the ***state variables***: i.e., *any variable with a "dot" is defined as a state variable.*

An examination of the right-hand side of Eq. (1.3) shows that it can be written in vector form as

$$\boldsymbol{f}(\boldsymbol{x}, \boldsymbol{u}) := \begin{bmatrix} v \sin\theta \\ v \cos\theta \\ g \cos\theta \end{bmatrix} \tag{1.4}$$

where, because g is a constant, we take the remaining variable θ as the ***control variable***,

$$\boldsymbol{u} := \theta$$

In other words, the angle the surface makes with respect to the horizontal naturally emerges as our control; see Fig. 1.4. *We can choose this any way we want; hence it is a control variable.*

In any given problem, a formulation of the dynamics automatically facilitates the identification of the state and control variables. Furthermore, as in the Brachistochrone problem, we can write a dynamical model for a given problem in a ***standard form*** called the ***state space form***,

$$\dot{\boldsymbol{x}} = \boldsymbol{f}(\boldsymbol{x}, \boldsymbol{u}) \tag{1.5}$$

where

$$\boldsymbol{x} \in \mathbb{R}^{N_x}, \qquad \boldsymbol{u} \in \mathbb{R}^{N_u}$$

and N_x and N_u denote the number of state and control variables, respectively. In the Brachistochrone problem, we have $N_x = 3$ and $N_u = 1$; hence, $\boldsymbol{x} \in \mathbb{R}^3$ and $\boldsymbol{u} \in \mathbb{R}$. We call

$$\mathbb{X} = \mathbb{R}^{N_x}$$

the ***state space***, and

$$\mathbb{U} = \mathbb{R}^{N_u}$$

the ***control space***. Thus, $\mathbb{X} = \mathbb{R}^3$ and $\mathbb{U} = \mathbb{R}$ are the state and control spaces for the Brachistochrone problem, respectively.

It is important to note the distinction between $\dot{\boldsymbol{x}}$ and $\boldsymbol{f}(\boldsymbol{x}, \boldsymbol{u})$. Although these two quantities are equal by virtue of Eq. (1.5), $\boldsymbol{f}(\boldsymbol{x}, \boldsymbol{u})$ *is a quantity independent of* $\dot{\boldsymbol{x}}$ as implied by Eq. (1.4). *Conflating these quantities is one of the main causes of confusion among beginners to optimal control theory.* The vector function $\boldsymbol{f}$ is a rule, a formula, or the name of a map that takes the pair $(\boldsymbol{x}, \boldsymbol{u}) \in \mathbb{R}^{N_x} \times \mathbb{R}^{N_u}$ as an input and outputs a vector in $\mathbb{R}^{N_x}$. This concept is written symbolically as

$$\begin{aligned} &\boldsymbol{f} : \mathbb{R}^{N_x} \times \mathbb{R}^{N_u} \to \mathbb{R}^{N_x} \\ \textit{or} \quad &\boldsymbol{f} : (\boldsymbol{x}, \boldsymbol{u}) \mapsto \mathbb{R}^{N_x} \end{aligned} \tag{1.6}$$

The ***map notation*** means the following: $\boldsymbol{f}$ *is a computational machine (such as a computer code) that takes two vector inputs, the first one in* $\mathbb{R}^{N_x}$ *(labeled* $\boldsymbol{x}$*) and the second one in* $\mathbb{R}^{N_u}$ *(labeled* $\boldsymbol{u}$*) and outputs a vector in* $\mathbb{R}^{N_x}$ *called* $\boldsymbol{f}(\boldsymbol{x}, \boldsymbol{u})$. The map notation of Eq. (1.6) is shorthand for the block diagram illustrated in Fig. 1.5.

Figure 1.5: A block diagram for the function $\boldsymbol{f}$. The input vectors are $\boldsymbol{x} \in \mathbb{R}^{N_x}$ and $\boldsymbol{u} \in \mathbb{R}^{N_u}$. The output is the vector, $\boldsymbol{f}(\boldsymbol{x}, \boldsymbol{u}) \in \mathbb{R}^{N_x}$. Note that $\boldsymbol{f}$ itself is the name of the computational machine, such as the name of the function file of a computer code.

For the Brachistochrone problem, $\boldsymbol{f}(\boldsymbol{x}, \boldsymbol{u})$ is given explicitly by Eq. (1.4). Thus, for instance, if we take

$$\boldsymbol{x}_1 = \begin{bmatrix} -1 \\ 0 \\ 1 \end{bmatrix}$$

and

$$\boldsymbol{u}_1 = \pi/8$$

we get (by setting $g = 1$),

$$\boldsymbol{f}(\boldsymbol{x}_1, \boldsymbol{u}_1) = \begin{bmatrix} 0.3827 \\ 0.9239 \\ 0.9239 \end{bmatrix} \in \mathbb{R}^3$$

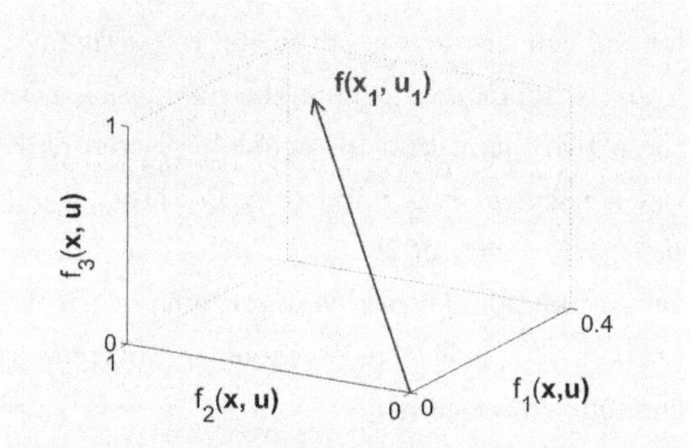

Figure 1.6: Illustration of $\boldsymbol{f}(\boldsymbol{x}, \boldsymbol{u})$ as a vector in $\mathbb{R}^3$ for the Brachistochrone problem.

as shown in Fig. 1.6. For different values of $\boldsymbol{x}$ and $\boldsymbol{u}$, we can generate a potentially large collection of $\boldsymbol{f}(\boldsymbol{x}, \boldsymbol{u})$-vectors. One way to visualize the management of this large collection of vectors is to move the origin of the arrow representing $\boldsymbol{f}(\boldsymbol{x}, \boldsymbol{u})$ to $\boldsymbol{x}$. That is, to every point $\boldsymbol{x} \in \mathbb{R}^3$, we associate a three-dimensional vector *emanating from* $\boldsymbol{x}$ and evaluated by the function $\boldsymbol{f}$ by fixing a value of $\boldsymbol{u}$. Hence, we can generate a ***vector field*** in $\mathbb{R}^3$, or more appropriately, a ***controlled vector field***, as shown in Fig. 1.7.

Study Problem 1.1

1. *Figures 1.6 and 1.7 were generated by using MATLAB® and the input-output structure indicated in Fig. 1.5. Using any computing language, regenerate these figures to understand firsthand the differences between the vector* $\boldsymbol{f}(\boldsymbol{x}, \boldsymbol{u})$ *and the vector field* $\boldsymbol{f}(\mathbb{X}, \boldsymbol{u})$.
2. *Next, modify the code and explore the meaning of* $\boldsymbol{f}(\boldsymbol{x}, \mathbb{U})$ *and* $\boldsymbol{f}(\mathbb{X}, \mathbb{U})$ *where* $\boldsymbol{f}$ *is reused as an overloaded function.*

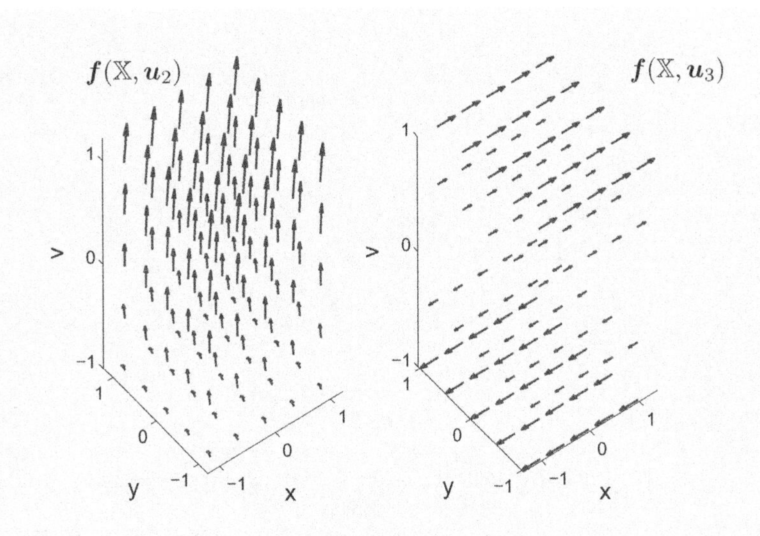

Figure 1.7: Two vector fields for the Brachistochrone problem generated by two values of control, $u_2 = \pi/4$ and $u_3 = \pi/2$. Appropriate symbols for the vector field are $f(\mathbb{X}, u_2)$ and $f(\mathbb{X}, u_3)$, where the symbol f is reused as an ***overloaded function*** to accept $\mathbb{X}$ as an input.

Note that this vector field is global; that is, it is a field over the entire state space, $\mathbb{X}$. When we change u, we change the entire field as shown in Fig. 1.7.

By definition, we are free to choose u because it is a control vector. In fact, we are not limited to a single value of u. We can choose an entire "shape function," $u(\cdot)$, called the ***control trajectory*** (see Fig. 1.8) to shape and control the vector field over time.

Thus, the two vector fields shown in Fig. 1.7 can be viewed as snapshots of the vector field taken at two instances of time, t_1 and t_2, corresponding to two values of controls, $u(t_1) = \pi/4$ and $u(t_2) = \pi/2$.

As implied in Fig. 1.8, we are free to choose the control trajectory with total abandonment and disregard for smoothness. This follows by definition as there is no such thing as $\dot{u}$. If there is a derivative on u, then, by definition, it is not a control but a state variable, as noted on page 6.

Similar to the notion of the control trajectory is the concept of a ***state trajectory***, denoted by $x(\cdot)$. A schematic of a state trajectory is shown in

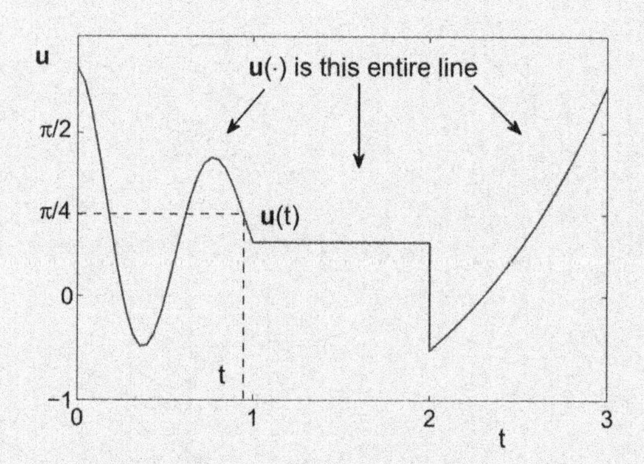

Figure 1.8: The control trajectory $u(\cdot)$ refers to the entire "shape function" while $u(t)$ means the value of $u(\cdot)$ at time t; hence the symbol $u(t)$ is a vector, not a function.

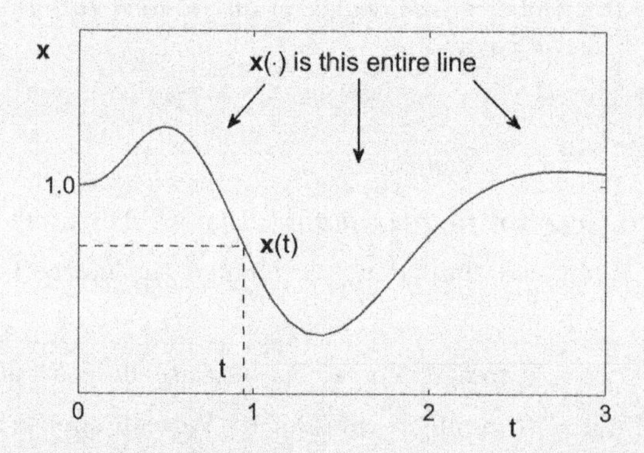

Figure 1.9: Similar to the control trajectory, the state trajectory is denoted by $x(\cdot)$. This symbol refers to the entire function; hence $x(t)$ means the value of $x(\cdot)$ at time t.

Fig. 1.9. Unlike a control trajectory, the state trajectory needs to be smoother so that one can perform the derivative operation $\dot{x}$.

Notation Alert: *Refer to page xxiv for nomenclature.*

1. The symbol $\boldsymbol{x} \in \mathbb{R}^{N_x}$ stands for the N_x-dimensional state vector. It is distinguished from $\boldsymbol{x}(\cdot)$, which is a *function*, meaning the entire ***state trajectory***; see Fig. 1.9. The function $\boldsymbol{x}(\cdot)$ is also denoted by the map notation $t \mapsto \boldsymbol{x}$. In the literature, the state trajectory is also called just a trajectory (or an "*arc*" in older texts).
2. Similar to $\boldsymbol{x}$, the symbol $\boldsymbol{u} \in \mathbb{R}^{N_u}$ is the N_u-dimensional control vector, while $\boldsymbol{u}(\cdot)$ is the ***control trajectory*** which is a *function*; see Fig. 1.8. The control function is also denoted by the map notation $t \mapsto \boldsymbol{u}$. In the literature, the control function is also called just a control, a control history or a control "program."
3. The symbols $\boldsymbol{x}(\cdot)$ and $\boldsymbol{u}(\cdot)$ that denote the state and control *functions* are used as a means to distinguish the use of $\boldsymbol{x}$ and $\boldsymbol{u}$ for the state and control *vectors*, respectively.
4. Note that we do not write $\boldsymbol{f}(\cdot)$ because the symbol $\boldsymbol{f}$ does not exist as a vector.

The totality of unknowns for the Brachistochrone problem, and for many optimal control problems, is $\boldsymbol{x}(\cdot)$, $\boldsymbol{u}(\cdot)$ and t_f. As the abstract quantity C in Eq. (1.1) on page 3 is the unknown, we can now define it explicitly as the triple

$$C := \{\boldsymbol{x}(\cdot), \boldsymbol{u}(\cdot), t_f\} \tag{1.7}$$

This implies that we can formalize the writing of the functional J in Eq. (1.1) as

$$J[\boldsymbol{x}(\cdot), \boldsymbol{u}(\cdot), t_f] = t_f \tag{1.8}$$

Because the arguments of J are themselves functions, it is customary to write such functionals using square brackets, as indicated in Eq. (1.8).

Study Problem 1.2

Explain why $J[\boldsymbol{x}(t), \boldsymbol{u}(t), t_f]$ is the wrong notation for the functional J.

The arguments of J also convey the concept that

- the state function $\boldsymbol{x}(\cdot) : t \mapsto \boldsymbol{x} \in \mathbb{R}^{N_x}$,

- the control function $\boldsymbol{u}(\cdot): t \mapsto \boldsymbol{u} \in \mathbb{R}^{N_u}$ and
- the number $t_f \in \mathbb{R}$

are all ***decision variables***. The pair

$$\{\boldsymbol{x}(\cdot), \boldsymbol{u}(\cdot)\}$$

is called a ***system trajectory***. A system trajectory is said to be ***dynamically feasible*** if, at each instant of time t it obeys the "law of motion":

$$\dot{\boldsymbol{x}}(t) = \boldsymbol{f}(\boldsymbol{x}(t), \boldsymbol{u}(t))$$

That is, the system trajectory satisfies the dynamics *pointwise*.

Collecting all the relevant equations, one particular formulation of Bernoulli's Brachistochrone problem can be cast in the following format:

$$\left.\begin{array}{ll} \mathbb{X} = \mathbb{R}^3 & \mathbb{U} = \mathbb{R} \\ \boldsymbol{x} = (x, y, v) & \boldsymbol{u} = \theta \end{array}\right\} \text{(preamble)}$$

$$\overbrace{(Brac:1)}^{\text{problem}} \left\{\begin{array}{lrl} \text{Minimize} & J[\boldsymbol{x}(\cdot), \boldsymbol{u}(\cdot), t_f] := t_f & \} \text{ (cost)} \\ \text{Subject to} & \dot{x} = v \sin\theta & \\ & \dot{y} = v \cos\theta & \text{(dynamics)} \\ & \dot{v} = g \cos\theta & \\ & t_0 = 0 & \\ & (x_0, y_0, v_0) = (0, 0, 0) & \text{(endpoints)} \\ & (x_f - x^f, y_f - y^f) = (0, 0) & \end{array}\right.$$

Casting an optimal control problem in this structured format preps it for both pencil/paper analysis (as described later in Chapter 2) as well as computer coding. This format is nearly identical to the actual format of a DIDO code. Details of this format are described next.

1.1.2 Introduction to Structured Optimization

A generic optimization problem (GOP) consists of *just three ingredients*:

- ***optimization*** variables (also known as ***decision*** variables) that need to be selected
- a selection criterion known as a ***cost functional*** or ***objective function*** or ***performance index*** and
- a set of ***constraints***.

These three ingredients are formatted as:

$$(GOP) \begin{cases} \text{Minimize} & \mathcal{C}(\boldsymbol{q}) \\ \text{Subject to} & \boldsymbol{c}^L \leq \boldsymbol{c}(\boldsymbol{q}) \leq \boldsymbol{c}^U \end{cases}$$

where $\boldsymbol{q}$ is a generic optimization variable, $\mathcal{C}$ is the cost functional and $\boldsymbol{c}$ is the constraint function whose values are required to be between two specified ***lower*** and ***upper bounds***, $\boldsymbol{c}^L$ and $\boldsymbol{c}^U$, respectively; see Fig. 1.10.

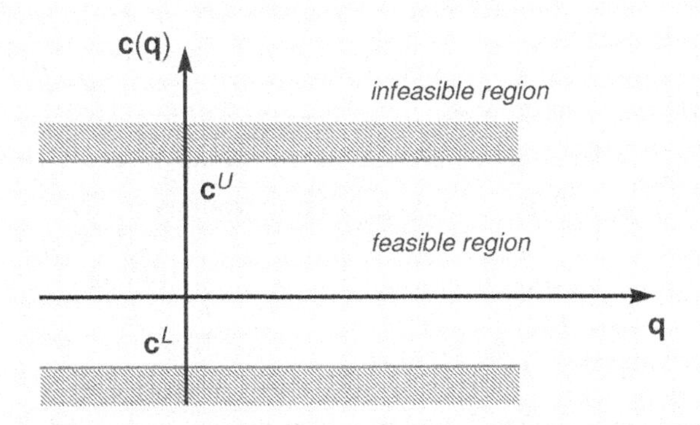

Figure 1.10: Feasible and infeasible regions in a generic optimization problem are given by lower and upper bounds on constraint functions.

An optimal control problem is an optimization problem with just such three ingredients; however, it is a *specifically structured* optimization problem. That is, the optimization variables are structured in terms of the state and control variables, and possibly final time. In addition, and this will be apparent later, the cost functional is also structured. Finally, the structured nature of the optimization variables in an optimal control problem lends itself to a special structuring of the constraint equations, as well.

The different structures involved in an optimal control problem affect the analysis (as well as the computer coding) of the problem in different ways. The Brachistochrone problem, as summarized and formatted on page 12, illustrates a typical structure in a ***standard optimal control problem*** formulation. We format an optimal control problem under two main parts:

- a preamble, that specifies the problem variables and sets; and,
- a problem definition by name; e.g., Brac:1

The preamble articulates the structure of the variables in the optimal control problem. It defines the state space (e.g., $\mathbb{X} = \mathbb{R}^3$), the control space (e.g., $\mathbb{U} = \mathbb{R}$) and the *names* of the state and control variables (e.g., $\boldsymbol{x} = (x, y, v)$ and $\boldsymbol{u} = \theta$). In addition to supporting the mathematical definition of the problem name that follows it, the preamble also serves as an "initialization" or preamble (!) to a computer code.

After the preamble, the problem is formatted by the cost statement, which contains the formulas for the computation of J, the cost functional. The structured arguments of J, namely $\boldsymbol{x}(\cdot), \boldsymbol{u}(\cdot)$ and t_f, supply the necessary inputs for computation. Deferring to §1.2 for more examples of J, the two sets of constraints identified in Brac: 1 on page 12 are grouped under

- dynamics, for the standard state-space dynamics and
- endpoint, for the boundary conditions.

As noted and discussed before, the system dynamics are a key element in formulating an optimal control problem. The boundary conditions are grouped as endpoints because they incorporate both the initial and the final point conditions. The initial point is chosen to be the origin of the coordinate system and the final point is given by the coordinates, (x^f, y^f). Refer back to Fig. 1.4 on page 5 for additional clarity.

A pseudocode that illustrates the usage of these concepts is presented next on page 15.

Illustrating the Concept: A Pseudocode for Brac:1 Dynamics

```
BEGIN preamble
    INPUT primal                        % structure for C; see Eq. (1.7)
        x(·) ← primal.states (1, ·)
        y(·) ← primal.states (2, ·)
        v(·) ← primal.states (3, ·)
        θ(·) ← primal.controls
          t ← primal.time
END preamble
BEGIN Dynamics
ẋ(t) ← v(t) * sin θ(t)
ẏ(t) ← v(t) * cos θ(t)
v̇(t) ← g * cos θ(t)                     % g is a constant
END Dynamics
% This pseudocode is fairly close to the actual format of a DIDO code.
```

Notation Alert: *Refer to page xxiv for nomenclature.* The pair (x_f, y_f) is a shorthand notation for $(x(t_f), y(t_f))$, while (x^f, y^f) represent their numerical values. See Fig. 1.4 on page 5. Similar notation is used for the initial-time conditions, as well, except that the values for the initial conditions in Brac: 1 are set to be zeros instead of some arbitrary numerical value. Additional details on the endpoints are described in Section 1.3 on page 37.

1.1.3 Avoiding Common Errors # 1

Once a problem is formulated, many beginners and even seasoned practitioners fall into the trap of making a series of totally avoidable early mistakes that have long-lasting unintended consequences.

1. The Bane of Non-Existence

One of the biggest mathematical questions is the problem of *existence of a solution.* Many practitioners dismiss this as a typical mathematician's obsession.

It turns out that this perception is not only ill-founded, but also that the question of existence of a solution is critical to addressing many practical problems such as the debugging of computer codes, autonomous operations of systems via optimal control theory and much more. See page 53 for further discussions on this topic related to flight safety. The existence problem is very important for practical applications because it is possible to easily formulate a problem that has no solution. This point is the subject of Study Problem 1.3.

Study Problem 1.3

Show that Brac:1 has no solution if the final point condition is set as either one of the following two cases:

1: $(x_f, y_f, v_f) = (1, 1, 0)$

2: $(x_f, y_f) = (1, -1)$

Hint: Show that $v(t)\dot{v}(t) - g\dot{y}(t) = 0 \ (\forall\ t)$.

The solutions to Study Problem 1.3 reveal that not all points in $\mathbb{R}^3$ can be arbitrarily stipulated for the target points. Armed with a basic knowledge of Newtonian dynamics, one might rightfully argue that the answer to Study Problem 1.3 is obvious; however, the point remains that in the absence of detailed knowledge of a given dynamical system, it is quite easy to formulate a problem with no solution.

A common error made by beginning students, and even some advanced practitioners, is the stipulation of arbitrary final-point conditions that are not "*reachable*." In simple terms, a point $\boldsymbol{x}^{Reach}$, is said to be ***reachable*** from a given point $\boldsymbol{x}^0$ if there exists a control function $\boldsymbol{u}(\cdot)$ that can steer $\boldsymbol{x}$ from $\boldsymbol{x}^0$ to $\boldsymbol{x}^{Reach}$. The set of all reachable points is called the ***reachable set***; see Fig. 1.11.

It is apparent that the reachability of a point may depend on its *start point* $\boldsymbol{x}^0$. Different start points may have different reachable sets. In general, reachable sets may also depend on the *start time* t^0. We call the pair

$$(\boldsymbol{x}, t) \in \mathbb{R}^{N_x} \times \mathbb{R} \tag{1.9}$$

an ***event***. Hence, in its *canonical form*, reachability is a property of events. For this reason, it is more appropriate to define the event $(\boldsymbol{x}^{Reach}, t^{Reach})$ as being reachable from the event $(\boldsymbol{x}^0, t^0)$.

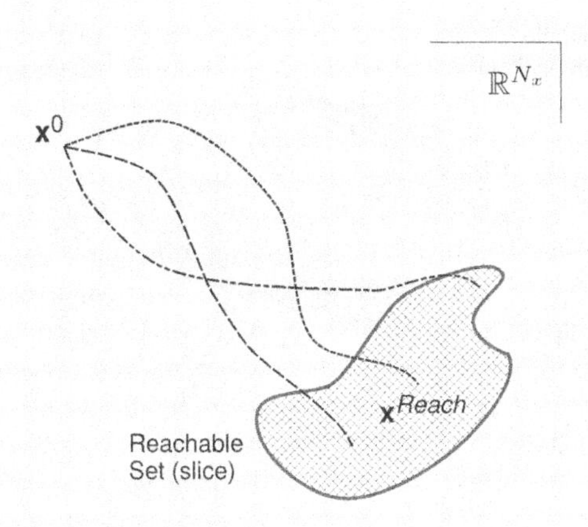

Figure 1.11: The reachable set from the event $(\boldsymbol{x}^0, t^0)$ is the set of all reachable points. The shaded region is a "slice" of the reachable set *at* some given time. The targeted event in an optimal control problem must be reachable.

In practical applications, it may be more useful to consider slices of reachable sets *at* a given time, t_f, as illustrated in Fig. 1.11. Because there is an infinite set of points that are not reachable, it is important to ensure that the events in an optimal control problem are reachable.

> **Computational Tip**: In DIDO, the endpoints are called events. If the target event is reachable, and DIDO outputs an infeasible display, the errors are likely elsewhere in the *problem formulation*, assuming, of course, that the problem was coded correctly in the first place.

In many applications, the collection of all reachable points is *implicitly* constrained by the dynamics of the problem to a more restrictive space than all of $\mathbb{X} = \mathbb{R}^{N_x}$. This concept is illustrated in Fig. 1.12 for Problem Brac:1. The space where all state trajectories evolve is called a *differentiable manifold.* Consequently, for a solution to exist, we cannot stipulate endpoints that are outside this differentiable manifold. *This is a typical error made by beginning students, and even some advanced practitioners, in using computer codes.*

As a means to illustrate this concept a little more simply, consider the fol-

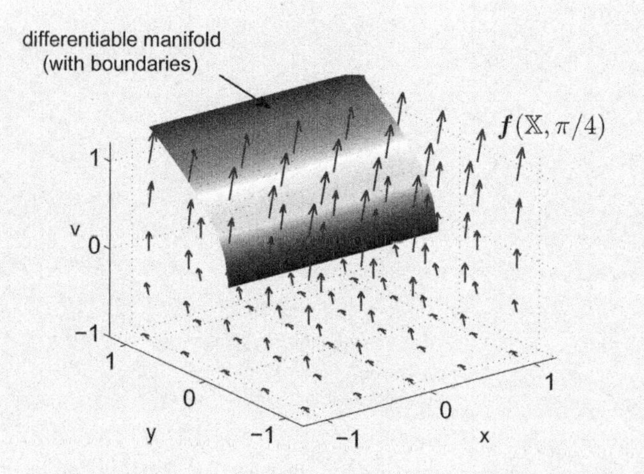

Figure 1.12: The dynamics of Brac:1, $\dot{\boldsymbol{x}} = \boldsymbol{f}(\boldsymbol{x}, \boldsymbol{u})$, naturally constrain the evolution of all state trajectories to a "surface" or manifold in $\mathbb{R}^3$. In addition, this manifold does not exist where $y < 0$. Refer to Study Problem 1.3.

lowing dynamical system,

$$\begin{aligned} \dot{x} &= uy \\ \dot{y} &= -ux \end{aligned} \tag{1.10}$$

where x, y and u are all real variables. Hence, $\mathbb{X} = \mathbb{R}^2$ and $\mathbb{U} = \mathbb{R}$. By simple multiplication, we get

$$x \times \dot{x} + y \times \dot{y} = x \times uy - y \times ux = 0 \qquad (\forall\, t) \tag{1.11}$$

Integrating Eq. (1.11) generates

$$x^2(t) + y^2(t) = \text{constant}$$

where the constant is completely determined by <u>any</u> pointwise condition, such as an initial condition. Hence, the dynamical system given by Eq. (1.10) has a ***hidden constraint*** that forces the evolution of its state trajectory to lie along a circle (a differentiable manifold). No amount of applied control, u, can make the state escape the "orbit" shown in Fig. 1.13. Thus, if the target points x_f

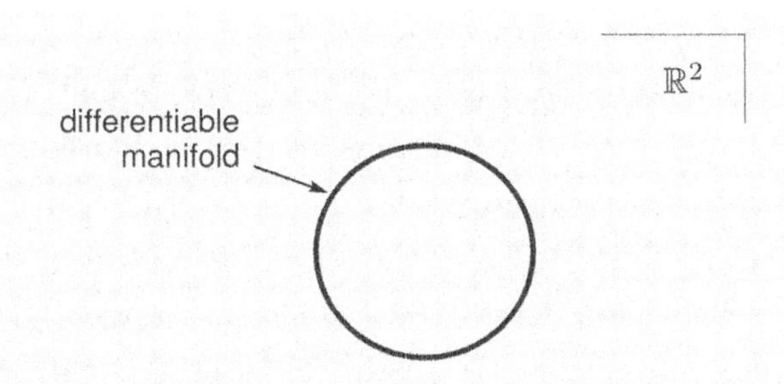

Figure 1.13: The state of the dynamical system given by Eq. (1.10) evolves along a circle. Any target point that is outside this circle is not reachable.

and y_f are not chosen to be such that

$$x_f^2 + y_f^2 = x_0^2 + y_0^2$$

then the problem will have no solution.

Situations where $\mathbb{X}$ is not all of $\mathbb{R}^{N_x}$ happen more often than not because dynamical systems tend to have "hidden" or "implicit" constraints known as *integrals of motion.* The integral of motion in the Brachistochrone problem is mechanical energy ($v^2(t)/2 - gy(t)$ = constant with respect to time). In mechanical systems, there is the possibility of several integrals of motion, like the conservation of angular momentum. In generic dynamical systems (like the one given by Eq. (1.10)) the motion may be implicitly constrained due to a whole host of factors that *include even the very choice of the state variables.* This implies that it is often easier than not (particularly in a computational setting) to formulate an optimal control problem that has no solution! In other words, problem formulation — the subject of this chapter — is indeed supreme!

2. Paralysis by Analysis

Based on the warnings of the preceding paragraphs, one might conclude that a full-blown reachability analysis is necessary before attempting to solve an optimal control problem. This is not true! Quite often, it is sufficient to perform a preliminary analysis to define target events that are reachable. Any and all knowledge of the specific dynamical system can be used to support this problem definition.

In many practical problems, it is hard or impossible to fully and mathematically characterize the entire set of integrals of motion that limit reachability. Even when it is possible to identify a few integrals of motion, it is generally advisable to use this information as part of a verification and validation (V&V) procedure. See, for instance, the *Tech Talk* comment on page 28.

In practical problem-solving, if it is suspected that a solution does not exist because of lack of reachability, it is possible to quickly diagnose the problem by solving another optimal control problem: one that minimizes the distance to the target. If the minimum is not zero, the target point is not reachable! See also Study Problem 1.9 on page 36 for additional discussions on reachability via a collection of optimal control problem formulations.

Although a detailed analysis of reachability is not essential (or practical) before attempting to solve a problem, the *projections* of reachable sets have immense practical value in many applications. See [10] and [25] for extensive details.

3. Scaling and Balancing

Scaling the problem variables and data are crucial for numerical solutions; they are also important for paper/pencil analysis as discussed next. Suffice it to say, bad scaling can make a good algorithm look terrible.

1.1.4 Scaling: How to Design Your Own Units!

Scaling is a critical element of analyzing the "problem of problems" resulting from a preliminary problem formulation.[§] At its basic level, it can be argued that scaling is really a clarification of the units we would like to choose to define the variables of the problem. To this end, let T be an arbitrary unit of time. That is, T need not be what we call a "second", a "minute" or an "hour", but just some agreed-upon quantity that we use to measure time. Similarly, let L and V be some agreed-upon quantity to measure length (distance) and speed, respectively. Then, using an overbar notation, we define our measured

[§]The dual concept of balancing[81] is a correspondingly crucial element for computation.

or "scaled" variables by

$$\bar{x} := \frac{x}{L}, \quad \bar{y} := \frac{y}{L}, \quad \bar{v} := \frac{v}{V}, \quad \text{and} \quad \bar{t} := \frac{t}{T} \tag{1.12}$$

Our goal is to write the entire problem in terms of $\bar{x}, \bar{y}, \bar{v}$ and $\bar{t}$. To achieve this, consider rewriting the differential equation for $\bar{x}$. The process is as follows:

$$\begin{aligned}
\frac{d\bar{x}}{d\bar{t}} &:= \frac{d(x/L)}{d(t/T)} \\
&= \dot{x}\left(\frac{T}{L}\right) \\
&= v\sin\theta\left(\frac{T}{L}\right) \\
&= \bar{v}V\sin\theta\left(\frac{T}{L}\right) \\
&= \left(\frac{VT}{L}\right)\bar{v}\sin\theta
\end{aligned} \tag{1.13}$$

Now, it seems reasonable, and even logical, to agree to define the units for measuring speed by

$$V := \frac{L}{T} \tag{1.14}$$

With this definition, Eq. (1.13) simplifies to

$$\frac{d\bar{x}}{d\bar{t}} = \bar{v}\sin\theta$$

In other words, the differential equation for x and $\bar{x}$ are identical; however, it was imperative that we use Eq. (1.14) to achieve this simplicity.

Similarly it can be shown that (do it!)

$$\frac{d\bar{y}}{d\bar{t}} = \bar{v}\cos\theta$$

Next, following the same process for the last equation, we have

$$\begin{aligned}\frac{d\overline{v}}{d\overline{t}} &:= \frac{d(v/V)}{d(t/T)} \\ &= \dot{v}\left(\frac{T}{V}\right) \\ &= \left(g\frac{T}{V}\right)\cos\theta \\ &= \overline{g}\cos\theta\end{aligned}$$

where we have deliberately defined a new constant, called $\overline{g}$, given by

$$\overline{g} := g\frac{T}{V}$$

If we now agree to choose T such that

$$g\frac{T}{V} = 1 \tag{1.15}$$

then we will have the "good fortune" of always having $\overline{g} = 1$. In other words, we do not need to know the planet where the Brachistochrone experiment is being conducted! Thus, the scaling process has revealed something more than mere numerics: It may be advantageous to use the planet's acceleration due to gravity as part of the constants that we choose to measure the problem variables in order to produce equations in their simplest, or ***canonical*** form. In furthering this *canonization*, suppose we choose

$$L = y^f$$

then, we will have

$$\overline{y}_f = 1$$

no matter what the "actual value" of y^f might be! The "trick" is to use the problem parameters themselves for judicious scaling. In doing so, we can always set $\overline{g} = 1$ and $\overline{y}_f = 1$. This implies that there is only one parameter, namely $\overline{x}_f$, that parameterizes the problem. Thus, by *canonically scaling* the problem, we have reduced the number of parameters and hence reduced the burden on parametric analysis to investigate the problem-of-problems. Suppose we had to investigate the effect of five values for each of the three parameters, g, x^f and

y^f. Then, in the absence of canonical scaling, a parametric analysis of $5^3 = 75$ solutions would be necessary. With scaling, the number of parameters are far fewer (how many?).

Study Problem 1.4

*Instead of Eq. (1.12), choose the following **designer units***

$$\overline{x} := \frac{x}{X}, \quad \overline{y} := \frac{y}{Y}, \quad \overline{v} := \frac{v}{V}, \quad \text{and} \quad \overline{t} := \frac{t}{T} \tag{1.16}$$

Redo the scaling procedure and discuss the impact of not using the same units to measure distances in x and y directions.

Do not read this problem and other similarly marked sections. Think of this as a hyperlink on a webpage that you want to skip at first read.

Study Problem 1.5

Show that canonical scaling leads to a numerically ill-conditioned problem if the final-time condition for the Brachistochrone problem is given by $(x_f, y_f) = (10^4, 1)$. Suggest an alternative scaling procedure to alleviate this problem.

This problem is solved in [81] as a "Bad Brachistochrone Problem" by using non-canonical scaling and exploiting some of the elements of constructing covectors in optimal control discussed in Section 2.2.

1.1.5 Alternative Problem Formulations

It should be no surprise that the Brachistochrone problem, or for that matter, any problem, takes on a different "flavor" if different choices were made for the state and/or control variables. This can be easily illustrated as follows. Suppose that instead of using $\boldsymbol{u} := \theta$, we decided to choose

$$\boldsymbol{u} = (u_x, u_y) \equiv (\sin\theta, \cos\theta) \equiv \mathbf{G}(\theta) \tag{1.17}$$

Then the dynamics can be reformulated as

$$\begin{aligned} \dot{x} &= v u_x \\ \dot{y} &= v u_y \\ \dot{v} &= g u_y \end{aligned}$$

From our previous choice of control, we had only one variable ($\boldsymbol{u} = [\theta]$); hence, we do not expect more decision variables simply by "changing the parameterization" for the control. From the trigonometric identities, we must obviously require

$$u_x^2 + u_y^2 = 1 \tag{1.18}$$

This equation essentially reduces our effective control to just one variable, as before. Nonetheless, in the language of optimal control theory, we not only have a different dynamical model (albeit the same physical problem), but also the number of control variables is indeed two: $\boldsymbol{u} \in \mathbb{R}^2$. In addition, we now have a ***control constraint***, given by Eq. (1.18). The control constraint limits $\mathbb{U}$ from all of $\mathbb{R}^2$ to just a circle of points, as illustrated in Fig. 1.14.

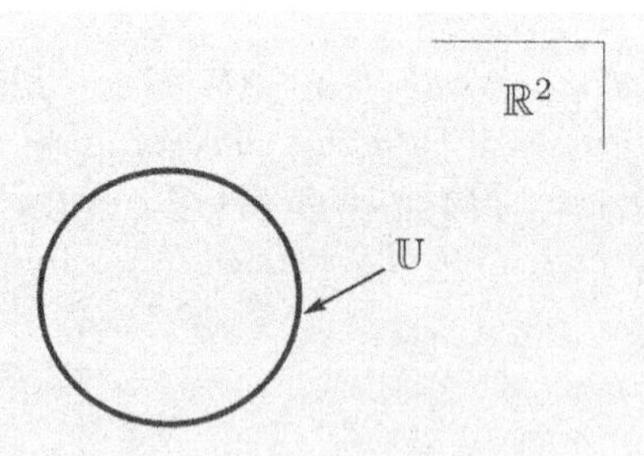

Figure 1.14: The control space, $\mathbb{U} \subset \mathbb{R}^2$, is a circle in $\mathbb{R}^2$ as specified by Eq. (1.19).

That $\mathbb{U}$ is a circle in $\mathbb{R}^2$ is mathematically expressed by the set notation,

$$\mathbb{U} = \left\{(u_x, u_y) \in \mathbb{R}^2 : \; u_x^2 + u_y^2 = 1\right\} \tag{1.19}$$

which is just shorthand for saying that $\mathbb{U}$ *is the set of all* u_x *and* u_y *such that* $u_x^2 + u_y^2 = 1$.

High-Brow Alert: The set $\mathbb{U}$ illustrated in Fig. 1.14, is called a one-dimensional sphere and is often denoted as S^1. Thus, we can think of $\mathbb{U} = S^1$ in a "coordinate-free" representation. In this view, $\boldsymbol{u}$ written either as a singleton θ, or as a pair (u_x, u_y) is two alternative "coordinatizations" of the same geometric object, namely the one-dimensional unit sphere, S^1.

Thus, we have a second formulation of the Brachistochrone problem that can be articulated as:

$$\left.\begin{array}{ll} \mathbb{X} = \mathbb{R}^3 & \mathbb{U} \subset \mathbb{R}^2 \\ \boldsymbol{x} = (x, y, v) & \boldsymbol{u} = (u_x, u_y) \end{array}\right\} \text{(preamble)}$$

$$\overbrace{(Brac:2)}^{\text{problem}} \left\{ \begin{array}{lrl} \text{Minimize} & J[\boldsymbol{x}(\cdot), \boldsymbol{u}(\cdot), t_f] = t_f & \} \ \text{(cost)} \\ \text{Subject to} & \dot{x} = v u_x & \\ & \dot{y} = v u_y & \text{(dynamics)} \\ & \dot{v} = g u_y & \\ & (t_0, x_0, y_0, v_0) = (0, 0, 0, 0) & \\ & (x_f - x^f, y_f - y^f) = (0, 0) & \text{(endpoints)} \\ & u_x^2 + u_y^2 = 1 & \} \ \text{(path)} \end{array} \right.$$

Besides the appropriate modifications to the preamble and the dynamics, $(Brac:2)$ differs from $(Brac:1)$ by the inclusion of the equation that defines $\mathbb{U}$ under a new label called path. It will be apparent later that the control constraint is part of a ***path constraint***. A path constraint is a more general constraint in an optimal control problem formulation that subsumes constraints on the control variables.

The problem formulated under $(Brac:2)$ shows that we can choose our control variables more generally than the implications suggested by $(Brac:1)$. This added freedom in choosing control variables can be exploited in many ways; however, in certain situations, it may generate a new problem called ***control allocation***, as illustrated in Fig. 1.15.

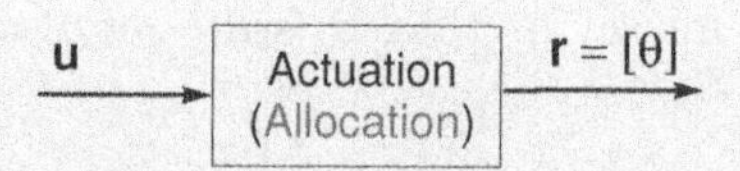

Figure 1.15: A trivial control allocation problem.

To understand the control allocation problem, suppose that we solved Brac:2, but wanted to extract θ from (u_x, u_y). This problem is quite trivial since to every $\boldsymbol{u} \in \mathbb{U}$, we can assign a unique $\theta \in [0, 2\pi]$ (using Eq. (1.17)). In many electromechanical systems the output of the allocator ($= \boldsymbol{r} =$ input to the plant) has a higher dimension than its input; hence, the inverse map, $\mathbf{G}^{-1}$, is one-to-many (i.e., a multifunction, $\mathbf{G}^{-1} : \mathbb{R}^{N_u} \twoheadrightarrow \mathbb{R}^{N_r}$). In such situations, we can define an optimization problem (static) where we pick the allocator output that minimizes some cost function (e.g., power).

Note that by choosing the control variables to be u_x and u_y, we can think of controls as some "internal" variables that need to be computed to generate the inputs for the plant in much the same way as we view state variables as being internal variables that are used to define the state of the system independent of the output.

In both problem formulations considered so far, we were apparently quite clever in writing Newton's dynamical equations along the velocity vector. Suppose we had chosen to write the equations along the more obvious directions of x and y; then, we would have obtained

$$\begin{aligned}
\dot{x} &= V_x \\
\dot{y} &= V_y \\
m\dot{V}_x &= N\cos\theta \\
m\dot{V}_y &= mg - N\sin\theta
\end{aligned}$$

where N is the normal force; see Fig. 1.4 on page 5. This means that we no longer have $\boldsymbol{x} \in \mathbb{R}^3$; rather, $\boldsymbol{x}$ is now an element of $\mathbb{X} = \mathbb{R}^4$. In the same spirit,

$$\boldsymbol{u} = \begin{bmatrix} N \\ \theta \end{bmatrix}$$

Thus, $\boldsymbol{u} \in \mathbb{R}^2$;¶ however, N and θ are not independent control variables. That the normal force is indeed normal to the surface (i.e., perpendicular to the velocity vector) is given by $\mathbf{N} \cdot \mathbf{v} = 0$, resulting in the "algebraic" equation

$$(N \cos\theta)V_x - (N \sin\theta)V_y = 0 \tag{1.20}$$

This equation is coupled in both the state (V_x and V_y) and control (N and θ) variables. It reveals that our new choice of state and control variables generates a new *path constraint* which, unlike the previous path constraint of Eq. (1.18), has a new twist in that Eq. (1.20) *jointly limits* our "space of allowable values" on both the state and control variables. In optimal control parlance, such constraints are known as ***mixed state-control constraints***.

Study Problem 1.6

Explain the ramifications of simplifying Eq. (1.20) to $V_x \cos\theta - V_y \sin\theta = 0$ by "canceling out" N.

In addition to Eq. (1.20), we also need to impose the constraint

$$N \geq 0 \tag{1.21}$$

if we insist that the particle is not allowed to "fly off" the surface. Equation (1.21) is also a path constraint (in the same spirit as Eq. (1.18)), except that we now have an *inequality*.

Collecting all the relevant equations, we now have a third formulation of the Brachistochrone problem:

¶In geek speak, $\boldsymbol{u} \in \mathbb{R}_+ \times S^1$.

$$
\left.\begin{array}{ll}
\mathbb{X} \subset \mathbb{R}^4 & \mathbb{U} \subset \mathbb{R}^2 \\
\boldsymbol{x} = (x, y, V_x, V_y) & \boldsymbol{u} = (N, \theta)
\end{array}\right\} \text{(preamble)}
$$

$$
\overbrace{(Brac:3)}^{\text{problem}}
\left\{\begin{array}{lll}
\text{Minimize} & J[\boldsymbol{x}(\cdot), \boldsymbol{u}(\cdot), t_f] = t_f & \text{(cost)} \\
\text{Subject to} & \dot{x} = V_x & \\
& \dot{y} = V_y & \\
& \dot{V}_x = \dfrac{N}{m}\cos\theta & \text{(dynamics)} \\
& \dot{V}_y = g - \dfrac{N}{m}\sin\theta & \\
& (t_0, x_0, y_0, V_{x,0}, V_{y,0}) = (0,0,0,0,0) & \text{(endpoints)} \\
& (x_f - x^f, y_f - y^f) = (0,0) & \\
& V_x N\cos\theta - V_y N\sin\theta = 0 & \text{(path)} \\
& N \geq 0 &
\end{array}\right.
$$

Study Problem 1.7

1. *Show that a fourth formulation of the Brachistochrone problem is possible by setting*

$$
u_x = \frac{N}{m}\cos\theta \quad \textit{and} \quad u_y = \frac{N}{m}\sin\theta
$$

   What is the control space $\mathbb{U}$ *for this formulation?*

2. *Let* $u \in \mathbb{R}$*. Show that by replacing* $(N/m)\cos\theta$ *by* uV_y *and* $(N/m)\sin\theta$ *by* uV_x *in Problem* $Brac:3$*, a fifth formulation of the Brachistochrone problem is achieved with* $\mathbb{U} = \mathbb{R}$*. Explain the legitimacy of this substitution. What is the meaning of* u *here?*

Tech Talk: An apparently simpler formulation of the problem is possible by using the principle of energy conservation. Since the system is conservative, we have

$$
\frac{1}{2}mv^2 = mgy
$$

This integral of motion implies that we should be able to reduce the dimension of the state space, which in fact, we can do by substituting $v = \sqrt{2gy}$ in the kinematical equations in Brac: 1

$$\begin{aligned}\dot{x} &= \sqrt{2gy}\sin\theta \\ \dot{y} &= \sqrt{2gy}\cos\theta\end{aligned}$$

Thus, the equation $\dot{v} = g\cos\theta$ is apparently unnecessary.

The problem with the above formulation is that the function $\boldsymbol{f}(\boldsymbol{x}, \boldsymbol{u})$, which describes the right-hand side of the differential equation, is not differentiable at $y = 0$ because the square-root function $y \mapsto \sqrt{y}$ is not differentiable at the origin

$$\left.\frac{\partial\sqrt{y}}{\partial y}\right|_{y=0} = \infty$$

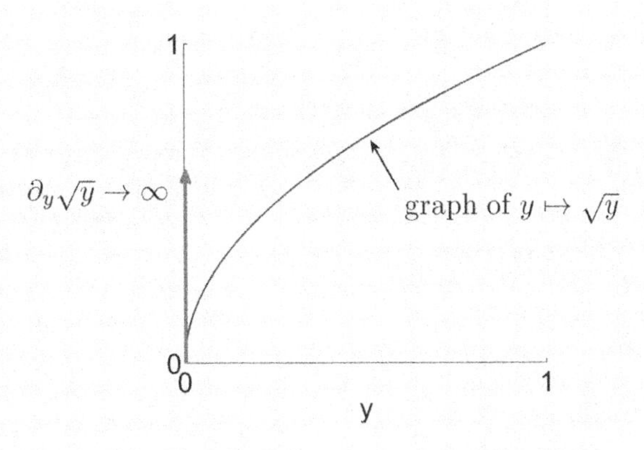

Figure 1.16: Differentiability problem with the square-root function.

Differentiability of $\boldsymbol{f}$ with respect to $\boldsymbol{x}$ is required for the "smooth" version of Pontryagin's Principle; otherwise, the more advanced nonsmooth versions are what is truly required. While the theoretical problem may be overcome by the nonsmooth version, differentiability is typically a necessary requirement in numerical computation (for computing gradients). Note also that even though this lack of differentiability is localized at $y = 0$, it is quite an important point: It is the initial condition!

In the absence of an *a priori* knowledge of the solution, the problem at $y = 0$ implies that if the optimal trajectory were to stay along $y = 0$ for some period of time, then the nondifferentiability problem is substantially worse than a problem at an isolated point. These concepts are not "merely" technical issues; such problems are pervasive in practical applications, particularly in engineering.

The take-away of this example is that model-reduction is not necessarily a good thing. What a modeler needs to ask is why model reduction is sought in the first place, and be cognizant of the new problems it creates. If these new problems are worse than the original problem, then it may be better to directly deal with the original problem or seek an alternative path for model reduction.

This material is marked with the double dangerous bend of Knuth. You must absolutely avoid reading such sections unless you are extremely comfortable and knowledgeable about advanced topics in optimal control that are not discussed in this book ... like a Sobolev space. If you insist on reading this subsection anyway, be forewarned that you are doing it at your own peril!

Study Problem 1.8

In formulating Problem Brac : 1, the equation of motion along the normal direction,

$$N - mg\sin\theta = mv\dot{\theta}$$

was ignored. Explain the ramifications of including this equation and justifications for excluding it. (Hint: Consider the appropriate Sobolev space for the function, $t \mapsto \theta$. Is θ a control or state variable? See also *§4.8.3 on page 298.)*

Note that the famous solvers of the Brachistochrone problem *did not* linearize it. It is quite possible they felt that linearization was an analyst's confession of defeat.

1.2 Changing the Paradigm

Many practical optimal control problems are initially formulated as "word" problems in much the same way the Brachistochrone problem was formulated on pages 2–4. A primary task of a practitioner is to translate the word problem to a mathematical problem. This translation is not unique as discussed in §1.1.5, pages 23–28. That is, the same problem can be formulated in several different ways. The question is, which problem formulation should a practitioner choose? The clever answer is to pick the "best one," but a better answer is to pick all of them for different purposes.

Studying different problem formulations provides different perspectives for the same problem. Such study has an inherent value in lending better understanding of the specific problem. More importantly, such study strengthens our knowledge of optimal control theory itself, which, in turn, can be used to confront new problems. In moving along this *dual* direction of better understanding optimal control theory, consider the following problem:

$$\left.\begin{array}{ll} \mathbb{X} = \mathbb{R}^3 & \mathbb{U} = \mathbb{R} \\ \boldsymbol{x} = (x, y, v) & \boldsymbol{u} = \theta \end{array}\right\} \text{(preamble)}$$

$$\overbrace{(DBrac:1)}^{\text{problem}} \left\{\begin{array}{lll} \text{Minimize} & J[\boldsymbol{x}(\cdot), \boldsymbol{u}(\cdot), t_f] := x_f & \text{(cost)} \\ \text{Subject to} & \dot{x} = v\sin\theta & \\ & \dot{y} = v\cos\theta & \text{(dynamics)} \\ & \dot{v} = g\cos\theta & \\ & (x_0, y_0, v_0, t_0) = (0,0,0,0) & \\ & (y_f - y^f, t_f - t^f) = (0,0) & \text{(endpoints)} \end{array}\right.$$

It is apparent that this problem formulation is *not* the Brachistochrone problem, but may have something to do with it. That is, the dynamics and initial conditions are the same as the Brachistochrone problem, but the cost functional and endpoint conditions are different.

Problem $DBrac:1$ is said to be dual to $Brac:1$. In $DBrac:1$, we have switched one of the endpoint conditions to a cost functional and vice versa. That is, instead of demanding $x_f = x^f$, we now seek to maximize the horizontal distance traveled by the point mass; see Fig. 1.17. Because we cannot ask to

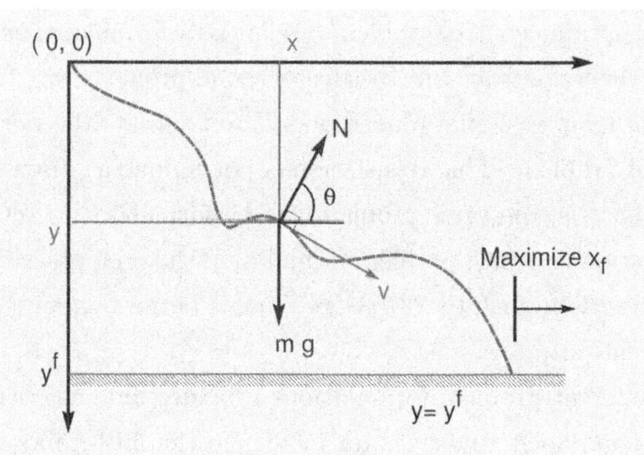

Figure 1.17: Schematic for a dual to the Brachistochrone problem.

maximize x_f and fix it at the same time, the condition that $x_f = x^f$ disappears from the original or *primal* formulation. If we let t_f go free, then $x_f \to \infty$ with $t_f \to \infty$.** Hence we constrain t_f to a fixed value of t^f.

The two functionals we have come across so far are

$$J[\boldsymbol{x}(\cdot), \boldsymbol{u}(\cdot), t_f] := t_f \tag{1.22}$$

and

$$J[\boldsymbol{x}(\cdot), \boldsymbol{u}(\cdot), t_f] := x_f \tag{1.23}$$

Defining a functional need not be limited to the final value of time or of some state variable. A problem designer is free to define an arbitrary scalar function E of the endpoint variables $\boldsymbol{x}_f \in \mathbb{R}^{N_x}$ and $t_f \in \mathbb{R}$:

$$E : \mathbb{R}^{N_x} \times \mathbb{R} \longrightarrow \mathbb{R} \tag{1.24}$$

and then set the cost functional as

$$J[\boldsymbol{x}(\cdot), \boldsymbol{u}(\cdot), t_f] := E(\boldsymbol{x}_f, t_f) \tag{1.25}$$

**Spoiler Alert: If this point is not apparent to a student who is not familiar with Newtonian dynamics, then an application of Pontryagin's Principle will actually show this! In other words, Pontryagin's Principle can be used to sniff out bad problem formulations!! See also the work-flow diagram on page 84 for further ideas in formulating the correct problem.

As noted before (see page 7), the map notation of Eq. (1.24) is simply a shorthand representation for clarifying the inputs and outputs of a function. A block diagram for Eq. (1.24) is shown in Fig. 1.18. Thus, in Eq. (1.22) the E-function

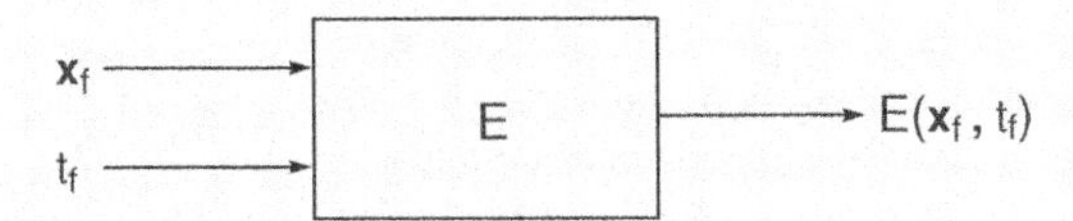

Figure 1.18: E is a user-supplied function that takes two inputs, $\boldsymbol{x}_f \in \mathbb{R}^{N_x}$ and $t_f \in \mathbb{R}$, and outputs a scalar $E(\boldsymbol{x}_f, t_f) \in \mathbb{R}$. Thus, E is the name of the computational machine.

is given by $E(\boldsymbol{x}_f, t_f) = t_f$. Likewise, $E(\boldsymbol{x}_f, t_f) = x_f$ in Eq. (1.23). Now, what transforms this process to a *functional*, J, is that in order to produce one of the inputs for E, namely $\boldsymbol{x}_f$, we need to have the entire function $\boldsymbol{x}(\cdot)$, so that $\boldsymbol{x}_f := \boldsymbol{x}(t_f)$ may be generated in the first place. This notion is illustrated in Fig. 1.19 along with a pseudocode for computing J.

Such cost functionals, that is, those that can be evaluated based solely on the final time conditions, are called ***endpoint costs***. A general form of an endpoint cost functional is given by Eq. (1.25); the symbol E stands for endpoint. Endpoint cost functionals occur quite often in many present-day optimal control problems. A less descriptive and somewhat older term for an endpoint cost is called a *Mayer cost*, and the cost functional is said to be in a *Mayer form*. This term is in honor of Adolph Mayer (1839–1907), who made contributions to the multiplier rule in the calculus of variations.

Cost functionals are not limited to endpoint forms; they can come in other different forms. To see this, observe the following:

$$x_f \equiv x(t_f) = \int_{t_0}^{t_f} \dot{x}\, dt \tag{1.26}$$

Substituting the differential equation $\dot{x} = v \sin\theta$ in Eq. (1.26), the cost functional given by Eq. (1.23) can be written equivalently as

$$J[\boldsymbol{x}(\cdot), \boldsymbol{u}(\cdot), t_f] := \int_{t_0}^{t_f} v(t) \sin\theta(t)\, dt \tag{1.27}$$

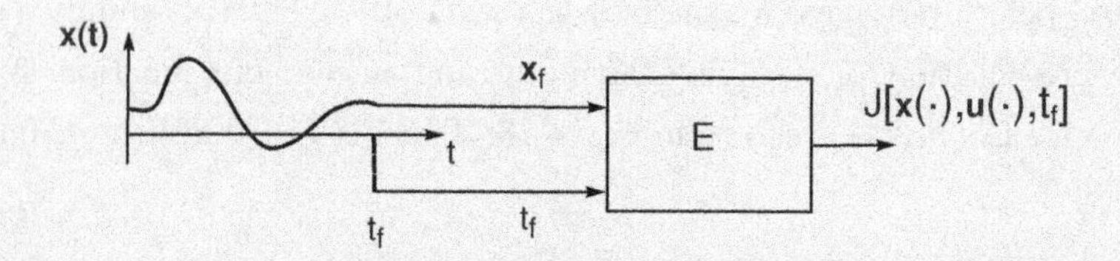

Figure 1.19: Conceptual block diagram for the computational machine J, given by a generic endpoint cost function E. The control function is not explicitly needed when J is given in the endpoint form E; however, if $\boldsymbol{x}(\cdot)$ is dynamically feasible, then the generation of $\boldsymbol{x}(\cdot)$ depends on $\boldsymbol{u}(\cdot)$. In addition, the general form of J depends on $\boldsymbol{u}(\cdot)$, as well.

A pseudocode for J can be constructed as:

BEGIN preamble

INPUT primal % *structure that contains* $\boldsymbol{x}(\cdot), \boldsymbol{u}(\cdot)$ *and* t_f

$\boldsymbol{x}(\cdot) \leftarrow$ *primal.states*

$\boldsymbol{u}(\cdot) \leftarrow$ *primal.controls*

$t_f \leftarrow$ *primal.time.final*

END preamble

BEGIN cost

$\boldsymbol{x}_f \leftarrow \boldsymbol{x}(t_f)$

cost $= E(\boldsymbol{x}_f, t_f)$ % E *is a user-supplied function*

END cost

This pseudocode is fairly close to the actual format of a DIDO code.

Obviously, this form of J is quite different from Eq. (1.25).

The integrand in Eq. (1.27) is a function of the control variable ($\boldsymbol{u} := \theta$) and one of the states ($x_3 := v$). A general form of such integrands can be written as $F(\boldsymbol{x}, \boldsymbol{u}, t)$; hence, a new class of cost functionals can be written as

$$J[\boldsymbol{x}(\cdot), \boldsymbol{u}(\cdot), t_f] := \int_{t_0}^{t_f} F(\boldsymbol{x}(t), \boldsymbol{u}(t), t)\, dt \tag{1.28}$$

where F is a scalar function of $\boldsymbol{x} \in \mathbb{R}^{N_x}$, $\boldsymbol{u} \in \mathbb{R}^{N_u}$ and t, written formally as

$$F : \mathbb{R}^{N_x} \times \mathbb{R}^{N_u} \times \mathbb{R} \longrightarrow \mathbb{R} \tag{1.29}$$

A block diagram for Eq. (1.29) is shown in Fig. 1.20.

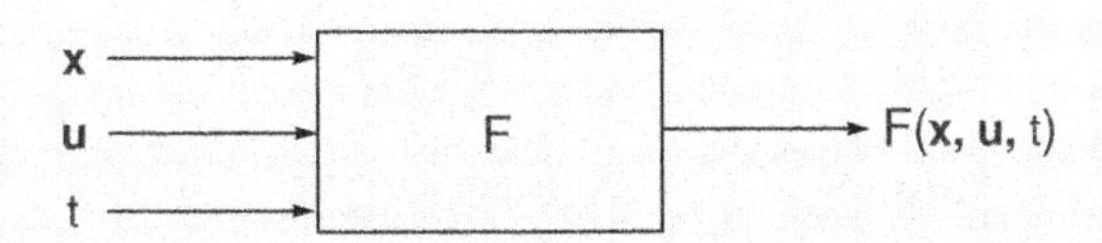

Figure 1.20: F is a user-supplied function that takes three inputs, $\boldsymbol{x} \in \mathbb{R}^{N_x}$, $\boldsymbol{u} \in \mathbb{R}^{N_u}$ and $t \in \mathbb{R}$, and outputs a scalar $F(\boldsymbol{x}, \boldsymbol{u}, t) \in \mathbb{R}$. Thus, F is the name of the computational machine.

For the example in Eq. (1.27) the F-function is given by

$$F(\boldsymbol{x}, \boldsymbol{u}) := v \sin \theta$$

There is no explicit dependence on t in $v \sin \theta$; hence, F is a function of $\boldsymbol{x}$ and $\boldsymbol{u}$ only.

The integrand in Eq. (1.28) is called a ***running cost*** and the cost functional is said to be in the *Lagrange form.* Such functionals can be visualized as generating the area under the function $t \mapsto F(\boldsymbol{x}(t), \boldsymbol{u}(t), t)$; see Fig. 1.21. For different system trajectories $(\boldsymbol{x}(\cdot), \boldsymbol{u}(\cdot))$, a new function is generated with new values for the area $J[\boldsymbol{x}(\cdot), \boldsymbol{u}(\cdot), t]$.

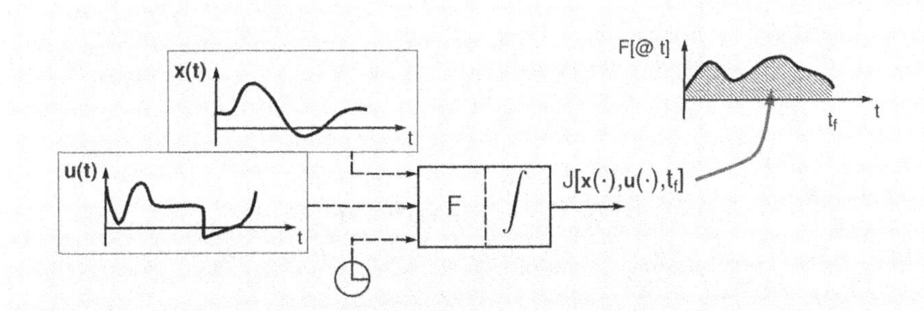

Figure 1.21: Conceptual block diagram for J given by the running cost function, F. Compare with Fig. 1.19. In DIDO, the integration required to compute J is automatically performed: A user only supplies the data function, F.

In any case, as a result of the equivalence between the Mayer and Lagrange cost functionals, Problem $DBrac : 1$ is equivalent to the following problem

$$
\left.\begin{array}{ll}
\mathbb{X} = \mathbb{R}^2 & \mathbb{U} = \mathbb{R} \\
\boldsymbol{x} = (y, v) & \boldsymbol{u} = \theta
\end{array}\right\} \text{(preamble)}
$$

$$
\overbrace{(DBrac:2)}^{\text{problem}}
\left\{
\begin{array}{lll}
\text{Maximize} & J[\boldsymbol{x}(\cdot), \boldsymbol{u}(\cdot), t_f] := \displaystyle\int_{t_0}^{t_f} v(t) \sin\theta(t)\, dt & \text{(cost)} \\
\text{Subject to} & \dot{y} = v\cos\theta & \text{(dynamics)} \\
 & \dot{v} = g\cos\theta & \\
 & t_0 = 0 & \text{(endpoints)} \\
 & (y_0, v_0) = (0, 0) & \\
 & y_f - y^f = 0 & \\
 & t_f - t^f = 0 &
\end{array}
\right.
$$

From one perspective, the reduction in the dimension of the state space in Problem $DBrac:2$ to two dimensions can be argued to be a better formulation than Problem $DBrac:1$.

Study Problem 1.9

1. *Explain how Problem $DBrac:1$ addresses part of the reachability problem discussed in Section 1.1.3; see Fig. 1.11 on page 17.*
2. *Refer to Fig. 1.17. Discuss the problem of minimizing x_f. Does this problem address reachability?*
3. *Does having the condition $y_f = y^f$ create a problem of existence of a solution?*
4. *Suppose we set y_f free in Problem $DBrac:1$. Is this a valid problem formulation? Does this problem support the production of reachable sets?*
5. *Discuss the problem of minimizing y_f.*

1.3 The Target Set

In all formulations of the (primal) Brachistochrone problem considered so far, the endpoint condition was a given point, $(x^f, y^f) \in \mathbb{R}^2$, in the physical space (see Fig. 1.4 on page 5). To facilitate the motivation and generation of a more standard framework for articulating target sets, observe that the target point (x^f, y^f) can be viewed as the intersection of two lines in $\mathbb{R}^2$:

$$\underbrace{(x_f - x^f) + 0 \cdot y_f}_{\tilde{e}_1(x_f, y_f)} = 0, \quad \underbrace{0 \cdot x_f + (y_f - y^f)}_{\tilde{e}_2(x_f, y_f)} = 0 \tag{1.30}$$

See Fig. 1.22. Each line is a set of "zeroes" of two *functions* $\tilde{e}_i : \mathbb{R}^2 \to \mathbb{R}$, $i = 1, 2$

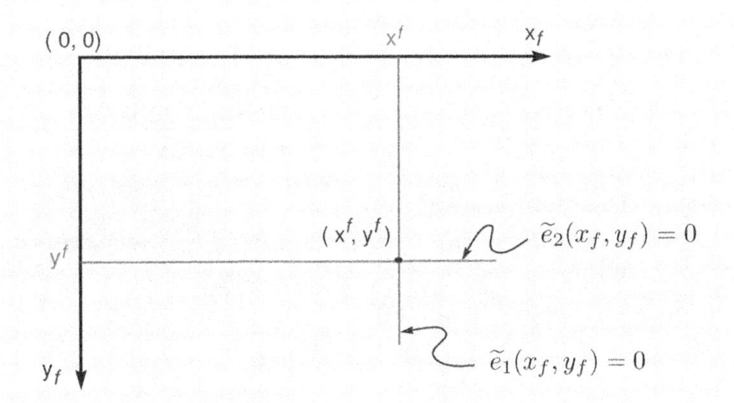

Figure 1.22: The given endpoint, (x^f, y^f), of Brac:1 is re-defined as the intersection point of two lines. Each line is given by the equation, $\tilde{e}_i(x_f, y_f) = 0$, $i = 1, 2$.

defined by

$$\tilde{e}_1(x_f, y_f) := x_f - x^f$$
$$\tilde{e}_2(x_f, y_f) := y_f - y^f$$

That is, $\tilde{e}_1(x_f, y_f) = 0$ is the *equation* of a vertical line passing through the point x^f, while $\tilde{e}_2(x_f, y_f) = 0$ is the *equation* of a horizontal line passing through the point y^f. In both equations, x_f and y_f are variables while x^f and y^f are the data points.

Function versus Equation: A common source of confusion among some beginners is mistaking a function for an equation and vice versa, particularly in understanding the stipulation of boundary conditions. We use Eq. (1.30) to help clarify some points at this juncture.

- The defining equation

$$\widetilde{e}_1(x_f, y_f) := x_f - x^f$$

  is a *formula* for evaluating the function $\widetilde{e}_1 : \mathbb{R}^2 \to \mathbb{R}$. The notation $\widetilde{e}_1(x_f, y_f)$ implies that $\widetilde{e}_1$ is a function of two variables, x_f and y_f.

- The statement

$$\widetilde{e}_1(x_f, y_f) = 0 \tag{1.31}$$

  is an *equation* (not a function) of two variables (x_f and y_f) whose solution is a set of points given by the vertical line in Fig. 1.22; hence, the equation to this vertical line is given by Eq. (1.31).

The vertical line in Fig. 1.22 is <u>not</u> the graph of the function $\widetilde{e}_1 : \mathbb{R}^2 \to \mathbb{R}$.

Study Problem 1.10

Graph the function $\widetilde{e}_1 : (x_f, y_f) \mapsto \mathbb{R}$. What is the connection between the graph of this function and the line given by the equation $\widetilde{e}_1(x_f, y_f) = 0$?

Because $\boldsymbol{x}_f := (x_f, y_f, v_f)$ is in $\mathbb{R}^3$, we can get a more accurate representation of the endpoint conditions by using $\mathbb{R}^3$ as the domain; hence, we rewrite Eq. (1.31) as

$$\underbrace{(x_f - x^f) + 0 \cdot y_f + 0 \cdot v_f}_{e_1(\boldsymbol{x}_f)} = 0 \qquad \underbrace{0 \cdot x_f + (y_f - y^f) + 0 \cdot v_f}_{e_2(\boldsymbol{x}_f)} = 0$$

where $e_i : \mathbb{R}^3 \to \mathbb{R}$, $i = 1, 2$. That is, the main difference between $\widetilde{e}_i$ and e_i is the domain: In the former, it is $\mathbb{R}^2$, whereas in the latter it is $\mathbb{R}^3$. As a result, the *target set* $\mathbb{E} \subset \mathbb{R}^3$ must be viewed as the set of points that lie at the intersection of the two planes defined by the equations

$$e_1(x_f, y_f, v_f) := x_f - x^f = 0$$
$$e_2(x_f, y_f, v_f) := y_f - y^f = 0$$

See Fig. 1.23. This target set is a line in $\mathbb{R}^3$, where the $\mathbb{R}^3$ coordinates are

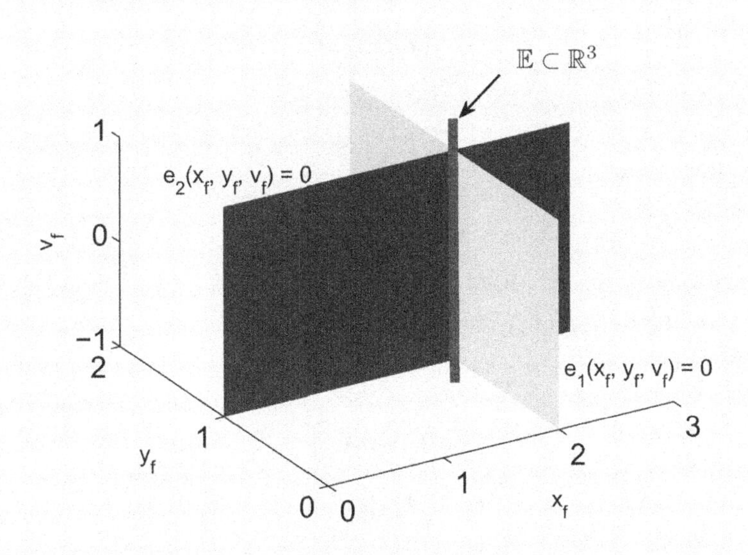

Figure 1.23: The target set $\mathbb{E}$ for Brac:1 is the manifold (line) defined by the intersection of the two planes.

given by the three variables (x_f, y_f, v_f), the components of $\boldsymbol{x}_f$. The target line is parallel to the v_f-axis and passes through the point $(x^f, y^f, 0)$. In vector notation, the equation

$$\boldsymbol{e}(\boldsymbol{x}_f) := \begin{bmatrix} e_1(\boldsymbol{x}_f) \\ e_2(\boldsymbol{x}_f) \end{bmatrix} = \mathbf{0} := \begin{bmatrix} 0 \\ 0 \end{bmatrix}$$

defines the target set $\mathbb{E}$.

Recall that (see Eq. (1.9) on page 16) the pair $(\boldsymbol{x}, t)$ is called an event. Hence, for Brac:1 the complete picture must be viewed in $\mathbb{R}^4$ coordinatized by the pair $(\boldsymbol{x}_f, t_f) \in \mathbb{R}^3 \times \mathbb{R}$. From this perspective, the final time variable t_f is the fourth coordinate and the endpoint set is a *hyperplane* in $\mathbb{R}^4$ obtained by the intersection of two other hyperplanes.

In a *standard optimal control problem*, the target set is given by a collection of N_e *nonlinear* equations:

$$\begin{aligned} e_1(\boldsymbol{x}_f, t_f) &= 0 \\ e_2(\boldsymbol{x}_f, t_f) &= 0 \\ &\vdots \\ e_{N_e}(\boldsymbol{x}_f, t_f) &= 0 \end{aligned}$$

Each equation can be visualized as a hypersurface in $\mathbb{R}^{N_x} \times \mathbb{R}$, and the target set $\mathbb{E}$ is the intersection of N_e hypersurfaces. This intersection is, in general, a hypersurface itself, and can be compactly written in terms of a vector equation:

$$\boldsymbol{e}(\boldsymbol{x}_f, t_f) := \begin{bmatrix} e_1(\boldsymbol{x}_f, t_f) \\ e_2(\boldsymbol{x}_f, t_f) \\ \vdots \\ e_{N_e}(\boldsymbol{x}_f, t_f) \end{bmatrix} = \boldsymbol{0} := \begin{bmatrix} 0 \\ 0 \\ \vdots \\ 0 \end{bmatrix}$$

where

$$\boldsymbol{e} : \mathbb{R}^{N_x} \times \mathbb{R} \to \mathbb{R}^{N_e} \tag{1.32}$$

is called the ***endpoint constraint function***. As noted several times before in this chapter, the map notation is simply a shorthand representation for clarifying the inputs and outputs of a function. A block diagram and a pseudocode for Eq. (1.32) are shown in Fig. 1.24.

A geometric interpretation of the ***endpoint constraint equation***

$$\boldsymbol{e}(\boldsymbol{x}_f, t_f) = \boldsymbol{0}$$

is illustrated in Fig. 1.25.

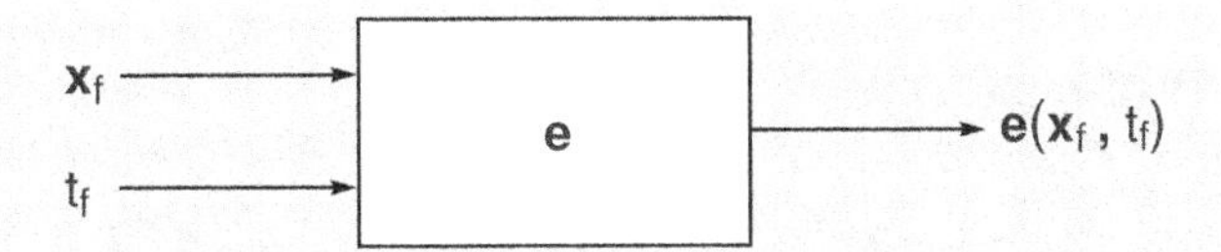

Figure 1.24: Block diagram for the endpoint constraint function $e: (\boldsymbol{x}_f, t_f) \mapsto \mathbb{R}^{N_e}$.

A pseudocode for e is as follows:

BEGIN preamble

INPUT primal   *% structure that contains $\boldsymbol{x}(\cdot), \boldsymbol{u}(\cdot)$ and t_f*

$\boldsymbol{x}(\cdot) \leftarrow$ *primal.states*

$\boldsymbol{u}(\cdot) \leftarrow$ *primal.controls*

$t_f \leftarrow$ *primal.time.final*

END preamble

BEGIN events

$\boldsymbol{x}_f \leftarrow \boldsymbol{x}(t_f)$

call function $e(\boldsymbol{x}_f, t_f)$ *% $\boldsymbol{e}$ is a user-supplied function*

END events

This pseudocode is fairly close to the actual format of a DIDO code.

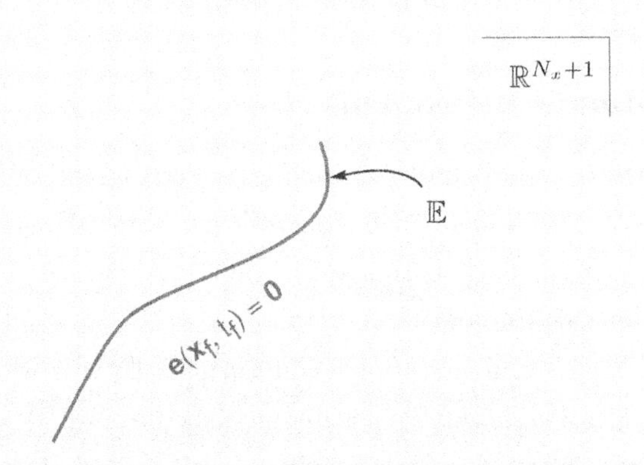

Figure 1.25: The target set $\mathbb{E}$ is the set of all points defined by the vector equation $e(\boldsymbol{x}_f, t_f) = \boldsymbol{0}$. It can be visualized as a hypersurface (manifold) in $\mathbb{R}^{N_x} \times \mathbb{R}$.

Study Problem 1.11

Suppose that the sliding surface in Brac:1 was subject to friction by a force, f, acting opposite to the velocity vector. Assuming a friction law given by $f = \mu N$, where μ is the coefficient of sliding friction, reformulate the Brachistochrone problem. Explain how this reality changes the scope of reachable sets. Revisit and revise your answers to Problem 1.3 discussed on page 16.

1.4 Guideposts for Problem Formulations

The various incarnations of the Brachistochrone problems discussed in the previous sections are all examples of a ***standard problem in optimal control*** that we call Problem B. In simple terms, this problem is to determine a dynamically feasible state-control function pair, $(\boldsymbol{x}(\cdot), \boldsymbol{u}(\cdot))$, that minimizes a cost functional while satisfying a collection of pointwise and path constraints. As already apparent in Sections 1.1, 1.2 and with many more examples to follow in Chapters 3 and 4, a vast number of problems in engineering, physics, economics, medicine, management and many other fields can be framed under this mathematical formulation — and in a very natural way.

1.4.1 The Standard Problem B

Each element of the standard problem was introduced in the previous sections as generalizations of Bernoulli's Brachistohcrone problem. We now collect these concepts to frame Problem B so that every new problem can be viewed as a specific "app" under this framework.

We begin by recalling some notational conventions. By $\boldsymbol{u} \in \mathbb{R}^{N_u}$, we mean an N_u-dimensional control variable. This is distinguished from $\boldsymbol{u}(\cdot)$, which means a ***control trajectory*** $t \mapsto \boldsymbol{u}$, the graph of which, with respect to time, is a curve. See Fig. 1.26. By definition, the control is allowed to "jump" from one point to another so long as it is contained in an allowable set, $\mathbb{U} \subset \mathbb{R}^{N_u}$, called the ***control space***. That is, the control trajectory is allowed to be discontinuous. If it is required to be limited by a rate, "$d\boldsymbol{u}/dt$", then $\boldsymbol{u}$ is not a control variable; such variables are called state variables. These are denoted by $\boldsymbol{x} \in \mathbb{R}^{N_x}$. The

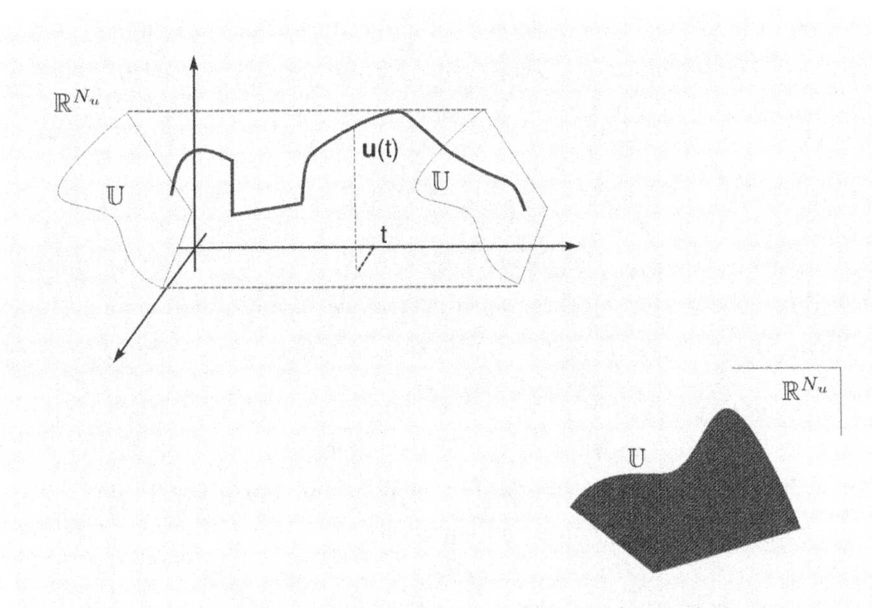

Figure 1.26: A schematic for the control variable $\boldsymbol{u}$, the control space $\mathbb{U}$, and the control trajectory $t \mapsto \boldsymbol{u}$. At each point in time, $\boldsymbol{u}(t)$ must be in $\mathbb{U} \subset \mathbb{R}^{N_u}$.

state vector is rate limited by a ***state velocity***, $\dot{\boldsymbol{x}}$. The limitations on the state velocity are articulated through a *standardized* differential equation called ***dynamics***:

$$\dot{\boldsymbol{x}} = \boldsymbol{f}(\boldsymbol{x}, \boldsymbol{u}, t) \tag{1.33}$$

Thus the ***state trajectory*** $t \mapsto \boldsymbol{x}$ is differentiable and smoother than the control trajectory. It begins at a given point $\boldsymbol{x}^0 \in \mathbb{R}^{N_x}$ at a given clock time $t^0 \in \mathbb{R}$. The pair

$$(\boldsymbol{x}^0, t^0) \in \mathbb{R}^{N_x} \times \mathbb{R}$$

is called the ***initial event***; all feasible trajectories $\boldsymbol{x}(\cdot)$ must satisfy the initial event condition:

$$\boldsymbol{x}(t_0) = \boldsymbol{x}^0 \tag{1.34a}$$

$$t_0 = t^0 \tag{1.34b}$$

In Eq. (1.34), the quantities on the left-hand side are variables while those on the right-hand side are symbols for numbers of unspecified values. Often, it becomes convenient to abbreviate $\boldsymbol{x}(t_0)$ to $\boldsymbol{x}_0$ so that we may use $\boldsymbol{x}_0$ as a variable in $\mathbb{R}^{N_x}$. In this context, we may write Eq. (1.34a) as

$$\boldsymbol{x}_0 = \boldsymbol{x}^0$$

Any confusion arising in the differences between $\boldsymbol{x}_0$ and $\boldsymbol{x}^0$ can be quickly dispelled by substituting $\boldsymbol{x}(t_0)$ for $\boldsymbol{x}_0$. Using similar notation for the final-time conditions, the pair

$$(\boldsymbol{x}_f, t_f)$$

denotes *variables* for the ***final event*** or the target set. As explained in Section 1.3, the target event in a standard problem formulation is specified in terms of a *standardized* endpoint constraint equation:

$$\boldsymbol{e}(\boldsymbol{x}_f, t_f) = \boldsymbol{0} \tag{1.35}$$

This equation defines a target set, $\mathbb{E}$, that can be visualized as a hypersurface in $\mathbb{R}^{N_x} \times \mathbb{R}$. See Fig. 1.27.

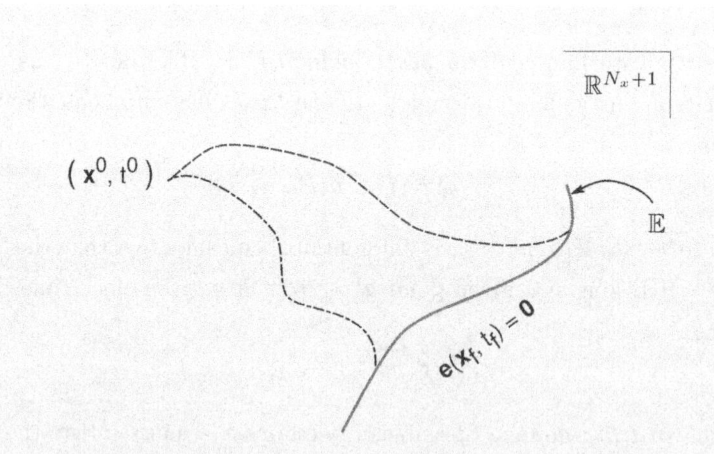

Figure 1.27: Schematic for the standard optimal control problem B.

With these preliminaries, we can now define the standard optimal control problem (B). The problem is to determine the state-control function pair $\{\boldsymbol{x}(\cdot), \boldsymbol{u}(\cdot)\}$ and possibly the final time t_f that steers the dynamical system,

Eq. (1.33), from its initial event, Eq. (1.34), to the target event set $\mathbb{E}$, while minimizing the ***standard cost functional***

$$J[\boldsymbol{x}(\cdot), \boldsymbol{u}(\cdot), t_f] := E(\boldsymbol{x}_f, t_f) + \int_{t_0}^{t_f} F(\boldsymbol{x}(t), \boldsymbol{u}(t), t)\, dt \tag{1.36}$$

The function E in Eq.(1.36) is called the ***endpoint cost*** or the *Mayer cost*, and the function F is called the ***running cost*** or the *Lagrange cost*. If $E = \Theta$, the zero function, then the cost functional is said to be in the Lagrange form, and if $F = \Theta$, the cost functional is said to be in the Mayer form. The standard form is also known as the *Bolza cost functional*.

Collecting all the relevant equations, the standard problem is summarized and structured as follows:

$$\boldsymbol{x} \in \mathbb{X} = \mathbb{R}^{N_x} \quad \boldsymbol{u} \in \mathbb{U} \subseteq \mathbb{R}^{N_u} \quad \text{(preamble)}$$

$$\overbrace{(B)}^{\text{problem}} \begin{cases} \text{Minimize} & J[\boldsymbol{x}(\cdot), \boldsymbol{u}(\cdot), t_f] = E(\boldsymbol{x}(t_f), t_f) \\ & \qquad + \displaystyle\int_{t_0}^{t_f} F(\boldsymbol{x}(t), \boldsymbol{u}(t), t)\, dt \quad \text{(cost)} \\ \text{Subject to} & \dot{\boldsymbol{x}}(t) = \boldsymbol{f}(\boldsymbol{x}(t), \boldsymbol{u}(t), t) \quad \text{(dynamics)} \\ & \boldsymbol{x}(t_0) = \boldsymbol{x}^0 \\ & t_0 = t^0 \\ & \boldsymbol{e}(\boldsymbol{x}_f, t_f) = \boldsymbol{0} \quad \text{(endpoints)} \end{cases}$$

We use uppercase for cost (or cost-like) functions and lowercase for the corresponding constraints. Thus, E and $\boldsymbol{e}$ go together, and F and $\boldsymbol{f}$ go together. That is, the domains of these function pairs are the same while their ranges (co-domains) are different:

$$E: \mathbb{R}^{N_x} \times \mathbb{R} \longrightarrow \mathbb{R} \qquad F: \mathbb{R}^{N_x} \times \mathbb{R}^{N_u} \times \mathbb{R} \longrightarrow \mathbb{R}$$

$$\boldsymbol{e}: \mathbb{R}^{N_x} \times \mathbb{R} \longrightarrow \mathbb{R}^{N_e} \qquad \boldsymbol{f}: \mathbb{R}^{N_x} \times \mathbb{R}^{N_u} \times \mathbb{R} \longrightarrow \mathbb{R}^{N_x}$$

For the application of Pontryagin's Principle, these functions need to be differentiable only with respect to their first argument (i.e., $\boldsymbol{x}_f$ or $\boldsymbol{x}$). In a computational environment, a majority of the software (including DIDO) requires differentiability of the functions with respect to all their arguments. These functions, in addition to the other quantities, like $\mathbb{U}$, $\boldsymbol{x}^0$ and so on, are called the ***problem data***. Thus, the term *problem data* refers not just to the numerical values associated with the problem but also to the given functions.

1.4.2 A Practical Problem Formulation

The standard problem formulation forms the basis (of this book) to understand and articulate the main elements of Pontryagin's Principle with the least amount of clutter. Pontryagin's Principle for this problem (B) is explained in detail in Chapter 2, and several example problems are worked out in Chapter 3. To better understand the standard problem for practical applications, we reformulate it under a broader umbrella. This formulation (called Problem P) allows a problem designer to rapidly generate a pseudocode, "plug" it in DIDO (or its clone) and "play" with the problem by solving the "problem of problems."

A schematic for Problem P is shown in Fig. 1.28. By comparing this figure with 1.27 it is readily apparent that Problems P and B are quite similar with $B \subset P$.

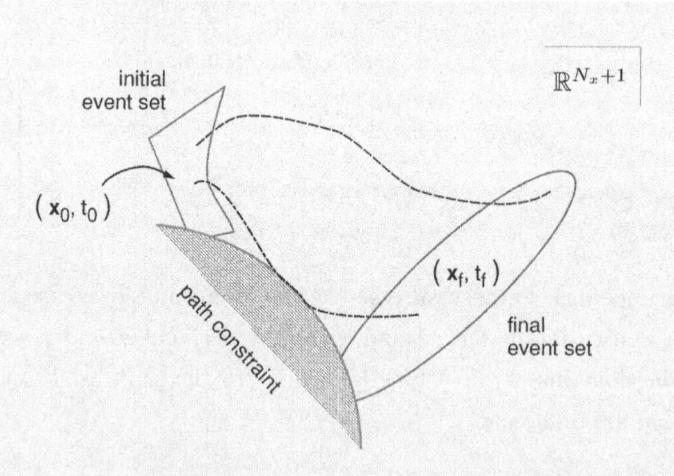

Figure 1.28: A schematic for Problem P. Note that the mathematical formulation is more general than what is apparent from this figure.

We define Problem P in the following format (whose syntax is nearly the same as in DIDO):

$$\left.\begin{array}{ll} \mathbb{X}_{search} = \mathbb{R}^{N_x} & \mathbb{U}_{search} = \mathbb{R}^{N_u} \\ \boldsymbol{x} = (x_1, \ldots, x_{N_x}) & \boldsymbol{u} = (u_1, \ldots, u_{N_u}) \end{array}\right\} \text{(preamble)}$$

$$\overbrace{(P)}^{\text{problem}} \left\{ \begin{array}{ll} \text{Minimize } J[\boldsymbol{x}(\cdot), \boldsymbol{u}(\cdot), t_0, t_f] := & \\ \qquad E(\boldsymbol{x}_0, \boldsymbol{x}_f, t_0, t_f) + \displaystyle\int_{t_0}^{t_f} F(\boldsymbol{x}(t), \boldsymbol{u}(t), t) & \} \text{ (cost)} \\ \text{Subject to} \qquad \dot{\boldsymbol{x}} = \boldsymbol{f}(\boldsymbol{x}(t), \boldsymbol{u}(t), t) & \} \text{ (dynamics)} \\ \qquad \boldsymbol{e}^L \leq \boldsymbol{e}(\boldsymbol{x}_0, \boldsymbol{x}_f, t_0, t_f) \leq \boldsymbol{e}^U & \} \text{ (events)} \\ \qquad \boldsymbol{h}^L \leq \boldsymbol{h}(\boldsymbol{x}(t), \boldsymbol{u}(t), t) \leq \boldsymbol{h}^U & \} \text{ (path)} \end{array} \right.$$

Notation Alert:

1. The symbols $\mathbb{X}_{search}$ and $\mathbb{U}_{search}$ in the preamble stand for "initial" search spaces for the state and control variables, respectively. The operational spaces are smaller than $\mathbb{X}_{search}$ and $\mathbb{U}_{search}$ due to the path constraints.
2. We use the symbol $\boldsymbol{h}$ for the path function because the corresponding inequality forms a constraint for an H-function called the Hamiltonian in Chapter 2. Thus, the pair $(H, \boldsymbol{h})$ go together in much the same way as $(F, \boldsymbol{f})$ and $(E, \boldsymbol{e})$ go together.

Computational Tip: In DIDO, the search spaces are bounded by box constraints:

$$\mathbb{X}_{search} := \left\{ \boldsymbol{x} \in \mathbb{R}^{N_x} : \boldsymbol{x}^L \leq \boldsymbol{x} \leq \boldsymbol{x}^U \right\}$$

$$\mathbb{U}_{search} := \left\{ \boldsymbol{u} \in \mathbb{R}^{N_x} : \boldsymbol{u}^L \leq \boldsymbol{u} \leq \boldsymbol{u}^U \right\}$$

Theoretically, the DIDO problem is equivalent to Problem P if the lower and upper bounds are selected to be $-\infty$ and ∞, respectively; however, in a computational environment it is generally inadvisable to have a humungous search space. It is assumed that the problem designer knows something about the problem.

It is helpful to note the similarities and differences between Problems B and P so that the power and limitations of Problem B can be appropriately gauged.

- As implied by the arguments of the cost functional J, the decision variables in Problem P include the initial clock time t_0. In addition, t_0 is constrained through the events function $\boldsymbol{e}$.

- The events function in Problem P combines the initial and final events. The fixed initial conditions of Problem B can be articulated under this construct by defining two functions, e_1 and $\boldsymbol{e}_2$:

$$\begin{aligned} e_1(\boldsymbol{x}_0, \boldsymbol{x}_f, t_0, t_f) &:= t_0 \\ \boldsymbol{e}_2(\boldsymbol{x}_0, \boldsymbol{x}_f, t_0, t_f) &:= \boldsymbol{x}_0 \end{aligned}$$

  that are bounded from below and above by the data:

$$\begin{aligned} e_1^L &:= t^0 \quad & e_1^U &:= t^0 \\ \boldsymbol{e}_2^L &:= \boldsymbol{x}^0 \quad & \boldsymbol{e}_2^U &:= \boldsymbol{x}^0 \end{aligned}$$

- Because the initial event is a decision variable in Problem P, it is incorporated as part of the argument of the endpoint cost function E.

- The path constraints in Problem P jointly limit the state and control spaces by the inequality $\boldsymbol{h}^L \leq \boldsymbol{h}(\boldsymbol{x}, \boldsymbol{u}, t) \leq \boldsymbol{h}^U$. Pure state ($\Rightarrow \boldsymbol{h}$ does not depend upon $\boldsymbol{u}$) and pure control ($\Rightarrow \boldsymbol{h}$ does not depend upon $\boldsymbol{x}$) constraints are automatically included in this formulation. The pure control constraints are included in B in the form of $\mathbb{U}$. From Problem Brac:3 on page 28, it is clear that ***mixed state-control path constraints*** can easily appear in a problem formulation simply because of the *choice* of a coordinate system.

1.4.3 Avoiding Common Errors # 2

Throughout this chapter, we have noted, in various ways, that the *totality of unknowns* in Problem B is given by the triple

$$(\boldsymbol{x}(\cdot), \boldsymbol{u}(\cdot), t_f)$$

Consequently, the functional J is always written as

$$J[\boldsymbol{x}(\cdot), \boldsymbol{u}(\cdot), t_f]$$

That is, the arguments of J are <u>*all*</u> of the decision variables. We emphasize this point further by noting that it is erroneous to oversimplify the collection of unknowns to just the control trajectory and possibly the final time by implying that once the pair $(\boldsymbol{u}(\cdot), t_f)$ is known, the state trajectory is determined; hence, $\boldsymbol{x}(\cdot)$ is not a "real unknown." In other words, the following argument is specious: Pick any control trajectory $t \mapsto \boldsymbol{u} \in \mathbb{U}$ and substitute it in the system dynamics, $\dot{\boldsymbol{x}} = \boldsymbol{f}(\boldsymbol{x}, \boldsymbol{u}, t)$ to generate the time-varying ordinary differential equation

$$\dot{\boldsymbol{x}} = \boldsymbol{g}(\boldsymbol{x}, t) \equiv \boldsymbol{f}(\boldsymbol{x}, \boldsymbol{u}(t), t)$$

Pick any time t_f, and using the given values of $\boldsymbol{x}(t_0) = \boldsymbol{x}^0$ at $t_0 = t^0$, propagate the differential equation up to $t = t_f$. This *algorithm* generates a dynamically feasible state trajectory $t \mapsto \boldsymbol{x}$; ergo, the argument goes, the real unknown is just the pair $(\boldsymbol{u}(\cdot), t_f)$ and not the triple $(\boldsymbol{x}(\cdot), \boldsymbol{u}(\cdot), t_f)$. In fact, one can make a seemingly reasonable point that it is just high-brow speak to repeatedly insist that the unknown includes the state trajectory, and that the functional J may be written quite simply as

$$J[\boldsymbol{u}(\cdot), t_f]$$

The reason the argument of the preceding paragraph is false can be best illustrated by considering the following simple example problem:

$$\left\{ \begin{array}{lll} \mathsf{Minimize} & f(x, y) & = x + y^2 \\ \mathsf{Subject\ to} & x + 2y & = 0 \end{array} \right. \tag{1.37}$$

where x and y are real numbers. Now, for any given y, we can always find a unique x by solving the constraint equation $x + 2y = 0$; in fact,

$$x = -2y$$

Hence, suppose we had argued that the "real unknown" is not the pair (x, y); rather, it is just y. Then, according to the arguments of the preceding paragraph, we can "simplify" Problem (1.37) and rewrite it as

$$\begin{cases} \textsf{Minimize} & f(y) = x + y^2 \\ \textsf{Subject to} & x + 2y = 0 \end{cases} \tag{1.38}$$

The statement of Problem (1.38) borders on the absurd. For starters, the notation

$$f(y) = x + y^2$$

implies x is a constant. Then, the constraint $x + 2y = 0$ implies x is a variable that depends on y, which contradicts the implication that x is a constant. The contradiction, one may argue, can be "fixed" by saying that x is a function of y, so that $f(y) = x + y^2$ really implies

$$f(y) = x(y) + y^2$$

Note that $x(y)$ is now not a real variable, but a function; this contradicts the original statement that x and y were real numbers. Furthermore, the constraint $x + 2y = 0$ must now be written as

$$x(y) + 2y = 0$$

This "fixing" is not entirely accurate because the constraint equation is no longer a constraint; rather, it is a *defining equation* for a function given by

$$x(y) := -2y$$

This implies that Problem (1.38) is not a constrained optimization problem as advertized; rather it is an unconstrained one that should be written as

$$\textsf{Minimize} \quad g(y) = -2y + y^2 \tag{1.39}$$

That is, Problem (1.37), which is an optimization problem of two variables, x and y, can indeed be written as an optimization problem of one variable; however, it must be written in the form of Problem (1.39). Thus, the statement and notation of Problem (1.38) are incorrect. Note also that the notation of Problem (1.39) allows us to reserve the original symbol f for the mapping:

$\mathbb{R} \times \mathbb{R} \to \mathbb{R}$ while introducing a new symbol g for a new map $\mathbb{R} \to \mathbb{R}$.

The discussion of the preceding paragraphs highlights why J in Problem B is indeed a functional that depends on the triple $(\boldsymbol{x}(\cdot), \boldsymbol{u}(\cdot), t_f)$ and not on just the pair $(\boldsymbol{u}(\cdot), t_f)$. If $\boldsymbol{x}(\cdot)$ is treated as "solved," then, it needs to be explicitly eliminated from the problem formulation as exemplified in Problem (1.39).

Some authors and textbooks avoid using the functional dependence of J altogether and write

$$J = E(\boldsymbol{x}_f, t_f) + \int_{t_0}^{t_f} F(\boldsymbol{x}(t), \boldsymbol{u}(t), t)\, dt \tag{1.40}$$

Technically, there is nothing wrong with this, but then, J is not a functional — it is an output. That is, writing J without its arguments implies it is a real number, not a functional, in much the same way as when we write $y = f(x)$.

Beyond creating totally avoidable confusion, the problem with such notation is that it lends itself to at least two other problems:

1. Writing J with no arguments specified is the equivalent of dropping (x, y) in Problem (1.37) leading to the following expression

   $$f = x + y^2$$

   In other words, if it is obvious that writing $f = x + y^2$ is confusing, then so is writing J without its arguments as in Eq. (1.40).

2. In many *practical applications*, we are not simply interested in an evaluation of the cost functional J over the optimal solution $(\boldsymbol{x}^*(\cdot), \boldsymbol{u}^*(\cdot), t_f^*)$, but also over alternative trajectories. Let $(\boldsymbol{x}^i(\cdot), \boldsymbol{u}^i(\cdot), t_f^i)$ be a collection of decision variables indexed by $i = 1, 2, \ldots$. Despite that it may seem elaborate, it is simpler and consistent to use the notation

   $$J[\boldsymbol{x}^*(\cdot), \boldsymbol{u}^*(\cdot), t_f^*]$$

to represent the optimal cost, and use symbols

$$J[\boldsymbol{x}^1(\cdot), \boldsymbol{u}^1(\cdot), t_f^1]$$
$$J[\boldsymbol{x}^2(\cdot), \boldsymbol{u}^2(\cdot), t_f^2]$$
$$J[\boldsymbol{x}^3(\cdot), \boldsymbol{u}^3(\cdot), t_f^3]$$
$$\vdots$$

to denote the evaluation of the *same cost functional* over alternative trajectories (indexed by $i = 1, 2, \ldots$).

In other words, the notation and writing of a cost functional are essentially no different that the notation used for writing an ordinary function in the sense of a clear usage of arguments!

The simple practice of not abusing notation is particularly important for beginners and even more so in solving real-world optimal control problems where human lives and/or large resources may be at stake. Both of these situations were at stake when ***pseudospectral optimal control theory**** debuted its historic flight in 2006[49]. Multiple solutions were found† for Bedrossian's celebrated "zero propellant maneuver." In this scenario, we had

$$J[\boldsymbol{x}^1(\cdot), \boldsymbol{u}^1(\cdot), t_f^1] = J[\boldsymbol{x}^2(\cdot), \boldsymbol{u}^2(\cdot), t_f^2] = J[\boldsymbol{x}^3(\cdot), \boldsymbol{u}^3(\cdot), t_f^3] = \cdots$$

The multiplicity of solutions were presented to NASA flight operators so that they could choose one based on the constraints of what they "liked better," where "like" was modeled through an iterative process that took into account many other "soft" system constraints.

In the practice of optimal control, we do not solve "the" optimal control problem; rather, multiple cost functionals for the same system are used to explore the trade space of implementation issues. This is part of the analysis process of solving the problem of problems. To illustrate this practice and, consequently, the need to use clear and consistent notation, let J_1 and J_2 be two different

*This concept encapsulates a functional analysis approach to optimal control that is separate from Pontryagin's Principle but connected to it through the covector mapping principle. See Section 2.9.2 on page 157 for an introduction to this concept and [87] for additional details.

†All solutions implemented onboard the International Space Station were computed using DIDO. Because DIDO is based on pseudospectral theory, the debut of pseudospectral optimal control is synonymous with the first flight implementation of DIDO.

cost functionals designed to explore the trade space for a given dynamical system. For example, J_1 may be fuel consumption and J_2 the time of travel. Let the triples $(\boldsymbol{x}_1^*(\cdot), \boldsymbol{u}_1^*(\cdot), t_{f1}^*)$ and $(\boldsymbol{x}_2^*(\cdot), \boldsymbol{u}_2^*(\cdot), t_{f2}^*)$ be the optimal solutions for minimizing J_1 and J_2, respectively. Then, by definition, this means that

$$J_1[\boldsymbol{x}_1^*(\cdot), \boldsymbol{u}_1^*(\cdot), t_{f1}^*] \leq J_1[\boldsymbol{x}_2^*(\cdot), \boldsymbol{u}_2^*(\cdot), t_{f2}^*]$$

and

$$J_2[\boldsymbol{x}_2^*(\cdot), \boldsymbol{u}_2^*(\cdot), t_{f2}^*] \leq J_2[\boldsymbol{x}_1^*(\cdot), \boldsymbol{u}_1^*(\cdot), t_{f1}^*]$$

See Fig. 1.29.

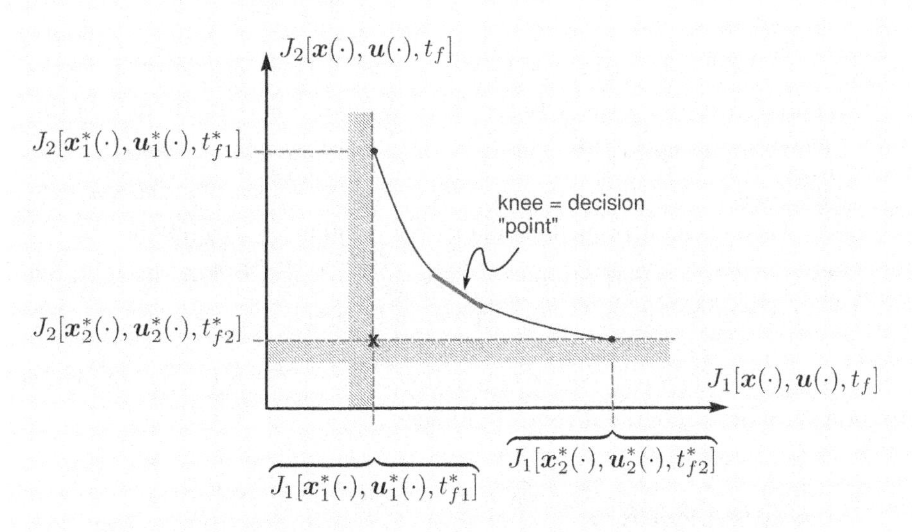

Figure 1.29: A typical trade space of cost functions. The knee of the curve represents typical decision points. The point marked "x" is usually an unobtainium.

Such trade-space analysis is routine in practical applications. For instance, in the design of the "optimal propellant maneuver" that was operationalized on the International Space Station in 2012[7], both J_1 and J_2 were fuel consumption. The difference was that in Problem "1," t_f was an optimization variable but in Problem "2," t_f was constrained to be less than one orbital period of the station (approximately 90 minutes). The time constraint for the second problem arose from dollar cost considerations as follows[‡]:

- The solution to Problem # 1 generated $J_1[\boldsymbol{x}_1^*(\cdot), \boldsymbol{u}_1^*(\cdot), t_{f1}^*] = 0$. Hence, the triple $(\boldsymbol{x}_1^*(\cdot), \boldsymbol{u}_1^*(\cdot), t_{f1}^*)$ is a zero propellant maneuver.
- The "price" of the zero propellant maneuver in terms of maneuver time (t_{f1}^*) was greater than $t_{2\pi}$, the orbital period.
- When a maneuver exceeds $t_{2\pi}$, NASA safety rules require an additional number of human-intensive system-level checks. The dollar costs of these system checks offset the dollar cost savings obtained by using no propellant. Thus, a cheaper zero-propellant maneuver would be one where $t_f^* < t_{2\pi}$.
- A zero-propellant maneuver may not exist if $t_f^* < t_{2\pi}$. Recall Section 1.1.3 on page 15 on the notion of existence of a solution. In view of this, Bedrossian designed Problem "2" with t_f constrained to be less than $t_{2\pi}$. This implies that $J_2[\boldsymbol{x}(\cdot), \boldsymbol{u}(\cdot), t_f] \geq 0$ with no assurance of $J_2[\boldsymbol{x}(\cdot), \boldsymbol{u}(\cdot), t_f] = 0$. The "optimal propellant maneuver" minimizes J_2 and the solution $(\boldsymbol{x}_2^*(\cdot), \boldsymbol{u}_2^*(\cdot), t_{f2}^*)$ may produce $J_2[\boldsymbol{x}_2^*(\cdot), \boldsymbol{u}_2^*(\cdot), t_{f2}^*] > 0$, but the total dollar cost may be (it was!) less than when $t_f > t_{2\pi}$.

It is apparent that simple "bookkeeping" practices are essential in formulating and solving practical optimal control problems.

Study Problem 1.12

In Sections 1.1 and 1.2, the Brachistochrone problems were assigned numerical labels for pedagogical reasons. Using the new perspective offered by the discussions in this subsection, produce new name tags and variable names for the various formulations of the Brachistochrone problem. For example, Brac:1 *may be called* Brac:main *while* Brac:2 *may be called* BracWithPath.

[‡]These insights and analysis are due to N. Bedrossian, the architect of both the zero-propellant- and the optimal-propellant maneuvers.

Computational Tip: In using DIDO, the more common problem is not in the production of solutions; instead, it is in the bookkeeping of various solutions generated from the problem of problems. Problem-solving is not the biggest challenge: Problem formulation is. Using problem names, variable names and file names that are similar to the notation used in this book will be extremely helpful in parsing and analyzing the "big data" generated by DIDO.

Geek Speak: Let $\mathcal{X}$ and $\mathcal{U}$ be the function spaces for $\boldsymbol{x}(\cdot)$ and $\boldsymbol{u}(\cdot)$, respectively. The cost functional J is defined by

$$J : \mathcal{X} \times \mathcal{U} \times \mathbb{R} \longrightarrow \mathbb{R}$$

For theoretical analysis, we take $\mathcal{X} = W^{1,1}$ and $\mathcal{U} = L^{\infty}$. In consideration of computational (and hence, practical) aspects of optimal control, we require $\mathcal{X} = W^{1,\infty}$ and $\mathcal{U} = L^{\infty}$; therefore, at the "intersection of theory and practice," we have

$$J : W^{1,\infty} \times L^{\infty} \times \mathbb{R} \longrightarrow \mathbb{R}$$

To be more precise, practical optimal control requires that

$$\boldsymbol{x}(\cdot) \in \mathcal{X} := W^{1,\infty}([t_0, t_f], \mathbb{X}) \qquad \boldsymbol{u}(\cdot) \in \mathcal{U} := L^{\infty}([t_0, t_f], \mathbb{U})$$

where $\mathbb{X} \subseteq \mathbb{R}^{N_x}$ is the state space and $\mathbb{U} \subseteq \mathbb{R}^{N_u}$ is the control space. Clearly, the problem is always stated as finding state-control function pairs $\{\boldsymbol{x}(\cdot), \boldsymbol{u}(\cdot)\}$ and not just $\boldsymbol{x}(\cdot)$ or $\boldsymbol{u}(\cdot)$. Note also the connection and differences between the symbols $\mathcal{X}$, $\mathbb{X}$ and $\mathcal{U}$, $\mathbb{U}$.

1.4.4 Transformational Tips and Tricks

Many practical optimal control problems can be posed under the constructs of Problem B. Oftentimes, a problem is more easily posed in a format that may not readily fit within these constructs. In such cases, it is useful to understand the various transformational "tricks" to determine if a given problem can be transformed to the format of Problem B.

1. Maximization Problems

The standard problem is a minimization problem. Often, we are more interested in maximizing a cost functional as apparent from $Brac : D_1$. In such cases, we can transform a maximization problem to a minimization problem by simply flipping the sign of the cost functional; that is,

$$\text{Maximize } J[\boldsymbol{x}(\cdot), \boldsymbol{u}(\cdot), t_f] \Rightarrow \text{Minimize } - J[\boldsymbol{x}(\cdot), \boldsymbol{u}(\cdot), t_f]$$

This result follows directly from the definition of a maximum and a minimum as follows:

$$\begin{aligned}&\text{Maximize } J[\boldsymbol{x}(\cdot), \boldsymbol{u}(\cdot), t_f] \Rightarrow \\ &\text{Find } \left\{\boldsymbol{x}^*(\cdot), \boldsymbol{u}^*(\cdot), t_f^*\right\} \text{ such that } J[\boldsymbol{x}^*(\cdot), \boldsymbol{u}^*(\cdot), t_f^*] \geq J[\boldsymbol{x}(\cdot), \boldsymbol{u}(\cdot), t_f]\end{aligned}$$

Multiplying the above inequality by -1 on both sides, we get

$$-J[\boldsymbol{x}^*(\cdot), \boldsymbol{u}^*(\cdot), t_f^*] \leq -J[\boldsymbol{x}(\cdot), \boldsymbol{u}(\cdot), t_f]$$

which proves that $-J[\boldsymbol{x}^*(\cdot), \boldsymbol{u}^*(\cdot), t_f^*]$ is the minimum.

2. L^1-Optimal Control

An L^1-optimal control problem has a cost functional given by

$$J_1[\boldsymbol{x}(\cdot), \boldsymbol{u}(\cdot), t_f] = \int_{t_0}^{t_f} |F_1(\boldsymbol{x}(t), \boldsymbol{u}(t), t)| \, dt \tag{1.41}$$

where $|\cdot|$ denotes the absolute value. Such problems arise in many applications that range from data-fitting to minimizing fuel consumption. The function

$$F(\boldsymbol{x}, \boldsymbol{u}, t) := |F_1(\boldsymbol{x}, \boldsymbol{u}, t)|$$

is non-differentiable with respect to its arguments, and more importantly, it is non-differentiable with respect to $\boldsymbol{x}$. The application of (the "smooth version") of Pontryagin's Principle requires differentiability at least with respect to $\boldsymbol{x}$, and many computational tools perform better if the data functions are differentiable with respect to all their arguments.

If F_1 is smooth, then the L^1-optimal control problem may be transformed to a smooth problem as follows: Let

$$z(t) := F_1(\boldsymbol{x}(t), \boldsymbol{u}(t), t)$$

Then we can split the function $t \mapsto z$ into a positive piece and a negative piece as shown in Fig. 1.30. Hence, we can write:

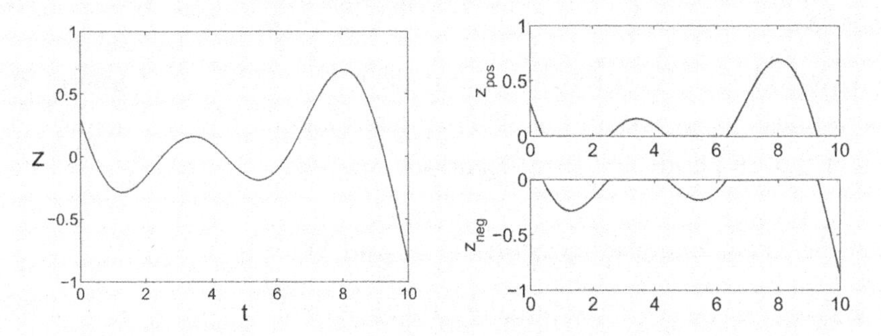

Figure 1.30: The function $t \mapsto z$ can be split into the sum of $t \mapsto z_{pos}$ and $t \mapsto z_{neg}$ as shown above; hence, we can write $|z(t)| = z_a(t) + z_b(t)$ as shown below.

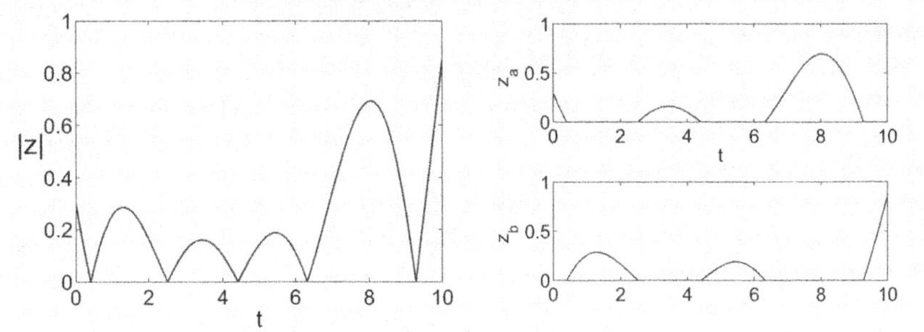

$$\begin{aligned}
z(t) &= z_a(t) - z_b(t) \\
|z(t)| &= z_a(t) + z_b(t) \\
z_a(t) &\geq 0 \\
z_b(t) &\geq 0
\end{aligned} \tag{1.42}$$

In applying these concepts to Eq. (1.41) we simply replace all occurrences of

$|F_1(\boldsymbol{x}, \boldsymbol{u}, t)|$ and $F_1(\boldsymbol{x}, \boldsymbol{u}, t)$ with $\big(F_a(\boldsymbol{x}, \boldsymbol{u}, t) + F_b(\boldsymbol{x}, \boldsymbol{u}, t)\big)$ and $\big(F_a(\boldsymbol{x}, \boldsymbol{u}, t) - F_b(\boldsymbol{x}, \boldsymbol{u}, t)\big)$, respectively. Thus, the L^1-optimal control problem transforms to

$$\left\{\begin{array}{lll} \textsf{Minimize} & J_1[\boldsymbol{x}(\cdot), \boldsymbol{u}(\cdot), t_f] & := \displaystyle\int_{t_0}^{t_f} F_a(\boldsymbol{x}(t), \boldsymbol{u}(t), t)\, dt \\ & & \quad + \displaystyle\int_{t_0}^{t_f} F_b(\boldsymbol{x}(t), \boldsymbol{u}(t), t)\, dt \\ \textsf{Subject to} & F_a(\boldsymbol{x}(t), \boldsymbol{u}(t), t) & \geq 0 \\ & F_b(\boldsymbol{x}(t), \boldsymbol{u}(t), t) & \geq 0 \end{array}\right. \tag{1.43}$$

Incidentally, as a general rule of thumb, it is possible to transform non-smooth functions to smooth ones through the addition of new functions; the new functions typically satisfy inequalities.

3. Other Transformations: Some Do's and Dont's

It is possible to transform the Standard Problem to other "equivalent" formats; however, there are some caveats.

Cost Functional Transformations: The standard cost functional can be transformed to an endpoint form by introducing an additional state variable, r, with dynamics given by

$$\dot{r} = F(\boldsymbol{x}, \boldsymbol{u}, t) \tag{1.44}$$

and an initial condition

$$r(t_0) = 0$$

It therefore follows that

$$\begin{aligned} J[\boldsymbol{x}(\cdot), \boldsymbol{u}(\cdot), t_f] &:= E(\boldsymbol{x}_f, t_f) + \int_{t_0}^{t_f} F(\boldsymbol{x}(t), \boldsymbol{u}(t), t)\, dt \\ &= E(\boldsymbol{x}_f, t_f) + \int_{t_0}^{t_f} \dot{r}(t)\, dt \\ &= E(\boldsymbol{x}_f, t_f) + r(t_f) \end{aligned}$$

Hence, the standard form of the cost functional is equivalent to an endpoint form via the introduction of the additional state variable r that satisfies the dynamics given by Eq. (1.44). The new state vector is of dimension $(N_x + 1)$:

$$\widetilde{\boldsymbol{x}} := \begin{bmatrix} \boldsymbol{x} \\ r \end{bmatrix} \in \mathbb{R}^{N_x+1}$$

Although this equivalence is mathematically sound, it is not necessarily computationally equivalent. This is because in the pre-transformation situation, the computational burden was on just the integration of $F[@t]$ — an easier task — while in the post-transformation case, the computational burden is on satisfying a differential equation $\dot{r}(t) = F(\boldsymbol{x}(t), \boldsymbol{u}(t), t)\ \forall t \in [t_0, t_f]$ — a more difficult task. These computational differences may eventually turn out to be minor depending on the accuracy of the integration, the accuracy of "differentiation" and machine precision; however, they are nonetheless different.

In the same vein, we can do a reverse operation: That is, an endpoint cost may be absorbed into a running cost by differentiating E:

$$\begin{aligned}
J[\boldsymbol{x}(\cdot), \boldsymbol{u}(\cdot), t_f] &:= E(\boldsymbol{x}_f, t_f) + \int_{t_0}^{t_f} F(\boldsymbol{x}(t), \boldsymbol{u}(t), t)\, dt \\
&= \int_{t_0}^{t_f} \left[\frac{d}{dt} E(\boldsymbol{x}(t), t) + F(\boldsymbol{x}(t), \boldsymbol{u}(t), t) \right] dt \\
&= \int_{t_0}^{t_f} \Big[\partial_{\boldsymbol{x}} E(\boldsymbol{x}(t), t) \cdot \dot{\boldsymbol{x}} + \partial_t E(\boldsymbol{x}(t), t) + F(\boldsymbol{x}(t), \boldsymbol{u}(t), t) \Big] dt \\
&= \int_{t_0}^{t_f} \Big[\partial_{\boldsymbol{x}} E(\boldsymbol{x}(t), t) \cdot \boldsymbol{f}(\boldsymbol{x}(t), \boldsymbol{u}(t), t) + \partial_t E(\boldsymbol{x}(t), t) \\
&\qquad\qquad + F(\boldsymbol{x}(t), \boldsymbol{u}(t), t) \Big] dt
\end{aligned}$$

Thus, the new running cost function, F_{new}, is given by

$$F_{new}(\boldsymbol{x}, \boldsymbol{u}, t) := \partial_{\boldsymbol{x}} E(\boldsymbol{x}, t) \cdot \boldsymbol{f}(\boldsymbol{x}, \boldsymbol{u}, t) + \partial_t E(\boldsymbol{x}, t) + F(\boldsymbol{x}, \boldsymbol{u}, t)$$

The caveat with this transformation is that E must be differentiable with respect to its arguments. In addition, Pontryagin's Principle requires that the running cost functional be differentiable with respect to $\boldsymbol{x}$; this implies that E must be twice-differentiable with respect to $\boldsymbol{x}$. In additional to all these "warnings," in the pre-transformation case, the computational burden is just on function evaluation (of E) while in the post-transformation case, many more additional operations (dot product, additional function evaluation and integration) are required.

Time Transformation: The time variable, t, in an optimal control problem, which is just a proxy for an independent variable, can be eliminated as a "third" argument, by introducing a "fake" state variable, $s \equiv t$, with dynamics given by

$$\dot{s} = 1 \tag{1.45}$$

Then, a new state vector, $\overline{\boldsymbol{x}}$, defined by

$$\overline{\boldsymbol{x}} := \begin{bmatrix} \boldsymbol{x} \\ s \end{bmatrix} \quad \in \mathbb{R}^{N_x+1}$$

transforms the standard problem to an "autonomous" problem (i.e., one that does not depend on time explicitly).

The mathematical caveat with this "trick" is that in the pre-transformation case, the problem data functions are merely required to be measurable with respect to t, whereas in the post-transformation situation, appropriate additional differentiability conditions are required. In addition, in the post-transformation case, a new computational burden of satisfying the differential equation $\dot{s}(t) = 1 \; \forall t \in [t_0, t_f]$ is imposed.

Study Problem 1.13

Rewrite the definition and meaning of Problem B discussed in Section 1.4.1 on page 42 by defining a new state variable

$$\underline{\boldsymbol{x}} := \begin{bmatrix} \boldsymbol{x} \\ r \\ s \end{bmatrix} \quad \in \mathbb{R}^{N_x+2}$$

where r and s satisfy the dynamics given by equations 1.44 and 1.45 respectively.

4. Infinite Horizon Optimal Control

A number of problems in economics, control theory, and other branches of engineering and mathematics can be framed as an infinite horizon optimal control problem; that is, an optimal control problem with t_f replaced by ∞. A standard infinite horizon problem can be posed as

$$\boldsymbol{x} \in \mathbb{X} = \mathbb{R}^{N_x}, \quad \boldsymbol{u} \in \mathbb{U} \subseteq \mathbb{R}^{N_u}$$

$$({}^{\infty}B) \begin{cases} \text{Minimize} & J[\boldsymbol{x}(\cdot), \boldsymbol{u}(\cdot), t_0] := \displaystyle\int_{t_0}^{\infty} F(\boldsymbol{x}(t), \boldsymbol{u}(t))\, dt \\ \text{Subject to} & \dot{\boldsymbol{x}}(t) = \boldsymbol{f}(\boldsymbol{x}(t), \boldsymbol{u}(t)) \\ & \boldsymbol{x}_0 = \boldsymbol{x}^0 \\ & t_0 = t^0 \end{cases}$$

Because weird things can happen "at infinity," the simple notion of using finite horizons and setting $t_f \to \infty$ has unintended consequences. Nonetheless, a class of problems can be analyzed by transforming the infinite horizon, $[t_0, \infty)$, to a finite horizon, $[\tau_a, \tau_b)$, by using a nonlinear time transformation. See Fig. 1.31. For example, the transform

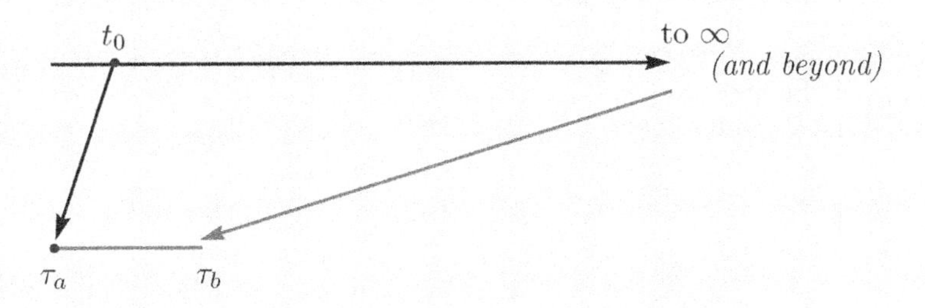

Figure 1.31: Linear and nonlinear domain transformation techniques help manage various horizons (fixed, free, infinite etc.) in optimal control problems[87].

$$\tau = \Gamma^{-1}(t; t_0) := \frac{(t - t_0) - 1}{(t - t_0) + 1}$$

maps $[t_0, \infty)$ to $[-1, 1)$. The advantage of this transformation is that the interval $[-1, 1)$ also happens to be the canonical computational domain of pseudospectral optimal control theory[29]. The inverse of this transformation is given by

$$t = \Gamma(\tau; t_0) := t_0 + \frac{1+\tau}{1-\tau}$$

from which it is apparent that as $\tau \to 1$, t tends to ∞. Hence the solutions obtained by solving the finite-horizon problem can be mapped back to the infinite horizon problem.

The bijective transforms are not unique; for instance, the transformations

$$\tau = \Gamma^{-1}(t; t_0) := \tanh(t - t_0) \qquad \Leftrightarrow \qquad t = \Gamma(\tau; t_0) := t_0 + \frac{1}{2}\ln\left(\frac{1+\tau}{1-\tau}\right)$$

map $[t_0, \infty)$ to $[0, 1)$.

Using any such domain transformation technique, Problem $({}^{\infty}B)$ transforms to

$$\boldsymbol{x} \in \mathbb{X} = \mathbb{R}^{N_x}, \quad \boldsymbol{u} \in \mathbb{U} \subseteq \mathbb{R}^{N_u}$$

$$(B) \left\{ \begin{array}{lrcl} \text{Minimize} & J[\boldsymbol{x}(\cdot), \boldsymbol{u}(\cdot), t_0] & := & \displaystyle\int_{\tau_a}^{\tau_b} F\Big(\boldsymbol{x}(@\tau), \boldsymbol{u}(@\tau)\Big)\, d\Gamma \\ \text{Subject to} & \dfrac{d\boldsymbol{x}}{d\tau} & = & \left(\dfrac{d\Gamma}{d\tau}\right) \boldsymbol{f}\Big(\boldsymbol{x}(@\tau), \boldsymbol{u}(@\tau)\Big) \\ & \boldsymbol{x}_0 & = & \boldsymbol{x}^0 \\ & t_0 & = & t^0 \end{array} \right.$$

where we have used the shorthand notation

$$\boldsymbol{x}(@\tau) \equiv \boldsymbol{x}(\Gamma(\tau; t_0)), \quad \boldsymbol{u}(@\tau) \equiv \boldsymbol{u}(\Gamma(\tau; t_0)), \quad \text{and} \quad d\Gamma \equiv \frac{d\Gamma(\tau; t_0)}{d\tau} d\tau$$

Study Problem 1.14

Pick or define a simple word problem from your field of expertise. Model it in at least three different ways. Discuss the pros and cons of the models.

1.5 Real-Time Optimal Control (RTOC)

Real-time optimal control, or RTOC, is a fundamental mathematical concept that is based on the instantaneous availability of an optimal control trajectory. To develop derivative concepts and applications, we simply assume that $t \mapsto \boldsymbol{u}$ is available instantly; see Fig. 1.32. Its practical implementation is realizable by

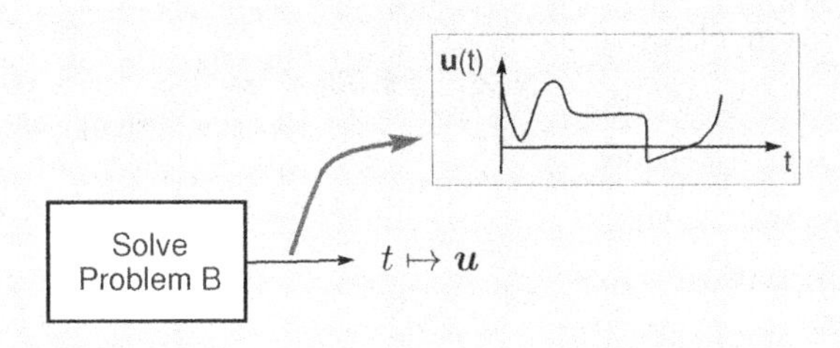

Figure 1.32: In RTOC, the entire optimal control signal $t \mapsto \boldsymbol{u}$ is available almost instantly.

properly accounting for the non-zero computational time of the optimal control trajectory. For instance, in feedback control applications (see Section 1.5.1) the control signal may be updated at a slow rate for slow dynamical systems. This intuition is mathematically captured by the ***time constant of the nonlinear dynamical system***[92] given by

$$\tau_c := \frac{1}{Lip_x \boldsymbol{f}} \tag{1.46}$$

where $Lip_x \boldsymbol{f}$ is the Lipschitz constant of $\boldsymbol{f}$ with respect to $\boldsymbol{x}$.

RTOC became technologically feasible in the early 2000s as a consequence of combining pseudospectral optimal control theory with digital computation[83]. Since then, it has been possible to produce faster than RTOC solutions to meet the demands of rapid footprint generation (projections of reachable sets; see Fig. 1.11 on page 17) for the next generation of aerospace vehicles[10].

The consequences of RTOC are far reaching: It can be applied to solve a number of challenging problems arising in the management of uncertainty as well as other practical problems in decision-making that need to take advantage of the entirety of the available information[40].

1.5.1 Feedback Control via RTOC

A ***system*** can be abstractly represented as an input-output system. In control theory, a system is commonly referred to as a ***plant*** or a ***process***. The outputs from a plant are measurement variables ($\boldsymbol{z}$) while its inputs are controllable signals ($\boldsymbol{a}$) and possibly additional ones that are exogenous in nature; see Fig. 1.33. A fundamental problem in control system design is to architect an input-output system called a ***controller*** that *autonomously* transforms a given ***task*** to inputs to the plant. The task is set by an operator. Other possible inputs in the design of the controller are the outputs from the plant ($\boldsymbol{z}$) and clock time, t; see Fig. 1.34. If the controller does not employ the output from the plant as its input, it is called an ***open-loop*** controller; otherwise, it is known as a ***closed-loop*** or ***feedback controller***. This terminology follows from Fig. 1.35.

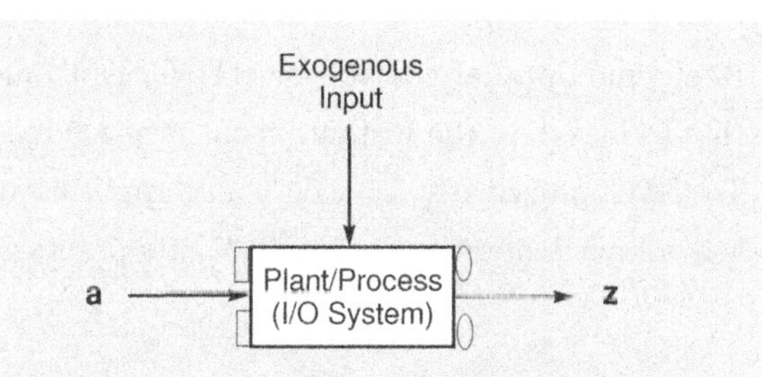

Figure 1.33: A plant or a process is an input-output system. Exogenous inputs to the plant are frequently regarded as "disturbances" in classical control.

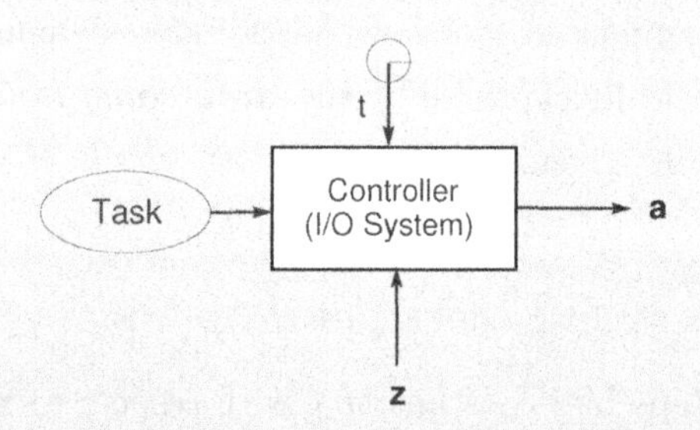

Figure 1.34: A controller is an input-output system. It can take three inputs: (i) a task, (ii) the clock time t and (ii) the output z from the plant.

As elaborated in the previous sections, the control $\boldsymbol{u}$ in an optimal control system can be any abstract quantity; e.g. a decision variable without a derivative. Similarly, the state variable $\boldsymbol{x}$ can also be an abstract quantity; e.g. a

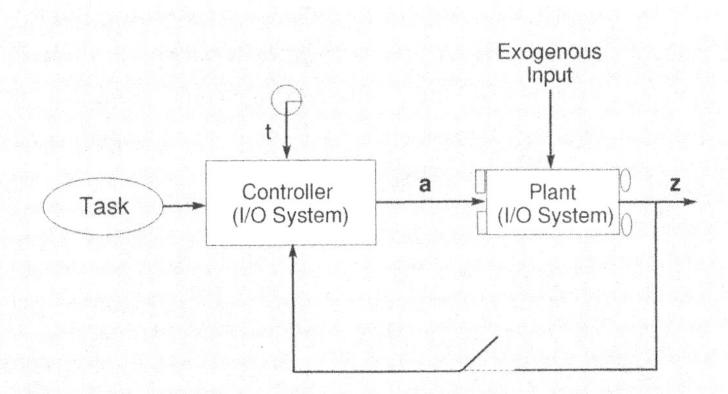

Figure 1.35: The main problem in control. The notional switch explains the difference between open- and closed-loop controls.

quaternion. Consequently, to map these abstractions to the design of the practical controller illustrated in Fig. 1.35, the subproblems of *allocation*, *control generation*, and *estimation* need to be addressed; see Fig. 1.36.

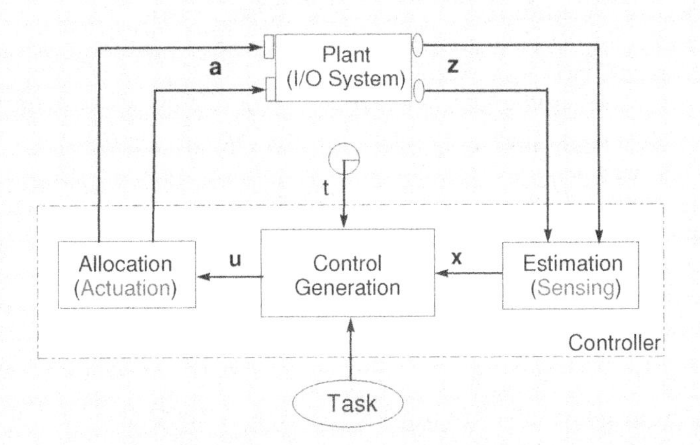

Figure 1.36: The generation of an "abstract" control can be practically implemented through additional processes of estimation and allocation. Control and state variables can be regarded as internal variables.

The control generation box in Fig. 1.36 can be designed using the principles of real-time optimal control as follows: Problem B can be considered to be parameterized by the initial event (x^0, t^0); hence, we write it explicitly as Problem $B(\boldsymbol{x}^0, t^0)$. By the principle of real-time optimal control, this implies we can produce a control signal $t \mapsto \boldsymbol{u}$ for all $t \geq t^0$ with initial condition $\boldsymbol{x}(t^0) = \boldsymbol{x}^0$. Feedback control is obtained by simply setting the incoming signal $\boldsymbol{x}$ at t to be $\boldsymbol{x}^0$ at t^0 in Problem $B(\boldsymbol{x}^0, t^0)$. This generates the $(\boldsymbol{x}^0, t^0)$-parameterized control signal $t \mapsto \boldsymbol{u}(t; \boldsymbol{x}^0, t^0)$, as shown in Fig 1.37.

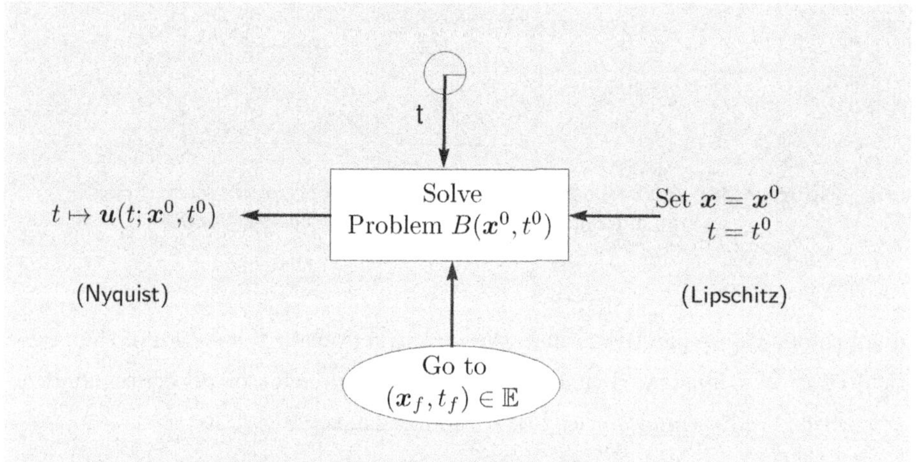

Figure 1.37: RTOC details for the control generation box of Fig. 1.36.

If $(\boldsymbol{x}^0, t^0)$ is not updated, $t \mapsto \boldsymbol{u}(t; \boldsymbol{x}^0, t^0)$ is an open-loop controller. If $(\boldsymbol{x}^0, t^0)$ is continuously updated, the result is a closed-loop controller. Thus, a solution to Problem B generates open- and closed-loop controls based on how it is implemented with optimality becoming a side benefit.

Practical Implementation: Nyquist and Lipschitz Sampling Rates

Intuitively, it is apparent that the continuous updates of $(\boldsymbol{x}, t) = (\boldsymbol{x}^0, t^0)$ can be relaxed to "discrete" updates provided the system is "sufficiently slow." This intuition is captured in Eq. (1.46) in terms of the time constant of the dynamics. Hence, for a practical closed-loop solution, the updates must happen at frequencies that are at or faster than the ***Lipschitz frequency***[92]:

$$\omega_{Lip} := \frac{1}{\tau_c} = Lip_x \boldsymbol{f}$$

Although the precise computation of the Lipschitz frequency is not easy, it can be practically estimated by holding the clock in a numerical simulation and varying the hold times; see [85] for details.

The Lipschitz frequency corresponds to the sampling of the input signal in RTOC. This sampling is independent of how the output signal (control) is implemented: digital or analog. In electromechanical systems, the control is commonly implemented via zero-order-holds. This digital sampling must be done at a frequency greater than or equal to the Nyquist frequency.

In RTOC, the entire control signal $[t_0, t_f] \ni t \mapsto \boldsymbol{u}$ is generated at each Lipschitz sample. This includes the possibility that t_f may be infinite[29]; see also page 60 for the concept of infinite horizon control. Thus, digital control at or greater than the Nyquist frequency is always available[92], even when the input signal is sampled only once!

Some of the principles of RTOC are illustrated in the example problems discussed in Chapter 3. As discussed in Section 3.3 (see page **??**) the classical notion of *PID controllers* can be considered as a simplified form of RTOC; hence, it should not be entirely surprising that a large number of *elementary set-point tasking* control systems (e.g., electric motor control, cruise control) perform extremely well with simple gain tuning.

> **Tech Talk**: Feedback control via RTOC is a generalized concept. It is based on the principle that *closed-loop* is not the same as *closed-form*. A closed-form solution is a sufficient (not necessary) condition for closed-loop control. On the other hand, real-time computation of control is a necessary condition for feedback control. Real-time computation — for the purposes of feedback control — can be defined universally as computation at Lipschitz rates (or higher). In most applications, Lipschitz rates are substantially smaller than Nyquist rates; see [85] and [92].

Guidance and Control

Because legacy control systems predate optimal control theory, the word control is often perceived more narrowly to a "low-level" set-point tasking system, as shown in Fig. 1.38. Consequently, PID controllers and their derivatives form the workhorse of industrial control systems. Such controllers work extremely well for the task they were designed to perform.

To establish distinctions and design philosophies, we define a ***guidance system*** as a continuous-time continuous tasker for an inner-loop control system; see Fig. 1.39. Thus, a guidance system is also a control system as defined in

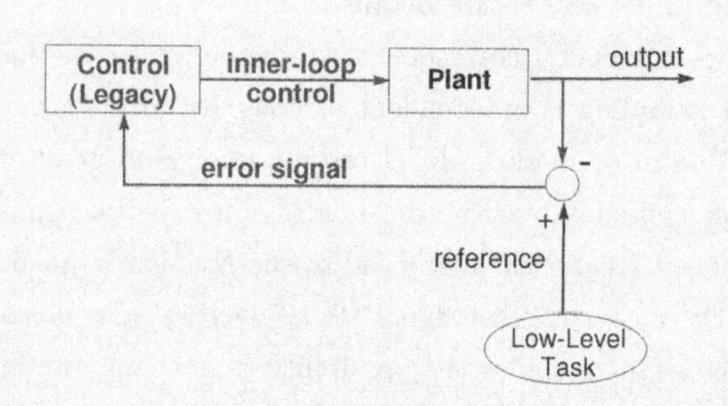

Figure 1.38: Historically, a control system has been defined narrowly to a low-level set-point tasker such as automobile's cruise control system.

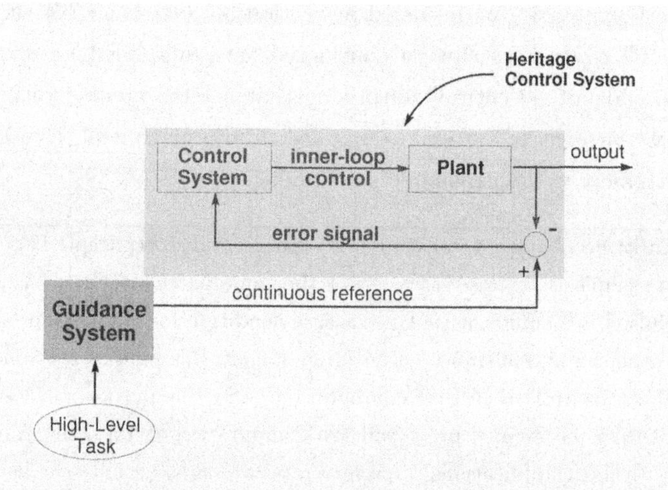

Figure 1.39: A guidance system is a continuous tasker for a heritage control system like the one shown in Fig. 1.38.

Figs. 1.34 and 1.35. It operates on a heritage control system and allows a higher level of automation than low-level set-point controllers.

The guidance system shown in Fig. 1.39 is called an open-loop guidance system. In aerospace engineering, open-loop guidance has been used since the 1960s; however, many guidance systems of this era were largely based on ad hoc techniques. Since the 1960s, optimal control theory has served as the foundation for designing a guidance system; hence, problems in open-loop guidance are called ***trajectory optimization*** problems.

In addition to providing a higher level of automation, a guidance system can also enhance the performance of a control system by providing an outer-loop ***feedforward control*** and/or closing the guidance loop via an outer-loop feedback system; see Fig. 1.40. It is apparent that a guidance system can be

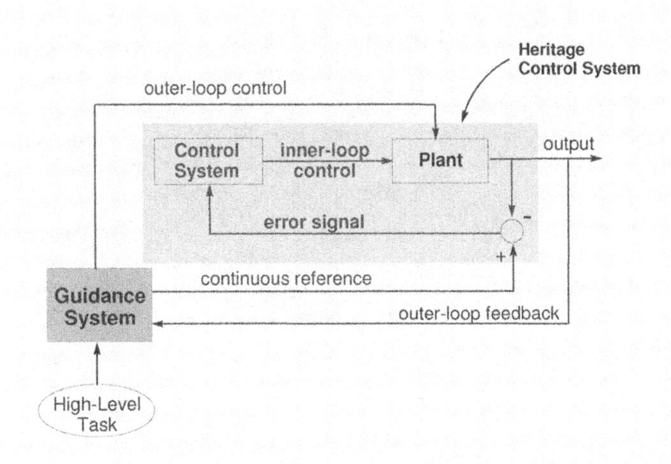

Figure 1.40: A guidance system operates as an outer-loop for an inner-loop control system. The "plant" for a guidance system is the plant plus its control system.

closed via RTOC. One of the earliest guidance algorithms closed by RTOC is the *powered-explicit guidance* (PEG) system. Details of PEG are discussed in Section 3.6 on page 231.

1.5.2 Towards Artificial Intelligence via RTOC

A guidance system concept allows an operator to perform high-level tasks while the low-level tasks are executed by the heritage control system; hence, it is a first step towards autonomous operations. Effective autonomous operations imply decision-making that fully exploit all available information. The RTOC concept meets this criterion by mapping information to problem data. That is, by considering Problem B to be parameterized by the entire data and not just $(\boldsymbol{x}^0, t^0)$ a large number of problems in autonomous operations and artificial intelligence (AI) become solvable. This concept is shown in Fig. 1.41 for the more general data incorporated in Problem P.

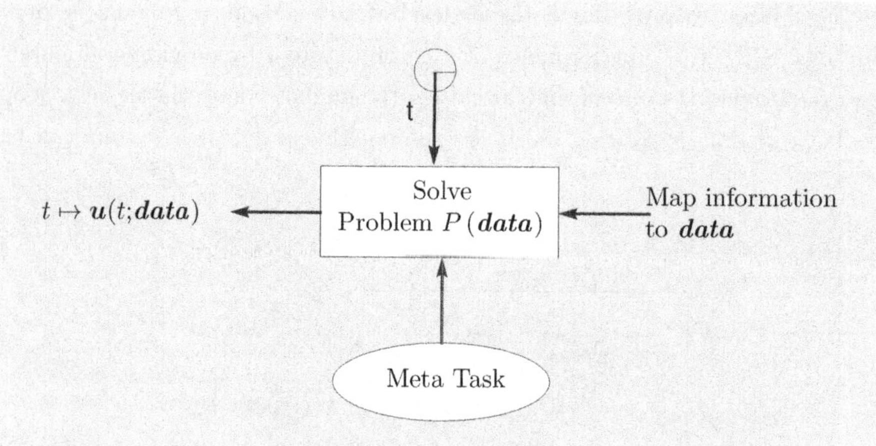

Figure 1.41: In RTOC, the control action is a function of t as well as the entire problem data. Compare with Fig. 1.37.

The data for Problem P are:

1. The initial event $(\boldsymbol{x}^0, t^0)$: mapping this to the current value of the state and time generates feedback control;

2. The target set (final event) given by the data function $\boldsymbol{e}$ and its bounds $(\boldsymbol{e}^L, \boldsymbol{e}^U)$: mapping this to new targets allows re-targeting and dynamic re-tasking;

3. The dynamics data function, $\boldsymbol{f}$: mapping in/ex situ sensor information to changes in the system dynamics (such as failures of actuation mechanisms, entry into new dynamical regimes, etc.) allows for adaptation to changes in the plant;

4. The path conditions given by the data function $\boldsymbol{h}$ and its bounds $(\boldsymbol{h}^L, \boldsymbol{h}^U)$: mapping in/ex situ sensor and environmental information (global) generates controls that avoid moving obstacles (for mobile robots), satisfy operational constraints (no-fly zones), etc.

As a result of all these functionalities, a much higher level of tasking is possible than what is provided by a more traditional guidance system. This implies that an operator can command an RTOC system to autonomously perform a collection of high-level tasks that constitute a ***meta task***. This structure is known as an ***operator on the loop*** in contrast to an operator <u>in</u> the loop. The etymology of this terminology is more apparent from Fig. 1.42.

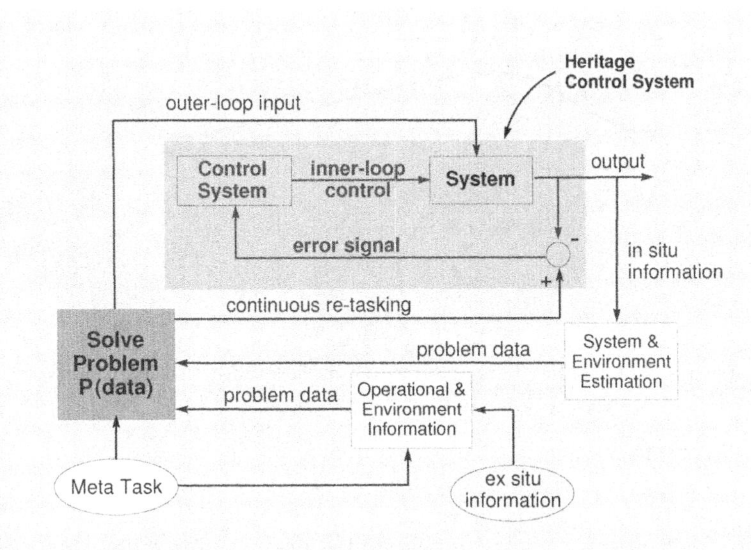

Figure 1.42: Incorporating the RTOC concept of Fig. 1.41 with a traditional control system facilitates meta tasking for an operator on the loop. Figure adapted from [59] and [80].

When RTOC is incorporated in a closed-loop manner as indicated in Fig. 1.42, it generates outcomes that mimic AI. Details are discussed extensively in [11, 13, 17, 18, 39, 40, 58, 59]. An example of an AI mimic generated by the architecture of Fig. 1.42 can be illustrated by Hurni's Sliding Door problem[38].

Hurni's Sliding Door

A mobile robot, with known dynamics $\dot{\boldsymbol{x}} = \boldsymbol{f}(\boldsymbol{x}, \boldsymbol{u})$, is tasked to move from point A to point B. At $t = t_0$, it has initial information of its environment specified in terms of a box constraint $(x^L, y^L) \leq (x, y) \leq (x^U, y^U)$ and a path constraint $\boldsymbol{h}_0^L \leq \boldsymbol{h}_0(\boldsymbol{x}, t_0) \leq \boldsymbol{h}_0^U$. The path constraint divides the box environment into two regions as shown in Fig. 1.43. At $t = t_0$, an optimal trajectory is computed and shown in Fig. 1.43. This trajectory was obtained by solving an optimal control "Problem P" that incorporates a cost functional based on time and safety criteria[38].

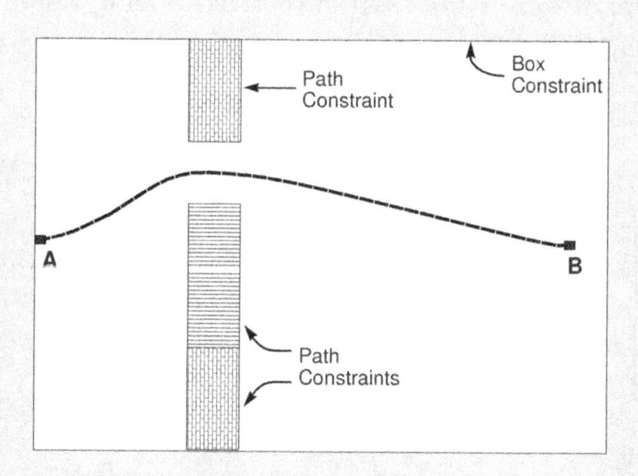

Figure 1.43: An optimal trajectory for a mobile robot tasked to go from point A to point B.

As the robot moves along its pre-computed path, a sliding door (i.e., one of the path constraints) blocks its path at $t = t_1$. This information maps to a new path constraint $\boldsymbol{h}_1^L \leq \boldsymbol{h}_1(\boldsymbol{x}, t_1) \leq \boldsymbol{h}_1^U$ for Problem P whose solution results in a new trajectory as shown in Fig. 1.44. Thus, RTOC with feedback produces a mobile robot that responds intelligently to a change in its environment while carrying out the meta task: Go to point B without hitting any obstacles that might be moving or popping.

While moving along its new path, a new obstacle pops up at $t = t_2$, rendering its current trajectory infeasible as shown in Fig. 1.45. Updating Problem P with an additional constraint $\boldsymbol{h}_2^L \leq \boldsymbol{h}_2(\boldsymbol{x}, t_2) \leq \boldsymbol{h}_2^U$ generates a new solution (see Fig. 1.46), leading to successful completion of the task.

Integrating RTOC in Categorical Space

Many more examples mimicking AI via RTOC are presented in [11]–[18], [38]–[40], [58, 59] and [84]. That RTOC in a feedback loop generates intelligent systems is not entirely surprising because the roots of AI go all the way back to the control-theory of the 1940s[93]. Modern AI is more closely aligned with RTOC; in fact, Russell and Norvig define AI (see page 15 in [93]) as "designing systems that behave optimally." Because optimal control theory is founded on real-valued functions in real space with notions of continuity, it cannot be

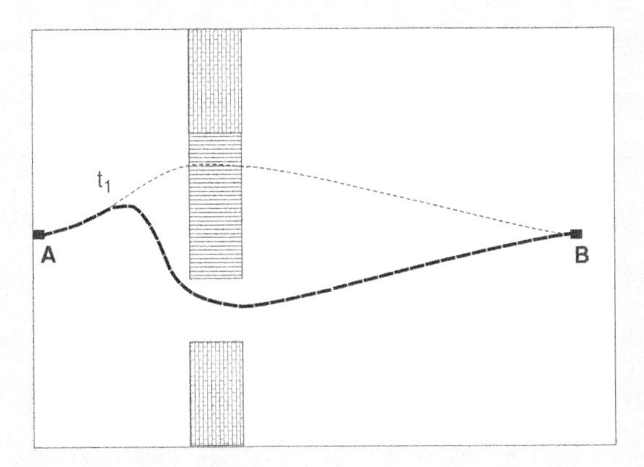

Figure 1.44: At $t = t_1$, RTOC generates a new path that incorporates the blocked passage information and makes the robot turn toward the new opening.

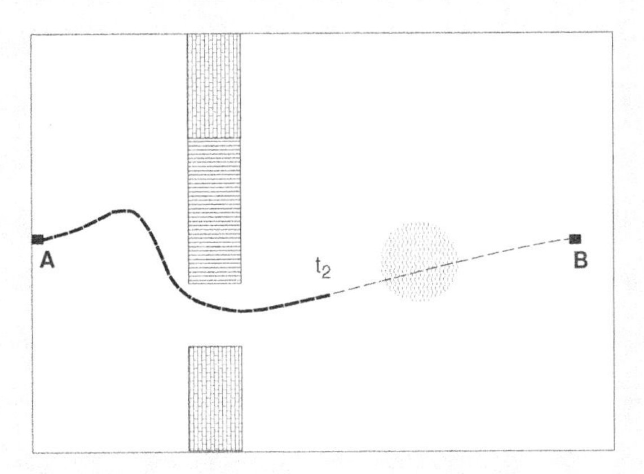

Figure 1.45: At $t = t_2$, new information of an impending obstacle is provided to the RTOC algorithm.

applied to systems that operate in discrete space[82]; however, the feedback principle can still be applied in categorical space as follows: An automaton accepts linguistic commands from an operator and generates a meta task based on the categorical state of the system generated by an inference system; see Fig. 1.47. In this three-degrees-of-freedom system, a human is further relieved

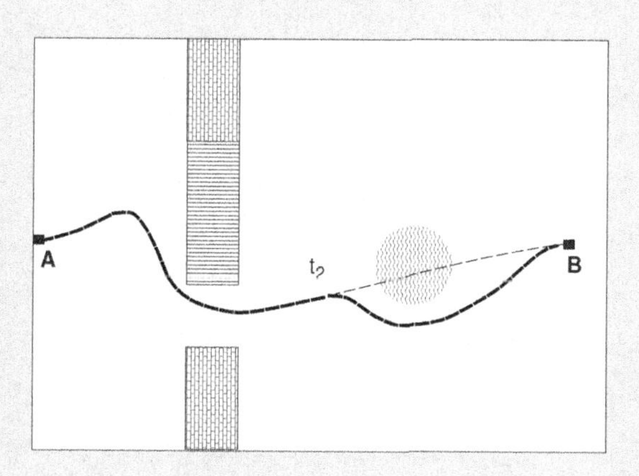

Figure 1.46: By incorporating the obstacle information at $t = t_2$, the RTOC algorithm swerves the robot to successful completion of its task.

of the burden of generating meta tasks and can operate the machine at a higher level of abstraction. This concept was used in [38, 40] to incorporate an expert system in the outer loop for mobile robots and in [98] for autonomously selecting the most beneficial targets for an imaging satellite.

Some Additional Comments About RTOC

Parallel implementations of RTOC can be used to produce real-time reachable sets. The main challenge here is an effective means to represent the reachable set, particularly for high dimensions.

To use RTOC, it is sufficient to produce "open-loop" solutions in real time. Feedback control and AI mimicking are obtained by closed-loop implementations. Much of the enabling technology for RTOC has been pseudospectral optimal control[87]; see also Section 2.9.2 on page 157. In principle, it is possible to produce the same results as those discussed in this section by solving the Hamilton-Jacobi-Bellman (HJB) equations; however, the HJB equations need to be solved in real time to incorporate the new information that enters Problem P.

As evident from the discussions in this section, RTOC manages uncertainty through the process of feedback. This management of uncertainty is implicit in the sense that no explicit information of uncertainty is used; the entire means to manage uncertainty is through a process of using all of the available deter-

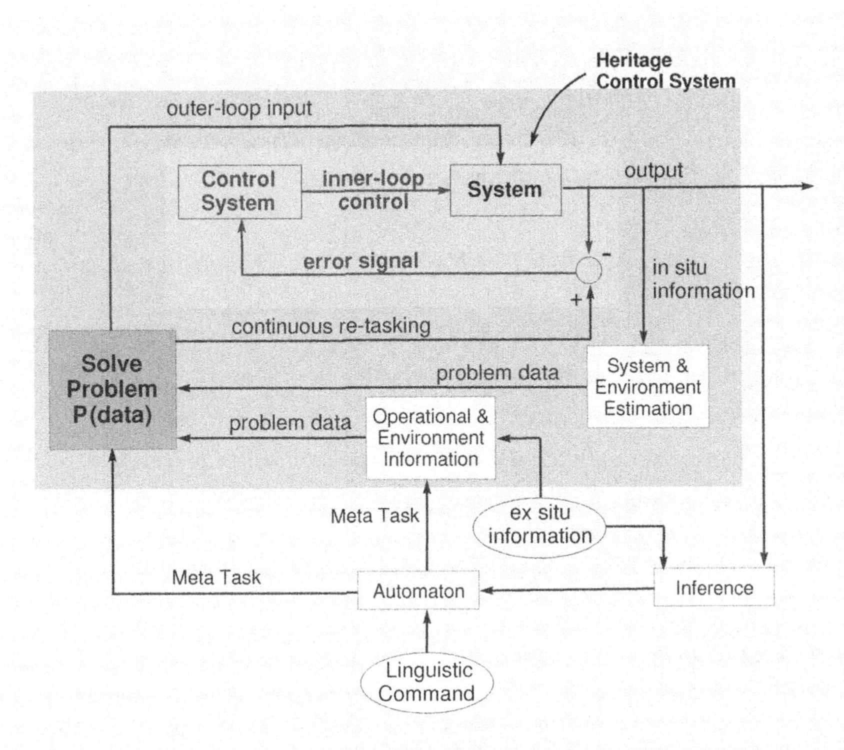

Figure 1.47: A categorical space feedback loop strapped to an RTOC system provides a framework for linguistic commands for an operator on the loop. Figure adapted from [82].

ministic information. In the next section, uncertainty is managed directly by formulating *tychastic* optimal control problems.

1.6 Tychastic Optimal Control

A *tychastic* differential equation is given by

$$\dot{\boldsymbol{x}} = \boldsymbol{f}(\boldsymbol{x}, \boldsymbol{u}, t; \boldsymbol{p}) \tag{1.47}$$

where $\boldsymbol{p} \in \mathbb{R}^{N_p}$ is an uncertain parameter such that $\boldsymbol{f} : (\boldsymbol{x}, \boldsymbol{u}, t, \boldsymbol{p}) \mapsto \mathbb{R}^{N_x}$ is deterministic if $\boldsymbol{p}$ is known. We would have called $\boldsymbol{p}$ a "random" parameter if, according to Aubin[3], the "terminology [was not] already confiscated by probability theory." Tyches means *chance* in classical Greek, from the Goddess

Tyche.

It will be apparent shortly that a tychastic optimal control is different from a stochastic optimal control because Eq. (1.47) is different from the Itô differential equation. Furthermore, a tychastic differential equation may be considered as a parametrization of a controlled differential equation

$$\dot{\boldsymbol{x}} \in \mathcal{F}(\boldsymbol{x}, \boldsymbol{u}, t) := \{\boldsymbol{f}(\boldsymbol{x}, \boldsymbol{u}, t; \boldsymbol{p}) : \ \boldsymbol{p} \subset supp(\boldsymbol{p})\} \tag{1.48}$$

where $supp(\boldsymbol{p})$ is the support of $\boldsymbol{p}$. Thus, the evolution $t \mapsto \boldsymbol{x}$ is set-valued, not "random." Furthermore, for a given value of $\boldsymbol{p}$, say $\boldsymbol{p}^0$, $\boldsymbol{f}(\boldsymbol{x}, \boldsymbol{u}, t; \boldsymbol{p}^0)$ is a ***deterministic selection*** of $\mathcal{F}$.

Now suppose we apply a control $t \mapsto \boldsymbol{u}$ to the uncertain system at $(\boldsymbol{x}^0, t^0)$; then, for a given value of $\boldsymbol{p}$, the state evolves such that

$$\partial_t \boldsymbol{x}(t, \boldsymbol{p}) := \dot{\boldsymbol{x}}(t, \boldsymbol{p}) = \boldsymbol{f}(\boldsymbol{x}(t, \boldsymbol{p}), \boldsymbol{u}(t), t; \boldsymbol{p}) \qquad \text{a.a. } t \tag{1.49}$$

where $\boldsymbol{x}(\cdot, \boldsymbol{p})$ denotes a particular trajectory for a given value of $\boldsymbol{p}$ and $\boldsymbol{x}(t, \cdot)$ is a slice of trajectories at t; see Fig. 1.48. Hence, we adopt the following notation:

$$\begin{aligned} \boldsymbol{x}(\cdot, \boldsymbol{p}) &: t \mapsto \mathbb{R}^{N_x} \\ \boldsymbol{x}(t, \cdot) &: \boldsymbol{p} \mapsto \mathbb{R}^{N_x} \\ \boldsymbol{x}(\cdot, \cdot) &: (t, \boldsymbol{p}) \mapsto \mathbb{R}^{N_x} \end{aligned} \tag{1.50}$$

Suppose we insert $\boldsymbol{x}(\cdot, \boldsymbol{p})$, $\boldsymbol{u}(\cdot)$ and t_f in a standard cost functional; then, we

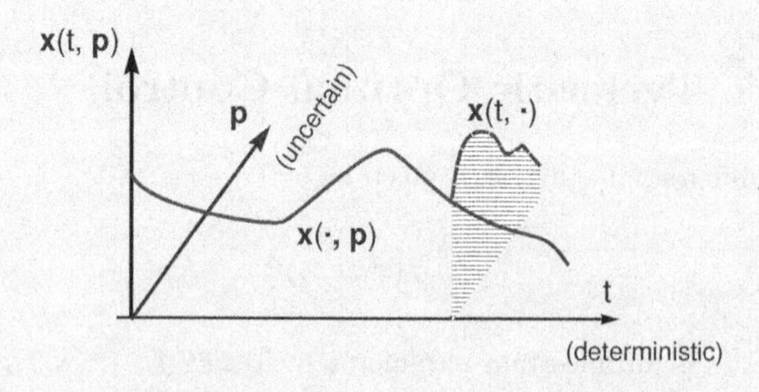

Figure 1.48: Schematic of state functions used in defining a tychastic optimal control problem.

get:

$$J[\boldsymbol{x}(\cdot, \boldsymbol{p}), \boldsymbol{u}(\cdot), t_f]$$

Obviously, this J is not deterministic; hence, it cannot be minimized. Nonetheless, we can use the standard cost functional as a process to generate a scalar function $\boldsymbol{p} \mapsto \mathbb{R}$ defined by

$$\boldsymbol{p} \mapsto J_{unc}[\boldsymbol{x}(\cdot, \boldsymbol{p}), \boldsymbol{u}(\cdot), t_f; \boldsymbol{p}] \in \mathbb{R} \tag{1.51}$$

where $J_{unc}[\boldsymbol{x}(\cdot, \boldsymbol{p}), \boldsymbol{u}(\cdot), t_f; \boldsymbol{p}]$ differs from $J[\boldsymbol{x}(\cdot, \boldsymbol{p}), \boldsymbol{u}(\cdot), t_f]$ through its explicit dependence on $\boldsymbol{p}$. We now use the function given by Eq. (1.51) to construct a new functional by integrating J_{unc} over $\boldsymbol{p}$

$$J[\boldsymbol{x}(\cdot, \cdot), \boldsymbol{u}(\cdot), t_f] := \int_{supp(\boldsymbol{p})} J_{unc}[\boldsymbol{x}(\cdot, \boldsymbol{p}), \boldsymbol{u}(\cdot), t_f; \boldsymbol{p}]\, dm(\boldsymbol{p}) \tag{1.52}$$

where $\int_{supp(\boldsymbol{p})}$ is an N_p-dimensional ***Lebesgue-Stieltjes integral*** with measure $m : \boldsymbol{p} \mapsto \mathbb{R}_+$.

Equation (1.52) defines a cost "***functional of a functional***"; it is minimizable because it is deterministic. In addition, it allows us to choose the measure function m to be the joint cumulative distribution function (CDF) of $\boldsymbol{p}$. Equations (1.52) and (1.47) constitute a basic ***tychastic or Lebesgue-Stieltjes optimal control problem***.

It is apparent that we have used the standard optimal control problem as a *generator* for a tychastic optimal control problem. We follow the same approach to discuss various formulations of tychastic target sets.

1.6.1 Specification of Tychastic Target Sets

In a standard optimal control problem, a "simple" specification of a target set is a point: $\boldsymbol{x}(t_f) = \boldsymbol{x}^f$. Suppose we port this idea to a tychastic problem and write:

$$\boldsymbol{x}(t_f, \boldsymbol{p}) = \boldsymbol{x}^f$$

Then, we need a "correction" because the left-hand-side is uncertain. A "simple" correction is to stipulate

$$\boldsymbol{x}(t_f, \boldsymbol{p}) = \boldsymbol{x}^f \qquad \forall\ \boldsymbol{p} \in supp(\boldsymbol{p}) \tag{1.53}$$

At first glance, this requirement seems like a setup for failure because Eq. (1.53) demands that the target set be a singleton despite uncertainties in the system. It turns out that there is indeed a class of dynamical systems where such severe requirements can be imposed. The tychastic Zermelo problem discussed in Section 4.7.2 (see page 287) demonstrates this surprising possibility. Generalizing Eq. (1.53) to a generic target set, we define a ***transcendental target condition*** as

$$\boldsymbol{e}\big(\boldsymbol{x}(t_f, \boldsymbol{p}), t_f; \boldsymbol{p}\big) \leq \boldsymbol{0} \qquad \forall\ \boldsymbol{p} \in supp(\boldsymbol{p}) \tag{1.54}$$

where $\boldsymbol{e} : \mathbb{R}^{N_x} \times \mathbb{R} \times \mathbb{R}^{N_p} \to \mathbb{R}^{N_e}$ is a given function. We would have called Eq. (1.54) *robust* target conditions; however, that term is used differently in feedback control theory and optimal control theory[14].

The challenge of using Eq. (1.54) as the standard bearer for "all" tychastic optimal control problems is, of course, the existence of a solution. To facilitate a reasonable chance of the existence of a solution, the transcendental nature of Eq. (1.54) must be abridged. The approach suggested in [91] is to impose the condition

$$\int_{supp(\boldsymbol{p})} \boldsymbol{e}\big(\boldsymbol{x}(t_f, \boldsymbol{p}), t_f; \boldsymbol{p}\big)\, dm(\boldsymbol{p}) \leq \boldsymbol{0} \tag{1.55}$$

The problem resulting from this relaxation is called an *ancillary* Lebesgue-Stieltjes optimal control problem[90]. Equation (1.55) is not the only way to relax the severity of Eq. (1.54); an alternative formulation is to take the integral in Eq. (1.55) inside the event-function; this generates the condition

$$\boldsymbol{e}\big(\boldsymbol{\mu}_{\boldsymbol{x}_f}, t_f; \boldsymbol{\mu}_{\boldsymbol{p}}\big) \leq \boldsymbol{0} \tag{1.56}$$

where $\boldsymbol{\mu}_{\boldsymbol{x}_f}$ is the mean of $\boldsymbol{x}(t_f, \boldsymbol{p})$ defined by

$$\boldsymbol{\mu}_{\boldsymbol{x}_f} := \int_{supp(\boldsymbol{p})} \boldsymbol{x}(t_f, \boldsymbol{p})\, dm(\boldsymbol{p})$$

with $\boldsymbol{\mu}_{\boldsymbol{p}}$ defined similarly. This approach is taken in [88, 89, 90] as a means to determine an initial feasibility of tychastic optimal control problems. *In fact, the problem of the existence of a solution is a main problem in tychastic optimal control.*

1.6.2 Chance-Constrained Optimal Control

Motivated by problems in management science (see Section 4.11 on page 316) where some variables are subject to uncertainties (weather, production schedules, etc.) Charnes and Cooper[19] formulated the notion of chance-constrained or stochastic programming. In adapting their ideas to tychastic optimal control, Eq. (1.54) is relaxed to the ***chance constraint***

$$Pr\Big\{\mathbf{e}\big(\boldsymbol{x}(t_f,\mathbf{p}),t_f;\mathbf{p}\big) \leq \mathbf{0}\Big\} \geq R \tag{1.57}$$

where $Pr\{A\}$ denotes the probability of event A, and $R \in (0,1]$ is a designer-chosen number (usually $0.95 - 0.99$) called ***reliability*** or *confidence coefficient* in [19]. When it satisfies Eq. (1.57), the endpoint condition is said to hold with reliability R.

The quantity

$$r := 1 - R \qquad (\geq 0) \tag{1.58}$$

is called ***risk***. There are other and better mathematical definitions of risk[76], but Eq. (1.58) is, conceptually, the simplest. Because $Pr\{A\} + Pr\{\neg A\} = 1$, Eq. (1.57) can also be written equivalently in terms of a risk constraint,

$$Pr\left\{\mathbf{e}\big(\boldsymbol{x}(t_f,\mathbf{p}),t_f;\mathbf{p}\big) > \mathbf{0}\right\} \leq r$$

If we choose $r = 0$ ($\Rightarrow$ $R = 1$), then Eq. (1.57) reduces to the transcendental condition given by Eq. (1.54). That is, the transcendental condition is equivalent to a requirement of zero risk or 100% reliability. For all other values of R, Eq. (1.57) generates a family of problems. Unfortunately, this family of problems is not necessarily "easier" than the transcendental formulation. To appreciate this point, note that the left-hand side of Eq. (1.57) can be expressed in terms of a Lebesgue-Stieltjes integral through the computational procedure

$$Pr\left\{\mathbf{e}\big(\boldsymbol{x}(t_f,\mathbf{p}),t_f;\mathbf{p}\big) \leq \mathbf{0}\right\} := \int_{\Omega\subset supp(\mathbf{p})} dm(\mathbf{p}) \tag{1.59}$$

where Ω is given by

$$\Omega := \{\boldsymbol{p} \in supp(\boldsymbol{p}) : \ \boldsymbol{e}(\boldsymbol{x}(t_f, \boldsymbol{p}), t_f; \boldsymbol{p}) \leq \boldsymbol{0}\}$$

Hence, the chance constraint can be written as the Lebesgue-Stieltjes constraint

$$\int_{\Omega \subset supp(\boldsymbol{p})} dm(\boldsymbol{p}) \geq R \tag{1.60}$$

The apparent simplicity of Eq. (1.60) belies its complexity: The representation of the Lebesgue-Stieltjes integral over $\Omega \neq supp(\boldsymbol{p})$ is a non-trivial task and forms a main problem in chance-constrained programming. In optimal control this difficulty is further exacerbated by at least two other problems:

1. $\boldsymbol{x}(t_f, \boldsymbol{p})$ depends implicity on $\boldsymbol{p}$, and
2. Even if Eq. (1.59) is computable, there is still no guarantee that the resulting tychastic optimal control problem, formulated through the constraint given by Eq. (1.60), will have a feasible solution for R near one (e.g., $R = 0.95$) if $R = 1$ has no solution.

Both of these problems are addressed in [88, 89, 90] by the procedure outlined in Fig. 1.49. Because Eq. (1.59) is computed a posteriori and compared to R, this approach produces a practical means to solve chance-constrained optimal control problems.

1.6.3 Connections to Quantum Control and Other Applications

It turns out that tychastic optimal control can be connected to many burgeoning problems in disparate fields: search theory, finance, quantum control etc.[91]. For instance, in one aspect of quantum control[15], the objective is to design an electromagnetic input, $t \mapsto \boldsymbol{u}$, to manipulate systems at the quantum level. The applications of controlling molecules are quite broad: from an ability to produce new materials to medical imaging using nuclear magnetic resonance spectroscopy. Because the dynamics of the quantum "particles" are on the scale of Avogadro's number ($\sim 6.022 \times 10^{23}$), it is far simpler to model the system as a continuum[61]; hence, in a quantum dynamical model $\dot{\boldsymbol{x}} = \boldsymbol{f}(\boldsymbol{x}, \boldsymbol{u}, t; p)$, the

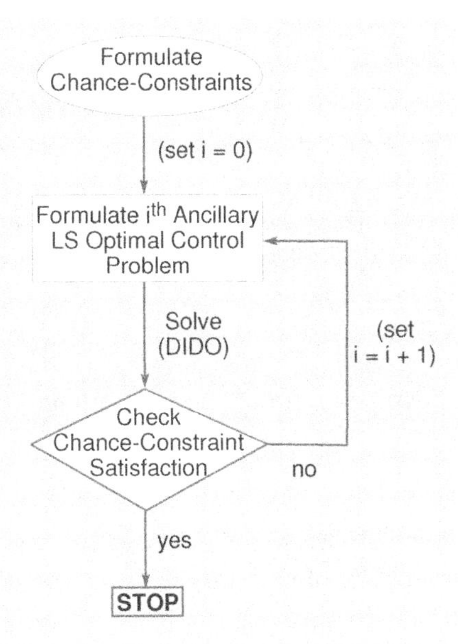

Figure 1.49: A procedure to solve chance-constrained optimal control problems. See Fig. 1.50 on page 84 for additional details on solving optimal control problems and their connections to the chapters in this book.

parameter p is a representation of this continuum. See Li et al[62] for further details.

In search theory[56], the problem is to find a moving target (missing boat, lost hiker, fugitive, etc.) whose location, $\boldsymbol{x}^T \in \mathbb{X}^T \subset \mathbb{R}^{N_{xT}}$, is unknown. We assume the motion of the target, $t \mapsto \boldsymbol{x}^T$, is conditionally deterministic: That is, the trajectory of the target conditioned on an uncertain vector, $\boldsymbol{p}$, is given by $\boldsymbol{x}^T(\cdot, \boldsymbol{p})$. A searcher (satellite, unmanned aerial vehicle, etc.) at $\boldsymbol{x}^S$ is equipped with a sensor suite whose effectiveness is modeled by an instantaneous probability density function called the search density function[101], $\psi : (\boldsymbol{x}^T, \boldsymbol{x}^S) \mapsto \mathbb{R}_+$. The probability that the searcher will find the target by searching along a given trajectory $\boldsymbol{x}^S(\cdot)$ over a time interval $(0, t]$ is given by[56, 101]

$$P(t, \boldsymbol{x}^T(t); \boldsymbol{p}) = 1 - \exp\left(-\int_0^t \psi(\boldsymbol{x}^S(\tau), \boldsymbol{x}^T(\tau, \boldsymbol{p}))\, d\tau\right)$$

Then, the probability that the searcher will detect the target over the time period $(0, t]$ is given by

$$\int_{supp(\boldsymbol{p})} P(t, \boldsymbol{x}^T(t, \boldsymbol{p}))d\alpha(\boldsymbol{p}) = 1 - \int_{supp(\boldsymbol{p})} \exp\left(-\int_0^t \psi(\boldsymbol{x}^S(\tau), \boldsymbol{x}^T(\tau, \boldsymbol{p}))\, d\tau\right) dm(\boldsymbol{p})$$

Thus, the quantity

$$\int_{supp(\boldsymbol{p})} \exp\left(-\int_0^t \psi(\boldsymbol{x}^S(\tau), \boldsymbol{x}^T(\tau, \boldsymbol{p}))\, d\tau\right) dm(\boldsymbol{p})$$

represents the probability of non-detection. Let the searcher's dynamics be given by the deterministic dynamics, $\dot{\boldsymbol{x}}^S = \boldsymbol{f}(\boldsymbol{x}^S, \boldsymbol{u})$. Then, we can define the optimal search problem as the problem of minimizing the Lebesgue-Stieltjes functional

$$J[\boldsymbol{x}(\cdot,\cdot), \boldsymbol{u}(\cdot), t_f] := \int_{supp(\boldsymbol{p})} \exp\left(-\int_0^{t_f} \psi(\boldsymbol{x}^S(t), \boldsymbol{x}^T(t, \boldsymbol{p}))\, dt\right) dm(\boldsymbol{p})$$

where $\boldsymbol{x} := (\boldsymbol{x}^T, \boldsymbol{x}^S)$.

The preceding examples are only two illustrative applications of the scope of Lebesgue-Stieltjes optimal control theory. See [91] for additional discussions and connections to many open problems in optimal control.

1.7 Endnotes

There are two aspects to problem formulation: One is its "invention" as an optimal control problem and the other is its transcription to a proper mathematical formulation. One of these two aspects is discussed in this chapter through several transcriptions of Bernoulli's invention of a "word problem." The invention component of many optimal control problems is discipline-specific. Knowledge of the discipline is a prerequisite to transcribing the word problem to a mathematical statement. A lack of knowledge of the discipline is one of the many reasons for incorrect formulation of an optimal control problem.

All of the topics covered in this chapter (including the sections marked with dangerous bends) have been used in some form or another in many real-world practical applications. As an example, in 2002, Bedrossian issued a $100B challenge[§] "word problem" that can be described as designing an ***attitude guidance*** solution for the International Space Station to "dump" accumulated momentum using no propellant consumption. Unlike Bernoulli, Bedrossian did not know the solution in advance. Discipline-specific knowledge was necessary to transcribe *Bedrossian's problem* to an optimal control formulation.

It is not necessary to transcribe a word problem to a "correct" optimal control problem right from the start. In fact, the first transcription of the word problem to an optimal control problem is likely going to be incorrect or incomplete. Many parts of the the problem statement can be corrected by exploring the initial mathematical formulation through a paper/pencil analysis via an application of Pontryagin's Principle — the main topic of this book, and introduced in Chapter 2. A systematic procedure for the paper/pencil analysis is illustrated in Chapter 3 through several example problems. The paper/pencil analysis will often provide clues on any missing details of the problem formulation; see, for example, Study Problem 3.16 on page 220. Additional missing details can be discovered by different mathematical formulations of the same or similar problem. Different mathematical formulations of the same problem, as illustrated in this chapter, have important, different and competing merits in terms of insights and analysis.

Post paper/pencil analysis, the problem can be analyzed by way of DIDO. As illustrated throughout this chapter, DIDO allows a problem designer to produce a code that looks nearly the same as its mathematical formulation. A typical work flow involved in the analysis of a present-day optimal control problem is illustrated in Fig. 1.50 along with its connections to specific chapters in this book.

Computer codes for solving some of the different formulations of the Brachistochrone problem discussed in this chapter can be downloaded from Elissar Global (`http://www.ElissarGlobal.com`).

The practice of optimal control fundamentally changed around the year 2000 (specifically, from about the late 1990s to the early 2000s). This is because, at the turn of the millennium, there was a fortuitous confluence of key mathematical breakthroughs juxtaposed with the availability of once-extraordinary computing power on ordinary computers[84]. The potent combination of new mathematics and new technology made solving of once-difficult problems easy.

[§] The price tag of the space station was $100B in 1999 US dollars[60].

This led to a fundamental intertwining of problem formulation with problem solving for new, inventive industrial problems. Around 2005, the work flow illustrated in Fig. 1.50 became routine in solving industrial-strength optimal control problems. This is how problem formulation became *the key* to solving practical real-world problems[87].

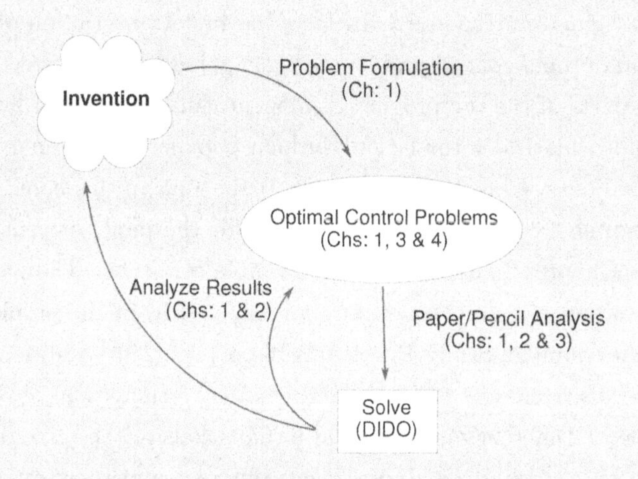

Figure 1.50: A typical work-flow in solving a present-day optimal control problem-of-problems. A free version of DIDO can be downloaded from http://www.ElissarGlobal.com.

Chapter 2

Pontryagin's Principle

You want proof?
You can't handle the proof!

In this chapter, we investigate and illustrate the optimality conditions for the standard optimal control problem developed in Chapter 1. The main result is called Pontryagin's Principle. It is also referred to as the Minimum Principle, the Maximum Principle,* Pontryagin's Minimum Principle or Pontryagin's Maximum Principle.

2.1 The Standard Problem

See Fig. 2.1. As discussed in §1.4, page 42, a standard optimal control problem can be defined in terms of finding a dynamically feasible state-control function pair, $\{\boldsymbol{x}(\cdot), \boldsymbol{u}(\cdot)\}$, that transfers the state of system, $\boldsymbol{x} \in \mathbb{R}^{N_x}$, from a given initial condition, $\boldsymbol{x}(t_0) = \boldsymbol{x}^0$, to a target condition, $\boldsymbol{e}(\boldsymbol{x}_f, t_f) = \mathbf{0}$, while minimizing a given cost functional, J. Following Chapter 1, this standard problem can be organized in a structured format† as

*See §2.5.5 for further discussions on minimum versus maximum.
†This format is nearly identical to the actual structure of a DIDO code.

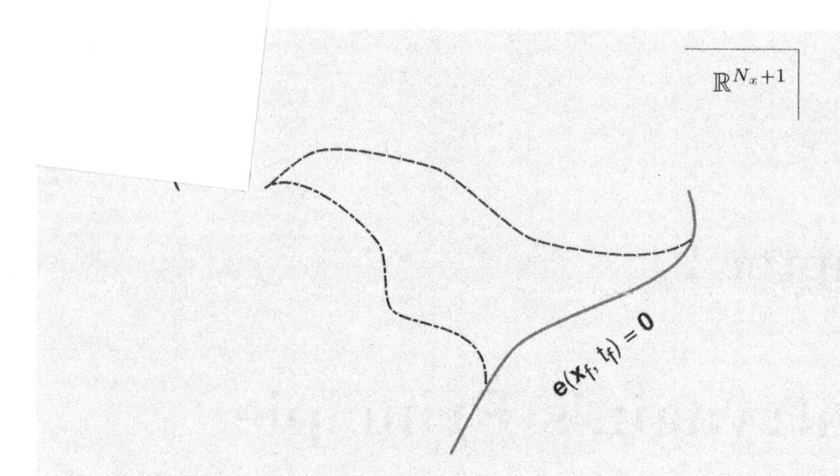

Figure 2.1: Schematic for the standard problem; same as Fig. 1.27, repeated here for quick reference.

$$\boldsymbol{x} \in \mathbb{X} = \mathbb{R}^{N_x} \quad \boldsymbol{u} \in \mathbb{U} \subseteq \mathbb{R}^{N_u} \qquad \Big\}\ \text{(preamble)}$$

$$\overbrace{(B)}^{\text{problem}} \left\{ \begin{array}{lrl} \text{Minimize} & J[\boldsymbol{x}(\cdot), \boldsymbol{u}(\cdot), t_f] = & E(\boldsymbol{x}(t_f), t_f) \\ & & + \displaystyle\int_{t_0}^{t_f} F(\boldsymbol{x}(t), \boldsymbol{u}(t), t)\, dt \\ \text{Subject to} & \dot{\boldsymbol{x}}(t) = & \boldsymbol{f}(\boldsymbol{x}(t), \boldsymbol{u}(t), t) \\ & \boldsymbol{x}(t_0) = & \boldsymbol{x}^0 \\ & t_0 = & t^0 \\ & \mathbf{e}(\boldsymbol{x}_f, t_f) = & \mathbf{0} \end{array} \right. \quad \begin{array}{l} \text{(cost)} \\ \\ \text{(dynamics)} \\ \\ \text{(endpoints)} \end{array}$$

In addition to $\mathbb{U}, \boldsymbol{x}^0$ and t^0, the four functions,

$$\begin{array}{ll} E : (\boldsymbol{x}_f, t_f) \mapsto \mathbb{R} & \text{(endpoint cost)} \\ \mathbf{e} : (\boldsymbol{x}_f, t_f) \mapsto \mathbb{R}^{N_e} & \text{(endpoint constraint)} \\ F : (\boldsymbol{x}, \boldsymbol{u}, t) \mapsto \mathbb{R} & \text{(running cost)} \\ \boldsymbol{f} : (\boldsymbol{x}, \boldsymbol{u}, t) \mapsto \mathbb{R}^{N_x} & \text{(dynamics)} \end{array}$$

are called the ***problem data***. The data functions are assumed to be continuously differentiable with respect to the state variable.

Notation Alert: By $\boldsymbol{x} \in \mathbb{R}^{N_x}$, we mean the N_x-dimensional state variable. This is distinguished from $\boldsymbol{x}(\cdot)$, which is a function (for example, $t \mapsto \boldsymbol{x}$), the graph of which, with respect to time, is a curve. The function $\boldsymbol{x}(\cdot)$ is called the ***state trajectory*** or *arc* (the latter term is somewhat antiquated). Similarly, $\boldsymbol{u} \in \mathbb{R}^{N_u}$ is the N_u-dimensional control variable, while $\boldsymbol{u}(\cdot)$ is the control function (for example, $t \mapsto \boldsymbol{u}$) called the ***control trajectory***; see Fig. 1.26. The pair $\{\boldsymbol{x}(\cdot), \boldsymbol{u}(\cdot)\}$ is called the ***system trajectory***. In addition, $\boldsymbol{x}_0$ is shorthand for $\boldsymbol{x}(t_0)$, while $\boldsymbol{x}^0$ means generic numerical data. Similar notation is used for the the final-time conditions, as well. Thus, for example, the set defined by the equation $\boldsymbol{e}(\boldsymbol{x}_f, t_f) = \boldsymbol{0}$ means that $\boldsymbol{x}_f$ and t_f are variables in $\mathbb{R}^{N_x} \times \mathbb{R}$, whereas $\boldsymbol{x}^f$ means a specific value of $\boldsymbol{x}(t_f) \equiv \boldsymbol{x}_f$; see Fig. 2.1. Finally, we use upper cases for cost (or cost-like) functions and lower cases for the corresponding constraints. Thus the pairs $(E, \boldsymbol{e})$ and $(F, \boldsymbol{f})$ go together. ***See page xix for additional nomenclature.***

In this standard problem formulation, the control space, $\mathbb{U}$, is quite an arbitrary set: it may be ***continuous*** (i.e. a subset of $\mathbb{R}^{N_u}$ with the power of the continuum), ***discrete*** (i.e. have finite cardinality), disjoint, non-convex ... and a host of other possibilities that are crucial for practical applications; see Fig. 2.2. The folklore that practical sets are "nice" is simply not true. A simple example of such a practical set is shown in Fig. 2.3.

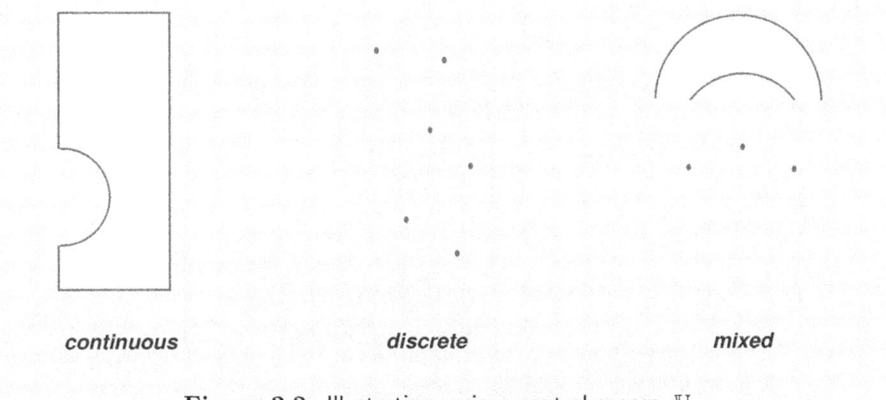

Figure 2.2: Illustrating various control spaces, $\mathbb{U}$.

Problem B was conceived by Pontryagin in 1955[33]. Today, on hindsight, it is a clear and natural way to formulate many problems in engineering, economics, physics, management, environment, social sciences, epidemiology and

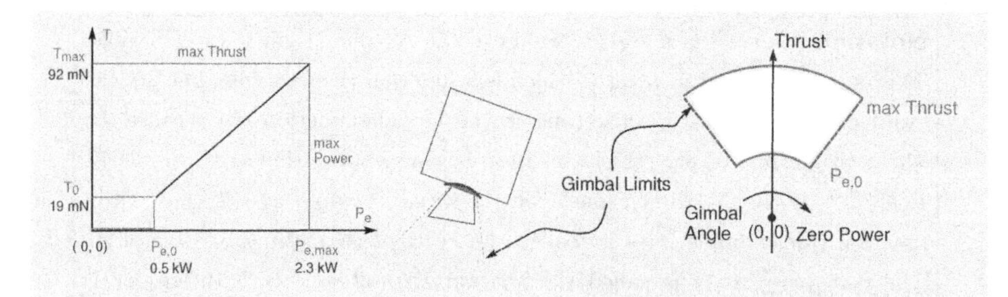

Figure 2.3: The control space for a practical electrically powered space thruster is nonconvex and disjoint: Zero power yields zero thrust; however, a minimum amount of nonzero power, $P_{e,0}$, is necessary to generate a nonzero minimum thrust, T_0.

other disciplines. In the 1950s, there was an absence of this clarity because the concepts in control and optimization were still evolving. The symbol $\boldsymbol{u}$ was adopted for control because, in Russian, ***upravlenie*** means control. The ***calculus of optimal control*** developed by Pontryagin and his students requires the concept of covectors.

Illustrating the Concept: Problem Formulation

As developed in Chapter 1, the Brachistochrone problem fits quite readily under the constructs of Problem B. Recall that the Brac:1 formulation (see page 12) is given by

$$\left.\begin{array}{ll} \mathbb{X} = \mathbb{R}^3 & \mathbb{U} = \mathbb{R} \\ \boldsymbol{x} = (x, y, v) & \boldsymbol{u} = \theta \end{array}\right\} \textit{(preamble)}$$

$$\underbrace{\text{problem}}_{(Brac:1)} \left\{ \begin{array}{lll} \textit{Minimize} & J[\boldsymbol{x}(\cdot), \boldsymbol{u}(\cdot), t_f] = t_f & \} \; \textit{(cost)} \\ \textit{Subject to} & \left.\begin{array}{r} \dot{x} = v \sin\theta \\ \dot{y} = v\cos\theta \\ \dot{v} = g\cos\theta \end{array}\right\} & \textit{(dynamics)} \\ & \left.\begin{array}{r} (t_0, x_0, y_0, v_0) = (0,0,0,0) \\ (x_f - x^f, y_f - y^f) = (0,0) \end{array}\right\} & \textit{(endpoints)} \end{array}\right.$$

In mapping the various symbols and notation of Problem B to Brac:1, we have $N_x = 3$, $N_u = 1$, $E(\boldsymbol{x}_f, t_f) = t_f$, $F = \Theta$, $f_1(\boldsymbol{x}, \boldsymbol{u}) = v\sin\theta$, $f_2(\boldsymbol{x}, \boldsymbol{u}) = v\cos\theta$, $f_3(\boldsymbol{x}, \boldsymbol{u}) = g\cos\theta$, $e_1(\boldsymbol{x}_f, t_f) = x_f - x^f$, *and* $e_2(\boldsymbol{x}_f, t_f) = y_f - y^f$.

2.2 Introduction to Covectors

By definition, the state vector $\boldsymbol{x}$ is just a medley of N_x variables stacked on top of one another:

$$\boldsymbol{x} := \begin{bmatrix} x_1 \\ x_2 \\ \vdots \\ x_{N_x} \end{bmatrix} \begin{matrix} x_1\text{-units} \\ x_2\text{-units} \\ \vdots \\ x_{N_x}\text{-units} \end{matrix} \tag{2.1}$$

We have purposefully written Eq. (2.1) with its units alongside to emphasize the point that $\boldsymbol{x}$ typically comprises a collection of "heterogeneous" scalars. Thus, our intuitive notion of vector as a quantity with magnitude and direction must be seriously revised.

Illustrating the Concept in $\mathbb{R}^3$

In (Brac:1), the state vector is defined by

$$\boldsymbol{x} := \begin{bmatrix} x \\ y \\ v \end{bmatrix} \begin{matrix} \textit{position-units} \\ \textit{position-units} \\ \textit{velocity-units} \end{matrix} \tag{2.2}$$

If the "magnitude" of $\boldsymbol{x}$ is computed by the standard Euclidean norm

$$\|\boldsymbol{x}\|_2 = \sqrt{x^2 + y^2 + v^2}$$

the result would be physically meaningless. In addition, we cannot legitimately add the square of position units to the square of velocity units.

If a vector is defined naïvely as a quantity with magnitude and direction, what, then, is the magnitude of this vector? What is its direction? What happens to the magnitude and direction if we change units?

These simple observations indicate that we need a potentially new approach to define a meaningful and natural way to measure vectors.

To motivate a new set of ideas, consider the following everyday example: Suppose we describe a sandwich that comprises the variables $x_1, x_2, x_3 \ldots$ where x_1 is bread, x_2 is mayonnaise, x_3 is lettuce, x_4 is meat, and so on. Then it is clear that we can describe the state of the sandwich in terms of a vector $\boldsymbol{x}$, of N_x ingredients, in much the same way as Eq. (2.1). Each component of this vector can be measured in terms of appropriate units. For instance, x_1 can be measured in terms of the number of slices of bread, x_2 in terms of ounces of mayonnaise, x_3 in terms of the number of lettuce leaves and so on. See Fig. 2.4. With these preliminaries, consider a two-dimensional model of a sandwich composed of just bread and meat

$$\boldsymbol{x} := \begin{bmatrix} b \\ m \end{bmatrix} \begin{matrix} \textit{slices} \\ \textit{ounces} \end{matrix}$$

Figure 2.4: State vector of a sandwich.

where b is bread measured in number of slices and m is meat measured in ounces. Suppose the numerical value of the state vector is given by

$$\boldsymbol{x} = \begin{bmatrix} 2 \\ 5 \end{bmatrix} \begin{matrix} \textit{slices} \\ \textit{ounces} \end{matrix}$$

Obviously, we would not compute the size of the sandwich using Euclidean ideas

$$\|\boldsymbol{x}\| = \sqrt{4 \text{ slices}^2 + 25 \text{ ounces}^2}$$

One common approach to measure the "size of the sandwich" is in terms of its calorie count. Suppose that each slice of bread is 80 *calories* and each ounce of (lean) meat is 55 *calories*; then, the calorie count of the sandwich is given by

$$\underbrace{80 \times 2}_{\textit{bread calories}} + \underbrace{55 \times 5}_{\textit{meat calories}} = \underbrace{435}_{\textit{total calories}}$$

We formalize this mathematics as follows: We agree to measure the "size" of the vector $\boldsymbol{x}$ (representing the sandwich) in terms of a ***common unit*** (CU) (which is calories for the sandwich). Then we let $\lambda_b = 80$ calories per slice and $\lambda_m = 55$ calories per slice and perform a ***linear measure*** of $\boldsymbol{x}$ using the (linear) operation

$$\begin{aligned} \omega(\boldsymbol{x}) &:= \lambda_b \times b + \lambda_m \times m \\ &= 80 \times 2 + 55 \times 5 \\ &= 435 \; calories \end{aligned} \tag{2.3}$$

The linear measure of a vector is not unique in the following sense: Instead of the calorie count, suppose we decided to measure the size of the sandwich in terms of its weight, a process adopted in many cafeterias. Then suppose we have $\widetilde{\lambda}_b = 1$ ounce per slice and $\widetilde{\lambda}_m = 1$ ounce per ounce; this generates a new linear measure of $\boldsymbol{x}$ that uses the same linear operation as before, but with new coefficients:

$$\begin{aligned} \omega(\boldsymbol{x}) &= \widetilde{\lambda}_b \times b + \widetilde{\lambda}_m \times m \\ &= 1 \times 2 + 1 \times 5 \\ &= 7 \; ounces \end{aligned}$$

Yet another way to measure the size of the sandwich is its dollar cost. Suppose $\widehat{\lambda}_b = 20$ cents per slice and $\widehat{\lambda}_m = 30$ cents per ounce; then, the new linear measure of $\boldsymbol{x}$ is given by

$$\begin{aligned} \omega(\boldsymbol{x}) &= \widehat{\lambda}_b \times b + \widehat{\lambda}_m \times m \\ &= 20 \times 2 + 30 \times 5 \\ &= 190 \; cents \; (= \$1.90) \end{aligned}$$

Thus, there are many ways to construct a linear measure of $\boldsymbol{x}$.

Study Problem 2.1

1. *Construct at least two additional linear measures for a sandwich. What are the numerical values of λ_b and λ_m for these new linear measures?*

2. *Pick any two items from your grocery bag that have an FDA label pasted on them. Identify at least two linear measures adopted by the FDA to inform a consumer.*

We formalize our sandwich example and declare that we will measure vectors in $\mathbb{R}^{N_x}$ by a linear scalar function, ω, defined by

$$\omega(\boldsymbol{x}) := \lambda_1 x_1 + \lambda_2 x_2 + \cdots + \lambda_{N_x} x_{N_x} \tag{2.4}$$

The collection of N_x quantities, $\lambda_1, \lambda_2, \ldots \lambda_{N_x}$, that *we choose to measure a vector* can be stacked up in a manner similar to $\boldsymbol{x}$

$$\boldsymbol{\lambda} := \begin{bmatrix} \lambda_1 \\ \lambda_2 \\ \vdots \\ \lambda_{N_x} \end{bmatrix} \begin{matrix} CU/x_1\text{-units} \\ CU/x_2\text{-units} \\ \vdots \\ CU/x_{N_x}\text{-units} \end{matrix} \tag{2.5}$$

to produce a new vector, $\boldsymbol{\lambda}$. This vector is called a ***covector***. Obviously, $\boldsymbol{\lambda} \in \mathbb{R}^{N_x}$; however, as evident from its units, ***its home-space is not*** $\mathbb{X}$; hence:

$$\boxed{\boldsymbol{\lambda} \in \mathbb{R}^{N_x} \quad \text{but} \quad \boldsymbol{\lambda} \notin \mathbb{X}}$$

That is, even though $\boldsymbol{\lambda}$ has the same dimension as $\boldsymbol{x}$, it does not occupy the same space as $\boldsymbol{x}$. *Hence,* $\boldsymbol{x} + \boldsymbol{\lambda}$ *is not a legal operation.* Obviously, $\boldsymbol{x}_1 + \boldsymbol{x}_2$ is a legal operation and so is $\boldsymbol{\lambda}_1 + \boldsymbol{\lambda}_2$. Thus, we have created a new vector space, out of thin air (!), that is different from $\mathbb{X}$. This space is called a ***dual space***; it is the home space of $\boldsymbol{\lambda}$. The dual space is said to be dual to the original or ***primal space***, $\mathbb{X}$. See Fig. 2.5.

Tech Talk: Strictly speaking, it is the *linear scalar function* ω, given by Eq. (2.4), that is called a covector. In fact, a dual vector space, or ***covector space***, is the space of all linear scalar functions, $\omega : \mathbb{X} \to \mathbb{R}$ where $\mathbb{X}$ can be any vector space. In our case, because $\mathbb{X}$ is finite-dimensional, the difference between ω and $\boldsymbol{\lambda}$ seems nuanced. The vector $\boldsymbol{\lambda}$ is called a ***representation*** of the linear function ω. Following the age-old practice of convenient abuse of terminology, we call $\boldsymbol{\lambda}$ a covector. It is fair warning to say that we have taken much more liberty than terminology in introducing the concept of a covector. A mathematician will undoubtedly cringe at this presentation in much the same way as science fictionados react to the phrase "Vulcan mind trick."

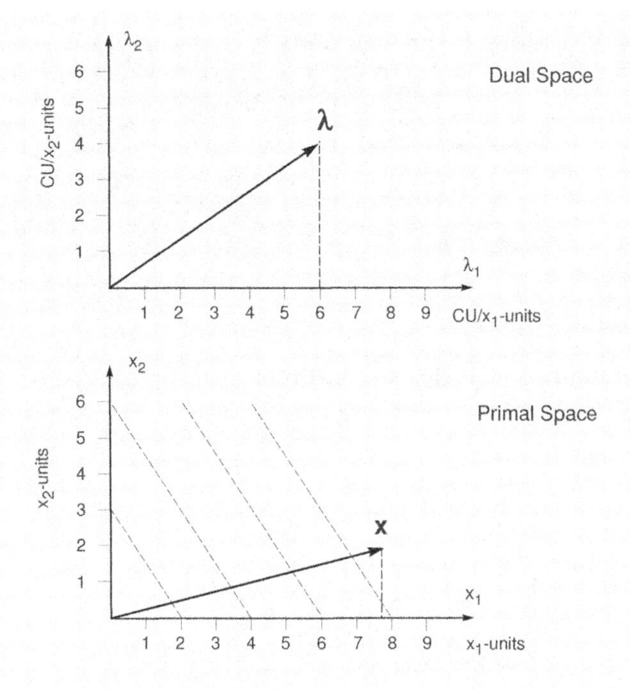

Figure 2.5: Primal and dual vector spaces. Also shown in $\mathbb{X}$ are the lines defined by $\omega(\boldsymbol{x})$ = constant for different values of the constant. Thus, the "true" covector ω grids the space $\mathbb{X}$ via lines (hyperplanes in N_x-dimensions) given by ω = constant.

2.3 The Covectors in Optimal Control

The *common unit* for measurement in an optimal control problem is the ***cost unit***; that is, the unit for measuring the value of J. On the basis of this *common unit*, we define a covector $\boldsymbol{\lambda}$ in exactly the manner as Eq. (2.5), where *CU* now stands for the *cost unit*! This particular covector is called the ***costate***.

Illustrating the Concept: Constructing the Costate

In Brac:1, $\boldsymbol{x} = (x, y, v) \in \mathbb{R}^3$. This implies that $\boldsymbol{\lambda}$ is in $\mathbb{R}^3$, as well. To aid its bookkeeping, we deliberately choose the subscripts of the components of $\boldsymbol{\lambda} \in \mathbb{R}^3$ to be λ_x, λ_y and λ_v so that it is properly tagged to the relevant *components of the state variables. Thus, the costate for Brac:1 is given by*

$$\boldsymbol{\lambda} := \begin{bmatrix} \lambda_x \\ \lambda_y \\ \lambda_v \end{bmatrix} \begin{matrix} TU/x\text{-units} \\ TU/y\text{-units} \\ TU/v\text{-units} \end{matrix} \tag{2.6}$$

where we have used the fact that the cost unit in Brac:1 is the same as the time unit, TU.

A quick examination of the data in Problem B reveals that

$$\boldsymbol{f}(\boldsymbol{x}, \boldsymbol{u}, t) \in \mathbb{R}^{N_x}$$

which can be elaborated to

$$\boldsymbol{f}(\boldsymbol{x}, \boldsymbol{u}, t) := \begin{bmatrix} f_1(\boldsymbol{x}, \boldsymbol{u}, t) \\ f_2(\boldsymbol{x}, \boldsymbol{u}, t) \\ \vdots \\ f_{N_x}(\boldsymbol{x}, \boldsymbol{u}, t) \end{bmatrix} \begin{matrix} x_1\text{-units}/t\text{-units} \\ x_2\text{-units}/t\text{-units} \\ \vdots \\ x_{N_x}\text{-units}/t\text{-units} \end{matrix} \quad \in \ \mathbb{R}^{N_x}$$

Illustrating the Concept: Constructing the Vector, $\boldsymbol{f}(\boldsymbol{x}, \boldsymbol{u}, t)$

In Brac:1, if we choose

$$\boldsymbol{x}_1 = \begin{bmatrix} -1 \\ 0 \\ 1 \end{bmatrix} \begin{matrix} m \\ m \\ m/s \end{matrix} \quad \text{and} \quad \boldsymbol{u}_1 = \pi/8$$

we get (by taking $g = 9.8\, m/s^2$)

$$\boldsymbol{f}(\boldsymbol{x}_1, \boldsymbol{u}_1) = \begin{bmatrix} 0.3827 \\ 0.9239 \\ 9.0540 \end{bmatrix} \begin{matrix} m/s \\ m/s \\ m/s^2 \end{matrix} \in \mathbb{R}^3$$

This vector is plotted in Fig. 2.6.

Clearly, the dimension of $\boldsymbol{f}(\boldsymbol{x}, \boldsymbol{u}, t)$ is always exactly equal to the dimension

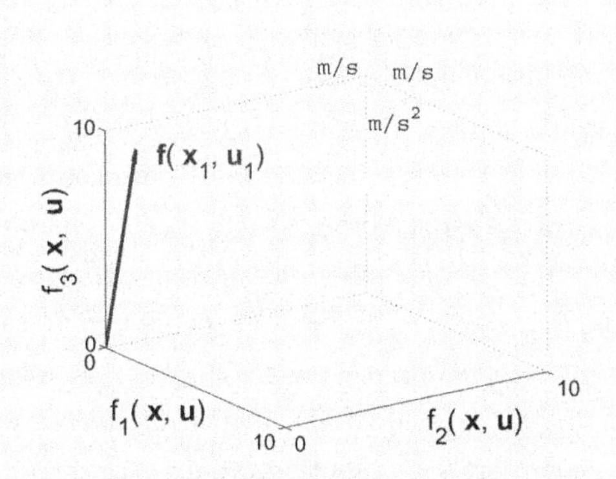

Figure 2.6: Illustration of $\boldsymbol{f}(\boldsymbol{x}, \boldsymbol{u})$ as a vector in $\mathbb{R}^3$ for *Brac:1*. Note the units and compare to Fig. 1.6 where g was scaled to 1.0.

of $\mathbb{X}$; however, as apparent by its units, note that

$$\boxed{\boldsymbol{f}(\boldsymbol{x}, \boldsymbol{u}, t) \in \mathbb{R}^{N_x} \quad \text{but} \quad \boldsymbol{f}(\boldsymbol{x}, \boldsymbol{u}, t) \notin \mathbb{X}}$$

By direct substitution, it can be easily verified that the "dot product"

$$\begin{aligned} &\boldsymbol{\lambda} \cdot \boldsymbol{f}(\boldsymbol{x}, \boldsymbol{u}, t) \\ &:= \sum_{i=1}^{N_x} \lambda_i f_i(\boldsymbol{x}, \boldsymbol{u}, t) \\ &\textit{(in CU/TU units)} \end{aligned}$$

is a legal operation and generates a scalar in CU/TU-units where TU is the unit of time or the independent variable. That is, we can use $\boldsymbol{\lambda}$ to measure $\boldsymbol{f}(\boldsymbol{x}, \boldsymbol{u}, t)$.

Recall from the discussions on page 89 that the dot product, $\boldsymbol{x} \cdot \boldsymbol{x}$, is not a legal operation and neither is $\boldsymbol{f}(\boldsymbol{x}, \boldsymbol{u}, t) \cdot \boldsymbol{f}(\boldsymbol{x}, \boldsymbol{u}, t)$; but $\boldsymbol{\lambda} \cdot \boldsymbol{f}(\boldsymbol{x}, \boldsymbol{u}, t)$ is indeed legal. This point further reinforces the fact that $\boldsymbol{\lambda}$ is a *representation of a linear scalar function* (called a ***linear form***) and that the space $\mathbb{X}$ is not necessarily equipped with a legal dot product.

Computational Tip: In a computational environment, a majority of software (including DIDO) compute quantities like $(\boldsymbol{x} \cdot \boldsymbol{x})$, $(\boldsymbol{u} \cdot \boldsymbol{u})$, etc. with ruthless and total disregard for units. This computational policy is used near universally in many software products as a unified means to measure distances, lengths and so on. Hence, it is imperative that the designer of a *specific* computational problem choose judicious units that do not mask the relative importance of one component (or variable) against another. In Fig. 2.6, the numerical value of $f_3(\boldsymbol{x}_1, \boldsymbol{u}_1)$ dominates over the values of $f_1(\boldsymbol{x}_1, \boldsymbol{u}_1)$ and $f_2(\boldsymbol{x}_1, \boldsymbol{u}_1)$ but this is not the case in Fig. 1.6. Refer back to §1.1.4, page 20, for further discussions on scaling. DIDO also uses a myriad of covectors for computing various other quantities as discussed in much of this chapter; hence, the scaling procedures discussed in §1.1.4 must be balanced with this perspective. Additional concepts and details on ***scaling and balancing*** are discussed in [81] and [84].

The unit of measurement for the running cost, F, is CU/TU; hence, we can legally add $F(\boldsymbol{x}, \boldsymbol{u}, t)$ to the scalar, $\boldsymbol{\lambda} \cdot \boldsymbol{f}(\boldsymbol{x}, \boldsymbol{u}, t)$. The sum of these two quantities,

$$\boxed{H(\boldsymbol{\lambda}, \boldsymbol{x}, \boldsymbol{u}, t) := F(\boldsymbol{x}, \boldsymbol{u}, t) + \boldsymbol{\lambda}^T \mathbf{f}(\boldsymbol{x}, \boldsymbol{u}, t)} \tag{2.7}$$

in CU/TU units is called the ***Hamiltonian for Problem B***.

Illustrating the Concept: Constructing the Hamiltonian

There is no running cost in Brac:1; hence the Hamiltonian is given by "dotting" $\boldsymbol{\lambda}$ to $\boldsymbol{f}(\boldsymbol{x}, \boldsymbol{u}, t)$ where $\boldsymbol{\lambda}$ is given by Eq. (2.6). Performing this simple operation, we have

$$H(\boldsymbol{\lambda}, \boldsymbol{x}, \boldsymbol{u}) := \lambda_x v \sin\theta + \lambda_y v \cos\theta + \lambda_v g \cos\theta \tag{2.8}$$

The Hamiltonian here is not an explicit function of time; hence, it is written as $H(\boldsymbol{\lambda}, \boldsymbol{x}, \boldsymbol{u})$ and not as $H(\boldsymbol{\lambda}, \boldsymbol{x}, \boldsymbol{u}, t)$. Note that the unit of this particular Hamiltonian is given by TU/TU; hence, H is dimensionless.

Tech Talk: The Hamiltonian is also known by other names in the literature: the Pontryagin-Hamiltonian, the Pontryagin H-function, the "unminimized Hamiltonian," the pseudo-Hamiltonian (older literature) or the control Hamiltonian. Later, when it is necessary to use other Hamiltonians, we will use other adjectives to clarify the appropriate reference. In this book, the adjective-free Hamiltonian will always be the ***control Hamiltonian*** given by Eq. (2.7).

The costate $\boldsymbol{\lambda}$ is in $\mathbb{R}^{N_x}$; hence, in principle, we can use it to measure any appropriate vector in $\mathbb{R}^{N_x}$. In the construction of the Hamiltonian, we used it to measure the vector $\boldsymbol{f}(\boldsymbol{x}, \boldsymbol{u}, t)$. We can use it to measure $\boldsymbol{x}$, as well, but it doesn't generate anything useful. This is because, it turns out, that what matters in optimal control is not the measurement of "input" vectors like $\boldsymbol{x}$ and $\boldsymbol{u}$, but their "outputs" or combined effects given in the form of the problem data.

By inspection, Problem B has four data functions, $E, F, \boldsymbol{f}$ and $\boldsymbol{e}$:

$$
\begin{aligned}
E &: (\boldsymbol{x}_f, t_f) \mapsto \mathbb{R} && \text{(CU)} \\
\boldsymbol{e} &: (\boldsymbol{x}_f, t_f) \mapsto \mathbb{R}^{N_e} && (e\text{-units}) \\
F &: (\boldsymbol{x}, \boldsymbol{u}, t) \mapsto \mathbb{R} && \text{(CU/TU)} \\
\boldsymbol{f} &: (\boldsymbol{x}, \boldsymbol{u}, t) \mapsto \mathbb{R}^{N_x} && (\boldsymbol{f}\text{-units} = \boldsymbol{x}\text{-units/TU})
\end{aligned}
$$

The functions E and F are scalar-valued. Only two of the data functions, $\boldsymbol{f}$ and $\boldsymbol{e}$, are vector functions. We applied the covector $\boldsymbol{\lambda}$ to measure the vector $\boldsymbol{f}(\boldsymbol{x}, \boldsymbol{u}, t)$. We cannot apply $\boldsymbol{\lambda}$ to measure $\boldsymbol{e}$ because it does not generate a legal operation either in terms of its units or dimensions. To see this, examine the vector, $\boldsymbol{e}(\boldsymbol{x}_f, t_f) \in \mathbb{R}^{N_e}$, more closely:

$$
\boldsymbol{e}(\boldsymbol{x}_f, t_f) := \begin{bmatrix} e_1(\boldsymbol{x}_f, t_f) \\ e_2(\boldsymbol{x}_f, t_f) \\ \vdots \\ e_{N_e}(\boldsymbol{x}_f, t_f) \end{bmatrix} \begin{matrix} e_1\text{-units} \\ e_2\text{-units} \\ \vdots \\ e_{N_e}\text{-units} \end{matrix} \quad \in \mathbb{R}^{N_e}
$$

In general $N_e \neq N_x$; hence, the operation $\boldsymbol{\lambda} \cdot \boldsymbol{e}(\boldsymbol{x}_f, t_f)$ cannot be performed. Even in the special case of $N_e = N_x$, the dot product $\boldsymbol{\lambda} \cdot \boldsymbol{e}(\boldsymbol{x}_f, t_f)$ does not generate a legal operation in terms of units. This suggests that we need to define an ***endpoint covector***, in a manner similar to $\boldsymbol{\lambda}$, and given by

$$\boldsymbol{\nu} := \begin{bmatrix} \nu_1 \\ \nu_2 \\ \vdots \\ \nu_{N_e} \end{bmatrix} \quad \begin{matrix} CU/e_1\text{-units} \\ CU/e_2\text{-units} \\ \vdots \\ CU/e_{N_e}\text{-units} \end{matrix} \tag{2.9}$$

This covector legalizes the "dot product"

$$\boldsymbol{\nu} \cdot \boldsymbol{e}(\boldsymbol{x}_f, t_f) := \sum_{i=1}^{N_e} \nu_i e_i(\boldsymbol{x}_f, t_f)$$

and produces a measurement mechanism for the data given by the $\boldsymbol{e}$-function. It can be easily verified that the unit of the scalar resulting from this operation is CU. This is exactly the same as the unit of E; hence, we can legally add E to the scalar product $\boldsymbol{\nu} \cdot \boldsymbol{e}(\boldsymbol{x}_f, t_f)$, and generate the quantity,

$$\boxed{\overline{E}(\boldsymbol{\nu}, \boldsymbol{x}_f, t_f) := E(\boldsymbol{x}_f, t_f) + \boldsymbol{\nu}^T \boldsymbol{e}(\boldsymbol{x}_f, t_f)} \tag{2.10}$$

The function $\overline{E}$ is called the ***Endpoint Lagrangian***.

Illustrating the Concept: Constructing the Endpoint Lagrangian

For Problem Brac:1, $N_e = 2$; hence $\boldsymbol{\nu} \in \mathbb{R}^2$. This implies that we can define

$$\boldsymbol{\nu} := \begin{bmatrix} \nu_1 \\ \nu_2 \end{bmatrix} \quad \begin{matrix} TU/x\text{-units} \\ TU/y\text{-units} \end{matrix} \tag{2.11}$$

Then, the Endpoint Lagrangian is given by

$$\overline{E}(\boldsymbol{\nu}, \boldsymbol{x}_f, t_f) := t_f + \nu_1(x_f - x^f) + \nu_2(y_f - y^f), \quad \textit{in } TUs \tag{2.12}$$

Thus, through a construction of two covectors, $\boldsymbol{\lambda}$ and $\boldsymbol{\nu}$, we have essentially packed all four data functions ($E, F, \boldsymbol{f}$ and $\boldsymbol{e}$) of Problem B in just two bags of scalar functions, H and E. These scalar functions form the bases for generating a collection of conditions known as Pontryagin's Principle.

2.4 Introducing Pontryagin's Principle

By definition, the cost functional J is the prescribed means to measure performance. That is, we say $\boldsymbol{u}^1(\cdot)$ is a better control trajectory than $\boldsymbol{u}^2(\cdot)$ if the cost for executing the former is lower than the latter. Hence, we can say that J is the chosen means to measure the "length" of $\boldsymbol{u}(\cdot)$, and that the problem is to find the "smallest" $\boldsymbol{u}(\cdot)$ measured in terms of the values of J. Recall that we are not interested in $\boldsymbol{u}(\cdot)$ only, and that the decision or independent variables are the triple[‡]

$$\{\boldsymbol{x}(\cdot), \boldsymbol{u}(\cdot), t_f\}$$

That is, J measures the entire triple by some formula

$$J : \{\boldsymbol{x}(\cdot), \boldsymbol{u}(\cdot), t_f\} \mapsto \mathbb{R}$$

chosen by the problem designer. See Fig. 2.7. In this context, it is the problem

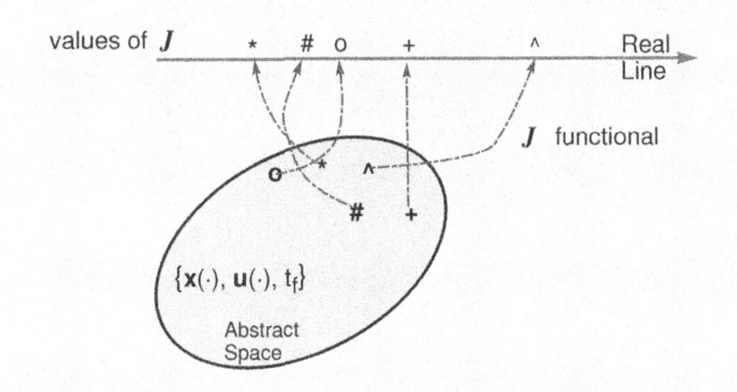

Figure 2.7: The functional J is "implottable." It can be visualized as a quantity that maps the triple $\{\boldsymbol{x}(\cdot), \boldsymbol{u}(\cdot), t_f\}$ to a real number. The leftmost point (*) on $\mathbb{R}$ is the optimal value.

designer who stipulates that J is the mechanism to measure the system trajectory; hence, we declare the decision variables, $\boldsymbol{x}^*(\cdot), \boldsymbol{u}^*(\cdot)$ and t_f^*, to be optimal if

$$J[\boldsymbol{x}^*(\cdot), \boldsymbol{u}^*(\cdot), t_f^*] \leq J[\boldsymbol{x}(\cdot), \boldsymbol{u}(\cdot), t_f] \tag{2.13}$$

[‡]Refer back to Section 1.4.3 (page 48) on avoiding common errors # 2.

for all $\boldsymbol{x}(\cdot), \boldsymbol{u}(\cdot)$ and t_f that are feasible. Pontryagin's Principle is based on the idea that this ***defining inequality*** can be transferred to the *instantaneous minimization of the Hamiltonian function* provided that the covectors are chosen appropriately. In other words, we have, so far, defined $\boldsymbol{\lambda}$ and $\boldsymbol{\nu}$ only in terms of their units and dimensions, and nothing else. As these covectors are the measurement devices, Pontryagin's Principle is based on translating the definition given by Eq. (2.13) to additional requirements on $\boldsymbol{\lambda}$ and $\boldsymbol{\nu}$. These concepts can be visualized as shown in Fig. 2.8.

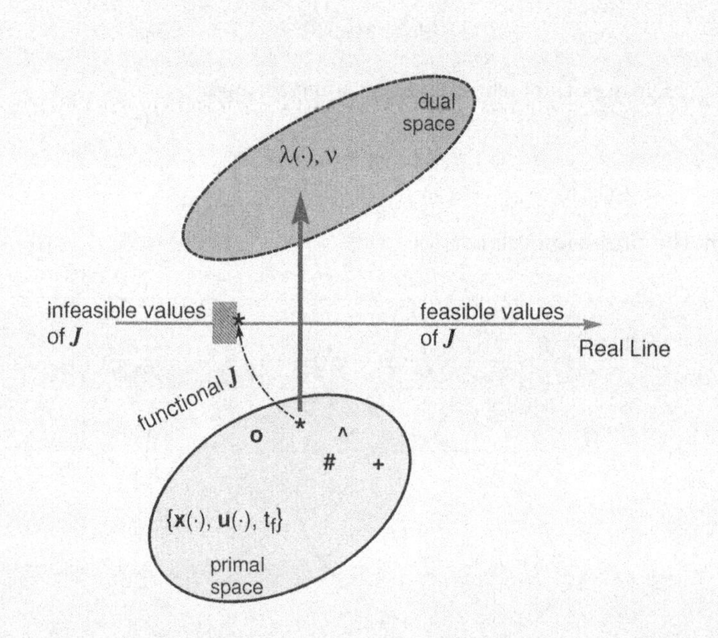

Figure 2.8: Pontryagin's Principle is based on the idea of transforming the defining equation (2.13) to computable conditions in the dual (covector) space.

2.4.1 The Basics

The Hamiltonian packs an enormous amount of information on the optimal control problem. Recall that

$$H(\boldsymbol{\lambda}, \boldsymbol{x}, \boldsymbol{u}, t) := F(\boldsymbol{x}, \boldsymbol{u}, t) + \boldsymbol{\lambda}^T \mathbf{f}(\boldsymbol{x}, \boldsymbol{u}, t) \qquad \textit{(in CU/TU units)}$$

From the elementary observation

$$\frac{\partial H}{\partial \boldsymbol{\lambda}} = \boldsymbol{f}(\boldsymbol{x}, \boldsymbol{u}, t)$$

we can immediately write:

$$\dot{\boldsymbol{x}} = \frac{\partial H}{\partial \boldsymbol{\lambda}} \tag{2.14}$$

That is, we can recover the state dynamical equations from the Hamiltonian.

Study Problem 2.2

Show that the units of $\partial H / \partial \boldsymbol{\lambda}$ are the same as that of $\dot{\boldsymbol{x}}$.

Next, observe that

$$\text{units of } \frac{\partial H}{\partial \boldsymbol{x}} = \frac{CU/TU}{\boldsymbol{x}\text{-units}} = \frac{CU/\boldsymbol{x}\text{-units}}{TU} = \frac{\boldsymbol{\lambda}\text{-units}}{TU} = \text{units of } \dot{\boldsymbol{\lambda}}$$

This is not a coincidence! It turns out that the costate satisfies the adjoint differential equation or, simply, the ***adjoint equation***

$$-\dot{\boldsymbol{\lambda}} = \frac{\partial H}{\partial \boldsymbol{x}} \tag{2.15}$$

where the minus sign is part of the mathematical technicality that makes Eq. (2.15) into an "adjoint": The adjoint of the differential operator, d/dt is given by $-d/dt$. This is why the costate is also known as the ***adjoint covector***.

Illustrating the Concept: Constructing the Adjoint Equations

From Eq. (2.15), the adjoint equations for Brac:1 are given by

$$\begin{aligned} -\dot{\lambda}_x &:= \partial_x H(\boldsymbol{\lambda}, \boldsymbol{x}, \boldsymbol{u}) &&= 0 \\ -\dot{\lambda}_y &:= \partial_y H(\boldsymbol{\lambda}, \boldsymbol{x}, \boldsymbol{u}) &&= 0 \\ -\dot{\lambda}_v &:= \partial_v H(\boldsymbol{\lambda}, \boldsymbol{x}, \boldsymbol{u}) &&= \lambda_x \sin\theta + \lambda_y \cos\theta \end{aligned} \tag{2.16}$$

The H-function is called a Hamiltonian because the state-costate pair satisfies the same differential equation as the one encountered in ***Hamiltonian*** *mechanics*

$$\begin{aligned} \dot{q} &= \frac{\partial \mathcal{H}}{\partial p} \\ -\dot{p} &= \frac{\partial \mathcal{H}}{\partial q} \end{aligned} \tag{2.17}$$

where q is a generalized coordinate and p is the momentum.

The adjoint covector trajectory can be visualized in several different ways. In one approach that parallels the concept illustrated in Fig. 2.5 on page 93, it can be viewed as a "shadow trajectory" that shadows the state trajectory in dual space; see Fig. 2.9.

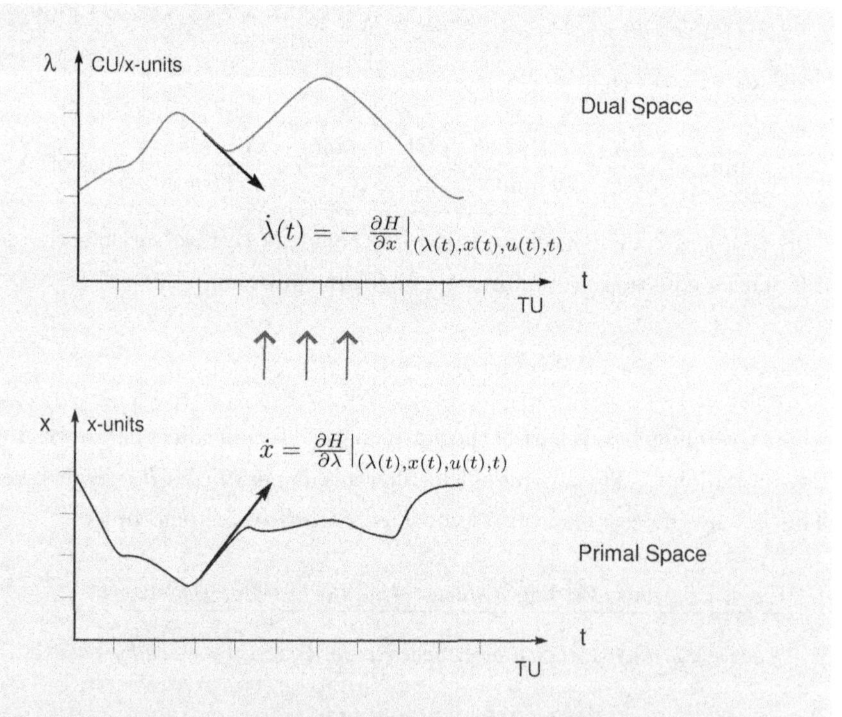

Figure 2.9: The costate trajectory in dual space can be viewed as a "shadow trajectory" of the state. In this context, Fig. 2.5 on page 93 is an instantaneous snapshot of x and λ in their respective spaces.

For the control function $u(\cdot)$ to be optimal, Pontryagin's Principle requires that u ***globally minimize the Hamiltonian for every*** $t \in [t_0, t_f]$. In other words, minimizing the Hamiltonian with respect to u (while holding λ and x

constant) yields a candidate solution for the optimal control. Hence, solving the ***pointwise static*** (finite dimensional) optimization problem in the parameter $\boldsymbol{u}$

$$(\text{HMC}) \quad \begin{cases} \underset{\boldsymbol{u}}{\text{Minimize}} & H(\boldsymbol{\lambda}, \boldsymbol{x}, \boldsymbol{u}, t) \\ \text{Subject to} & \boldsymbol{u} \in \mathbb{U} \end{cases}$$

at each instant of time t, yields a candidate function for the optimal control. This simple statement is called the ***Hamiltonian Minimization Condition*** (HMC) and is the heart of Pontryagin's Principle.

Illustrating the Concept: Constructing Problem HMC

Problem HMC for Brac:1 is quite simply formulated as the unconstrained minimization problem

$$\begin{cases} \underset{\theta}{\textit{Minimize}} & H(\boldsymbol{\lambda}, \boldsymbol{x}, \boldsymbol{u}) := \lambda_x v \sin\theta + \lambda_y v \cos\theta + \lambda_v g \cos\theta \\ \textit{Subject to} & \theta \in \mathbb{R} \end{cases}$$

where we have taken the usual liberty of calling θ a real number rather than its more appropriate representation, $\theta \in S^1$.

Tech Alert: The minimization in Problem HMC is to be performed only with respect to $\boldsymbol{u}$, and hence the notation, $\underset{\boldsymbol{u}}{\text{Minimize}}$ (i.e., H is to be regarded as a function of $\boldsymbol{u}$ ***only***, in Problem HMC).

Let us further clarify the meaning of Problem HMC by denoting its objective function as $R(\boldsymbol{u}, t)$:

$$(\boldsymbol{u}, t) \xrightarrow{R} H(\boldsymbol{\lambda}, \boldsymbol{x}, \boldsymbol{u}, t)$$

That is, let $R(\boldsymbol{u}, t)$ be the same as $H(\boldsymbol{\lambda}, \boldsymbol{x}, \boldsymbol{u}, t)$ with $\boldsymbol{\lambda}$ and $\boldsymbol{x}$ held constant. Then Problem HMC can be written without the clutter of $\boldsymbol{\lambda}$ and $\boldsymbol{x}$ as:

$$\begin{cases} \underset{\boldsymbol{u}}{\text{Minimize}} & R(\boldsymbol{u}, t) \\ \text{Subject to} & \boldsymbol{u} \in \mathbb{U} \end{cases} \qquad (\text{for each } t)$$

This concept is illustrated in Fig. 2.10 where a snapshot of the Hamiltonian as a function of $\boldsymbol{u}$ only (i.e., R) is taken at discrete times.

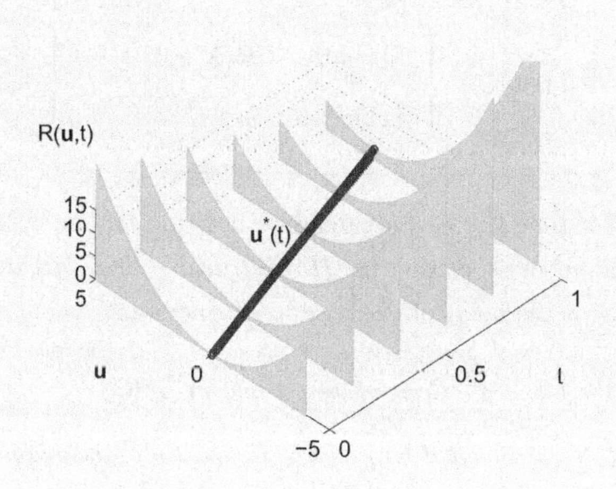

Figure 2.10: An illustration of the Hamiltonian Minimization Condition where $R(\boldsymbol{u},t)$ is H as a function of $\boldsymbol{u}$ only at each time stamp t.

Problem HMC is a finite-dimensional optimization problem; hence, all tools from finite dimensional optimization theory can be used to solve this problem.

Illustrating the Concept: Solving Problem HMC - 1/2

Problem HMC for Brac:1 can be solved quite readily. Since the problem is unconstrained, we can set $\partial_\theta H(\boldsymbol{\lambda}, \boldsymbol{x}, \boldsymbol{u}) = 0$ at each instant of time t. This immediately generates the equation:

$$\lambda_x v \cos\theta - \lambda_y v \sin\theta - \lambda_v g \sin\theta = 0 \tag{2.18}$$

According to Pontryagin's Principle, this equation must hold at each instant of time; hence, we can write:

$$\lambda_x(t)v(t)\cos\theta(t) - \lambda_y(t)v(t)\sin\theta(t) - \lambda_v(t)g\sin\theta(t) = 0$$

As a sidebar, note that $\partial_\theta^2 H(\boldsymbol{\lambda}, \boldsymbol{x}, \boldsymbol{u}) = -H(\boldsymbol{\lambda}, \boldsymbol{x}, \boldsymbol{u})$. Thus, the convexity condition $\partial_\theta^2 H(\boldsymbol{\lambda}, \boldsymbol{x}, \boldsymbol{u}^*) \geq 0$ requires that $H(\boldsymbol{\lambda}, \boldsymbol{x}, \boldsymbol{u}^*) \leq 0$ over the entire trajectory.

Deferring a discussion on the details of solving it to Section 2.5, assume, for the moment, that Problem HMC can be solved. Then, its solution $\boldsymbol{u}^*$ is given by some function $\boldsymbol{g}$:

$$(\boldsymbol{\lambda}, \boldsymbol{x}, t) \xrightarrow{g} \boldsymbol{u}^* \tag{2.19}$$

This control function $\boldsymbol{u}^* = \boldsymbol{g}(\boldsymbol{\lambda}, \boldsymbol{x}, t)$ is known as the Pontryagin-extremal control or simply the ***extremal control***.

Illustrating the Concept: Solving Problem HMC - 2/2

The extremal control for Problem Brac:1 is obtained from Eq. (2.18). This equation can be solved quite readily as:

$$\underbrace{\theta^*}_{u^*} = \underbrace{\tan^{-1}\left(\frac{\lambda_x v}{\lambda_y v + \lambda_v g}\right)}_{g(\lambda, x)} \tag{2.20}$$

As simple as this equation seems, note that inverses of trigonometric functions are multi-valued.

A solution to Problem HMC generates an extremal control as some function $\boldsymbol{g}$ of the costate, state and time:

$$\boldsymbol{u}^* = \boldsymbol{g}(\boldsymbol{\lambda}, \boldsymbol{x}, t) \tag{2.21}$$

Substituting this function in the dynamcis, $\dot{\boldsymbol{x}} = \boldsymbol{f}(\boldsymbol{x}, \boldsymbol{u}, t)$, eliminates $\boldsymbol{u}$ and generates a new differential equation for the states:

$$\begin{aligned}\dot{\boldsymbol{x}} &= \boldsymbol{f}(\boldsymbol{x}, \boldsymbol{u}, t) \\ &= \boldsymbol{f}(\boldsymbol{x}, \boldsymbol{g}(\boldsymbol{\lambda}, \boldsymbol{x}, t), t) \\ &:= \widetilde{\boldsymbol{f}}_1(\boldsymbol{x}, \boldsymbol{\lambda}, t)\end{aligned} \tag{2.22}$$

The state trajectory $\boldsymbol{x}^*(\cdot)$ resulting from Eq. (2.22) is called an an ***extremal state trajectory*** or an extremal *arc* (in some classical texts), and the state-control function pair, $\{\boldsymbol{x}^*(\cdot), \boldsymbol{u}^*(\cdot)\}$, is called an ***extremal*** solution.

Illustrating the Concept: Generating ODEs for the Extremal States

The extremal states for Brac:1 satisfy the differential equations:

$$\dot{\boldsymbol{x}} := \begin{bmatrix} \dot{x} \\ \dot{y} \\ \dot{v} \end{bmatrix} = \underbrace{\begin{bmatrix} v \sin\left(\tan^{-1}\left(\frac{\lambda_x v}{\lambda_y v + \lambda_v g}\right)\right) \\ v \cos\left(\tan^{-1}\left(\frac{\lambda_x v}{\lambda_y v + \lambda_v g}\right)\right) \\ g \cos\left(\tan^{-1}\left(\frac{\lambda_x v}{\lambda_y v + \lambda_v g}\right)\right) \end{bmatrix}}_{\tilde{\boldsymbol{f}}_1(\boldsymbol{x},\boldsymbol{\lambda})}$$

It is apparent that even though we may have found a candidate optimal control $\boldsymbol{u}^*$ by solving Problem HMC, a production of the extremal states is not complete because we need a quantitative knowledge of the costates. This knowledge can be obtained by substituting $\boldsymbol{u}^*$ in the adjoint equation in much the same way as was done in producing Eq. (2.22). This procedure produces a new differential equation for the costates that does not depend on $\boldsymbol{u}$. That is, substituting Eq. (2.21) in the adjoint equation, $\dot{\boldsymbol{\lambda}} = -\partial_{\boldsymbol{x}} H(\boldsymbol{\lambda}, \boldsymbol{x}, \boldsymbol{u}, t)$, generates a new differential equation for the costates:

$$\begin{aligned} -\dot{\boldsymbol{\lambda}} &= \partial_{\boldsymbol{x}} H(\boldsymbol{\lambda}, \boldsymbol{x}, \boldsymbol{u}, t) \\ &= \partial_{\boldsymbol{x}} H(\boldsymbol{\lambda}, \boldsymbol{x}, \boldsymbol{u}, t)\big|_{\boldsymbol{u} = \boldsymbol{g}(\boldsymbol{\lambda}, \boldsymbol{x}, t)} \\ &:= \tilde{\boldsymbol{f}}_2(\boldsymbol{x}, \boldsymbol{\lambda}, t) \end{aligned} \tag{2.23}$$

Illustrating the Concept: Generating ODEs for the Extremal Costates

The extremal costates for Brac:1 must satisfy the differential equations:

$$\begin{aligned} -\dot{\lambda}_x &= 0 \\ -\dot{\lambda}_y &= 0 \\ -\dot{\lambda}_v &= \underbrace{\lambda_x \sin\left[\tan^{-1}\left(\frac{\lambda_x v}{\lambda_y v + \lambda_v g}\right)\right] + \lambda_y \cos\left[\tan^{-1}\left(\frac{\lambda_x v}{\lambda_y v + \lambda_v g}\right)\right]}_{\tilde{\boldsymbol{f}}_2(\boldsymbol{x},\boldsymbol{\lambda})} \end{aligned} \tag{2.24}$$

From the general equation of (2.23) and the example of Eq. (2.24) it is evident that we need a quantitative knowledge of the states to produce a quantitative knowledge of the costates and vice versa. That is, solving Problem HMC leads to a pair of of coupled differential equations:

$$\begin{bmatrix} \dot{\boldsymbol{x}} \\ -\dot{\boldsymbol{\lambda}} \end{bmatrix} = \begin{bmatrix} \widetilde{\boldsymbol{f}}_1(\boldsymbol{x}, \boldsymbol{\lambda}, t) \\ \widetilde{\boldsymbol{f}}_2(\boldsymbol{x}, \boldsymbol{\lambda}, t) \end{bmatrix}$$

In other words, we need to solve a system of $2N_x$ state-costate differential equations *simultaneously*. To achieve this task, we need $2N_x$ point conditions: preferably N_x initial conditions on the states and N_x initial conditions on the costates. We also need the value of the initial (clock) time t_0 to propagate the initial condition as well the value of the final time t_f to stop the propagation. In all, we need $(2N_x + 2)$ numbers or point conditions. Collecting all of the available point data for Problem B, we have:

$$\begin{aligned} t_0 &= t^0 && \text{(1 equation)} \\ \boldsymbol{x}(t_0) &= \boldsymbol{x}^0 && (N_x \text{ equations}) \\ \boldsymbol{e}(\boldsymbol{x}(t_f), t_f) &= \boldsymbol{0} && (N_e \text{ equations}) \end{aligned}$$

Illustrating the Concept: Available Point Conditions

The totality of point conditions for Brac:1 are:

$$\begin{aligned} t_0 &= 0 && (1 \textit{ equation}) \\ (x_0, y_0, v_0) &= (0, 0, 0) && (3 \textit{ equations}) \\ (x_f - x^f, y_f - y^f) &= (0, 0) && (2 \textit{ equations}) \end{aligned}$$

(Recall that $N_x = 3$ and $N_e = 2$.)

Clearly, not only do we not have any initial condition on $\boldsymbol{\lambda}$, we also do not have the requisite $(2N_x + 2)$ point conditions. The deficit of point conditions is given by

$$\underbrace{(2N_x + 2)}_{required} - \underbrace{(1 + N_x + N_e)}_{available} = \underbrace{N_x + 1 - N_e}_{deficit} \tag{2.25}$$

For the Brac:1 formulation of the Brachistochrone problem, this deficit of point conditions is $8 - 6 = 2$.

Study Problem 2.3

1. *Under what conditions is there a zero deficit of point conditions?*
2. *Is it possible to have $N_e > N_x + 1$?*

2.4.2 In Search of the Missing Boundary Conditions

A deficit of point conditions in the Hamiltonian system of $2N_x$ differential equations implies that there are additional optimality conditions that await discovery. This point should also be apparent from the fact that, so far, we have not used the endpoint cost function E in generating any of the optimality conditions discussed in the previous section.

In seeking the missing point conditions, we cannot just add additional conditions on $\boldsymbol{x}_0$ or $\boldsymbol{x}_f$ without changing the problem. This implies that the missing point conditions must be on $\boldsymbol{\lambda}(t_0)$ and/or on $\boldsymbol{\lambda}(t_f)$.

So far, we have only used the H function and hence the pair $(F, \boldsymbol{f})$. We have not used the pair $(E, \boldsymbol{e})$ and hence the function $\overline{E}$. As part of this analysis, consider the gradient of $\overline{E}$ with respect to $\boldsymbol{x}_f$:

$$\frac{\partial \overline{E}(\boldsymbol{\nu}, \boldsymbol{x}_f, t_f)}{\partial \boldsymbol{x}_f} := \begin{bmatrix} \partial_{x_{1f}} \overline{E}(\boldsymbol{\nu}, \boldsymbol{x}_f, t_f) \\ \partial_{x_{2f}} \overline{E}(\boldsymbol{\nu}, \boldsymbol{x}_f, t_f) \\ \vdots \\ \partial_{x_{N_x f}} \overline{E}(\boldsymbol{\nu}, \boldsymbol{x}_f, t_f) \end{bmatrix} \quad \begin{matrix} CU/x_1\text{-units} \\ CU/x_2\text{-units} \\ \vdots \\ CU/x_{N_x}\text{-units} \end{matrix} \tag{2.26}$$

Obviously, the units of $\partial_{\boldsymbol{x}_f} \overline{E}$ are the same as the units of $\boldsymbol{\lambda}$. Given that $\overline{E}$ is a function of the final time variables, it is not too hard to guess that

$$\boldsymbol{\lambda}(t_f) = \frac{\partial \overline{E}}{\partial \boldsymbol{x}_f} \tag{2.27}$$

is a "missing" boundary condition. This equation is called the ***terminal Transversality Condition***.

Illustrating the Concept: Terminal Transversality Condition

From Eq. (2.12) on page 98, the endpoint Lagrangian for Brac:1 *is given by*

$$\overline{E}(\boldsymbol{\nu}, \boldsymbol{x}_f, t_f) := t_f + \nu_1(x_f - x^f) + \nu_2(y_f - y^f)$$

Applying Eq. (2.27), we get:

$$\begin{aligned} \lambda_x(t_f) &= \frac{\partial \overline{E}}{\partial x_f} &&= \nu_1 \\ \lambda_y(t_f) &= \frac{\partial \overline{E}}{\partial y_f} &&= \nu_2 \\ \lambda_v(t_f) &= \frac{\partial \overline{E}}{\partial v_f} &&= 0 \end{aligned} \tag{2.28}$$

It is clear that the transversality condition for $\lambda_x(t_f)$ and $\lambda_y(t_f)$ provide no new information in the sense that it says that these two unknown quantities are equal to two other unknown quantities. The story on the transversality condition $\lambda_v(t_f) = 0$ is different; it provides a non-trivial boundary condition. Recall that we need 2 point conditions for Brac:1*; hence, we still need one more point condition to complete the circle.*

Study Problem 2.4

Show that the units of $\lambda_x, \lambda_y, \nu_1$ and ν_2 in Eq. (2.28) are all consistent with their definitions given by Eqs. (2.5) and (2.9).

Equation (2.27) provides us N_x point conditions for $\boldsymbol{\lambda}(t)$. From Eq. (2.25) we needed $(N_x + 1 - N_e)$ point conditions; hence, the new number of missing boundary conditions are

$$(N_x + 1 - N_e) - N_x = 1 - N_e$$

This number seems a little strange as we now seem to have an excess of point conditions when $N_e > 1$; however, as evident from Eq. (2.28), we do not actually have N_x point conditions from Eq. (2.27) but something less. This is because the N_x equations from the terminal transversality condition contain an additional set of N_e unknowns in terms of $\boldsymbol{\nu} \in \mathbb{R}^{N_e}$. In other words, the effective number of equations from Eq. (2.27) is not N_x but $(N_x - N_e)$. This why in Brac:1, we were able to extract only $3 - 2 = 1$ useful condition by applying the transversality condition. Thus, the correct number of missing boundary conditions is,

$$(N_x + 1 - N_e) - (N_x - N_e) = 1$$

This "last" missing point condition comes from the ***Hamiltonian Value Condition***[§]

$$H[@t_f] = -\frac{\partial \overline{E}}{\partial t_f} \tag{2.29}$$

where we have used a shorthand notation, $H[@t_f]$, for the value of the Hamiltonian at $t = t_f$; that is,

$$H[@t_f] \equiv H(\boldsymbol{\lambda}(t_f), \boldsymbol{x}(t_f), \boldsymbol{u}(t_f), t_f)$$

Illustrating the Concept: Hamiltonian Value Condition

For Brac:1, we have

$$\frac{\partial \overline{E}}{\partial t_f} = 1 \tag{2.30}$$

Hence, the Hamiltonian value condition is given by

$$\lambda_x(t_f)v(t_f)\sin\theta(t_f) + \lambda_y(t_f)v(t_f)\cos\theta(t_f) + \lambda_v(t_f)g\cos\theta(t_f) = -1 \tag{2.31}$$

Study Problem 2.5

Write down the missing units in Eq. (2.29). What are the units in Eq. (2.30)?

Equations (2.27) and (2.29) complete the set of missing boundary conditions necessary to obtain a *candidate solution* to the optimal control problem. The collection of all these equations constitutes a special ***boundary value prob-***

[§]The Hamiltonian value condition applies to the *lower* Hamiltonian discussed later in Section 2.5.5. No serious error results in mixing up these two Hamiltonians at the the present time.

lem (BVP). We denote the BVP generated from Problem B as Problem B^λ.

2.4.3 Formulating the Boundary Value Problem B^λ

A generic BVP is essentially two or more differential equations where some of the point conditions are specified at the initial time and the remainder at the final time; see Fig. 2.11. If all conditions are specified at the initial time, it is called

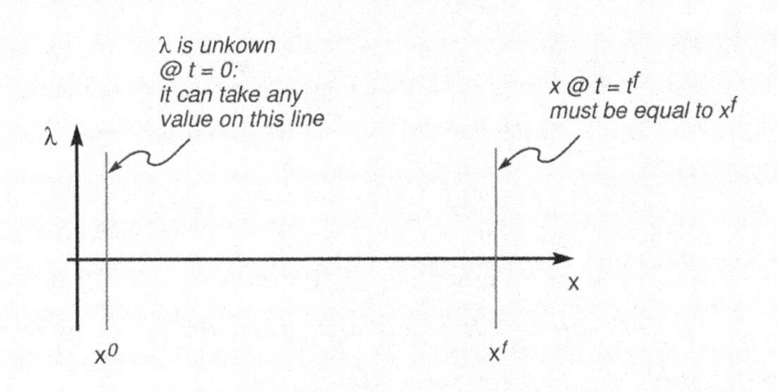

Figure 2.11: A fundamental two-dimensional BVP: x^0 is a given value of $x(t_0)$ while $\lambda(t_0)$ is unknown; instead, $x(t_f)$ is given by x^f.

an *initial value problem,* which is generally considered to be a solved problem as a consequence of good numerical integration techniques like Runge-Kutta methods.

The BVP developed in the preceding two subsections is not generic; it is ***structured in a special manner*** as follows:

- A pair of $2N_x$ (state-costate) differential equations

$$\begin{aligned} \dot{\boldsymbol{x}} &= \widetilde{\boldsymbol{f}}_1(\boldsymbol{x}, \boldsymbol{\lambda}, t) \qquad (N_x \text{ equations}) \\ -\dot{\boldsymbol{\lambda}} &= \widetilde{\boldsymbol{f}}_2(\boldsymbol{x}, \boldsymbol{\lambda}, t) \qquad (N_x \text{ equations}) \end{aligned}$$

- $2N_x$ effective boundary conditions given in the algebraic form

$$\begin{aligned} \boldsymbol{x}_0 &= \boldsymbol{x}^0 && (N_x \text{ equations}) \\ \boldsymbol{e}(\boldsymbol{x}_f, t_f) &= \boldsymbol{0} && (N_e \text{ equations}) \\ \boldsymbol{\lambda}(t_f) &= \frac{\partial \overline{E}}{\partial \boldsymbol{x}_f} && ((N_x - N_e) \text{ effective equations}) \end{aligned}$$

- Two clock conditions (to start and stop)

$$\begin{aligned} t_0 &= t^0 && (1 \text{ equation}) \\ H[@t_f] &= -\frac{\partial \overline{E}}{\partial t_f} && (1 \text{ equation}) \end{aligned}$$

Thus, the application of Pontryagin's Principle has not solved Problem B; instead, it has converted or "mapped" it to a BVP denoted by Problem B^λ; see Fig. 2.12. ***Note that this BVP is always given by an even number of differential equations (i.e., $2N_x$).***

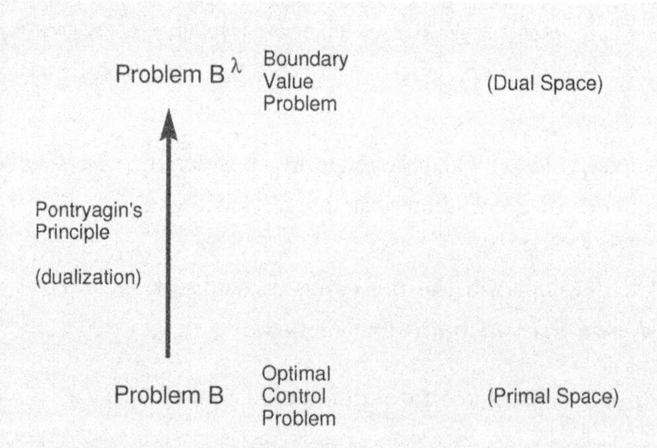

Figure 2.12: Pontryagin's Principle is a problem generator: it does not solve an optimal control problem; it maps Problem B to a boundary value problem by "lifting" it to a dual space.

Study Problem 2.6

Collect the complete set of differential and boundary conditions derived in the preceding pages for Brac:1. *Obtain, or discuss how to obtain, a solution to this BVP. Explore a numerical solution for this BVP in MATLAB or any other tool set. Discuss the pitfalls.*

2.4.4 Solving the BVP: Work, Plug, Pray

It is possible to "solve" the Brachistochrone problem by a simpler process than the procedure identified in Study Problem 2.6. One such process is semi-analytic; it is discussed later in Section 3.1, page 171. Semi-analytic procedures, often employed in academic-strength problems, usually use clever coordinate transformations and other mathematical "tricks" to avoid or beat down the BVP to something "manageable." While such ad hoc techniques are indeed useful and important for the analysis of specific problems, they are not portable to the broader Problem B. Furthermore, ad hoc techniques have the appearance of making "easy" problems look hard. On the other hand, the systematic process summarized in Fig. 2.12 is generic and extremely powerful. The generation of Problem B^{λ} might seem arduous at first, but note that the process is routine; hence, it can be algorithmized. The routineness of this process is demonstrated in Chapter 3 while an algorithm to solve Problem B^{λ} is encoded in DIDO.

Easy Problems Made Hard

Recall that the production of the differential equations

$$\dot{\boldsymbol{x}} = \widetilde{\boldsymbol{f}}_1(\boldsymbol{x}, \boldsymbol{\lambda}, t)$$
$$-\dot{\boldsymbol{\lambda}} = \widetilde{\boldsymbol{f}}_2(\boldsymbol{x}, \boldsymbol{\lambda}, t)$$

assumed that we can solve Problem HMC explicitly in an equation form:

$$\boldsymbol{u} = \boldsymbol{g}(\boldsymbol{\lambda}, \boldsymbol{x}, t)$$

Unfortunately, this is often not possible and is the subject of Section 2.5 on page 114. In other words, it is not easy or possible to produce an "unconstrained"

BVP; more often, it can be formulated only as a ***differential-algebraic inequality***. Standard methods for solving differential-algebraic inequalities are based on "optimal control techniques," such as *collocation methods*, which are briefly introduced later in Section 2.9.2, page 157. This, coming to a full circle, implies that Problem B^λ is not necessarily simpler than Problem B! While this realization may seem "obvious" today, note, however, that it is result of a culmination of research that took place from the 1960s to the year 2000.

Interestingly, if a problem is posed as a differential-algebraic inequality, it may be possible to "map it" down to an optimal control problem, that is, a down arrow in Fig. 2.12. In this case, the differential equations may be viewed as a manifestation of something more fundamental: an optimal control problem!

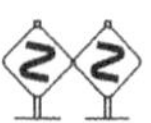

Symplectic BVPs

Problem B^λ is not a generic BVP: the $2N_x$ differential equations have a *symplectic* structure

$$\dot{\boldsymbol{y}} = \mathfrak{J}\nabla_{\boldsymbol{y}} H(\boldsymbol{y}, \boldsymbol{u}, t), \quad \mathfrak{J} := \begin{bmatrix} \mathbf{0} & \boldsymbol{I}_d \\ -\boldsymbol{I}_d & \mathbf{0} \end{bmatrix} \tag{2.32}$$

where $\boldsymbol{y} := (\boldsymbol{x}, \boldsymbol{\lambda}) \in \mathbb{R}^{N_x} \times \mathbb{R}^{N_x}$ and $\boldsymbol{I}_d$ is an $N_x \times N_x$ identity matrix. Hence, in principle, the entire power of symplectic vector space analysis can be brought to bear on analyzing Problem B^λ. Note, however, that the analysis of Problem B^λ is a little more complicated than those encountered in mathematical physics due to the presence of $\boldsymbol{u}$. Conversely, the entire power of optimal control theory can be brought to bear on the analysis of problems in mathematical physics. In this spirit, as physicists discover that more and more of Nature's laws are symplectic, it is apparent that Nature must be solving an optimal control problem (à la Hamilton's principle of "least" action). Thus, a grand physics Problem B awaits discovery.

2.5 Minimizing the Hamiltonian

It is evident from the previous section that the problem of minimizing the Hamiltonian with respect to the control $\boldsymbol{u}$ (i.e., Problem HMC) is a critical step in applying Pontryagin's Principle for any given problem. Problem HMC is a *static optimization problem* that must be performed at each instant of time. In many instances, $\mathbb{U}$ is a continuous set; then, Problem HMC reduces to a ***nonlinear programming (NLP)*** problem.

2.5.1 How Things Go Wrong

Too often, Problem HMC is erroneously oversimplified to the condition:[¶]

$$\left.\frac{\partial H}{\partial \boldsymbol{u}}\right|_{\boldsymbol{u}=\boldsymbol{u}^*} = \mathbf{0} \tag{2.33}$$

As sketched in Fig. 2.13, Eq. (2.33) would isolate points b and c as candidate solutions, neither of which are correct. Any claim that the *convexity condition*,

$$\left.\frac{\partial^2 H}{\partial \boldsymbol{u}^2}\right|_{\boldsymbol{u}=\boldsymbol{u}^*} \geq 0 \tag{2.34}$$

would isolate point c from point b, and hence would generate a *local* minimum to Problem B is also not true! *This is simply because there is no guarantee that*

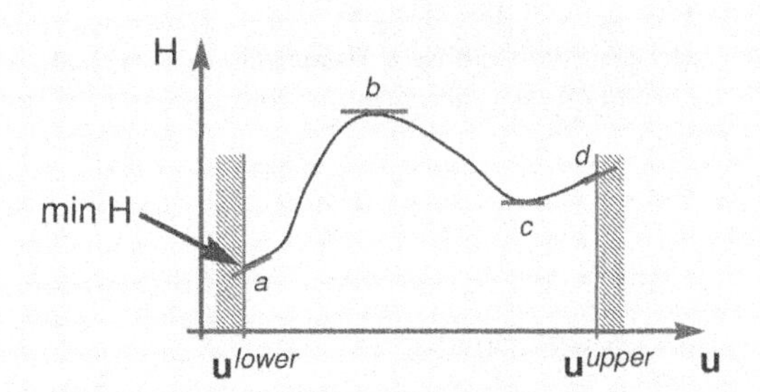

Figure 2.13: A snapshot of the graph of a possible Hamiltonian sketched as a function of $\boldsymbol{u}$ only; i.e., the graph of the function, $[\boldsymbol{u}^{lower}, \boldsymbol{u}^{upper}] \ni \boldsymbol{u} \mapsto H(\boldsymbol{\lambda}, \boldsymbol{x}, \boldsymbol{u}, t)$.

a local minimum to Problem HMC generates a local minimum for Problem B. Conversely, a global minimum to Problem HMC does not guarantee a global minimum for Problem B. The global minimum of H is only a part of the totality of necessary conditions for the local minimum of Problem B. In Fig. 2.13, this

[¶]We plead guilty to this charge in our illustration in Fig. 2.10; our defense is pedagogy.

occurs at point a which is quite far from point c; hence, *any claims of "sub optimality" for the point c would also be false.*

In the situation illustrated in Fig. 2.13, the solution to Problem HMC is the simple statement

$$\boldsymbol{u}^* = \boldsymbol{u}^{lower}$$

where $\boldsymbol{u}^{lower}$ is the ***lower bound*** of $\boldsymbol{u}$. Obviously, the solution to Problem HMC is not always going to be at a lower bound of $\boldsymbol{u}$; it might occur at its ***upper bound*** or at some point in the interior as illustrated in Fig. 2.14. Thus,

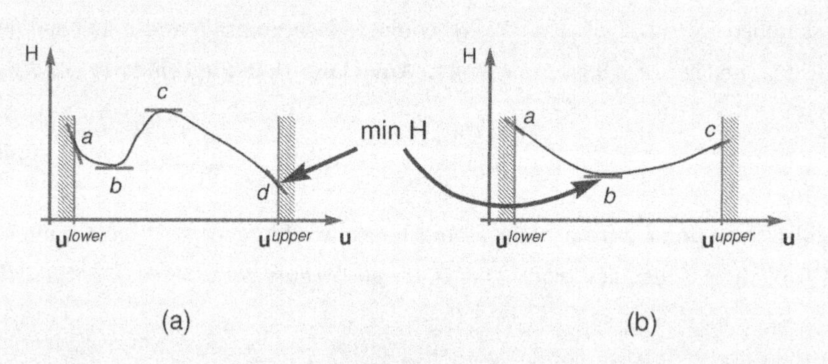

Figure 2.14: Two possible graphs of Hamiltonians sketched as a function of $\boldsymbol{u}$ only for consideration of solutions to Problem HMC.

the conditions given by Eqs. (2.33) and (2.34) are not always incorrect; they are just not inclusive of all possibilities.

One simple way to articulate the findings of the preceding discussion is to determine all points that satisfy Eq. (2.33) and compare the values of the objective function (H) at these points to those at the corners (lower and upper bounds). The point with the lowest value of H is the optimal one. This tedious process of isolation and enumeration can be substantially reduced by observing the following three properties:

1. If $\boldsymbol{u}^* = \boldsymbol{u}^{lower}$, then $\partial_{\boldsymbol{u}} H > \mathbf{0}$. This is the situation at point a in Fig. 2.13.

2. If $\boldsymbol{u}^* = \boldsymbol{u}^{upper}$, then $\partial_{\boldsymbol{u}} H < \mathbf{0}$. This is the situation at point d in Fig. 2.14 (a).

3. If $\boldsymbol{u}^{lower} < \boldsymbol{u}^* < \boldsymbol{u}^{upper}$, then $\partial_{\boldsymbol{u}} H = \mathbf{0}$. This is the situation at point b

in Fig. 2.14 (b).

Study Problem 2.7

Sketch figures to show the possibility of the following:

1. $\partial_{\boldsymbol{u}} H = \mathbf{0}$ *when* $\boldsymbol{u}^* = \boldsymbol{u}^{lower}$.
2. $\partial_{\boldsymbol{u}} H = \mathbf{0}$ *when* $\boldsymbol{u}^* = \boldsymbol{u}^{upper}$.

Based on these possibilities, revise the three conditions discussed in the paragraph preceding this problem.

The revised three conditions (resulting from the solution to Study Problem 2.7) can be unified through the following additional observations: When $\boldsymbol{u}^* = \boldsymbol{u}^{lower}$, we can rewrite the condition, $\partial_{\boldsymbol{u}} H \geq \mathbf{0}$, as

$$\partial_{\boldsymbol{u}} H + \boldsymbol{\mu} = \mathbf{0}, \qquad \boldsymbol{\mu} \leq \mathbf{0} \tag{2.35}$$

where $\boldsymbol{\mu}$ is the "equalizer"; that is, it cancels out $\partial_{\boldsymbol{u}} H$ exactly. This seemingly silly trick is wonderfully powerful. To see this, observe that

$$\text{units of } \boldsymbol{\mu} = \text{units of } \frac{\partial H}{\partial \boldsymbol{u}} = \frac{H\text{-units}}{\boldsymbol{u}\text{-units}} \quad \left(= \frac{CU/TU}{\boldsymbol{u}\text{-units}} \right)$$

Hence, $\boldsymbol{\mu}$ is a covector that can be used to measure $\boldsymbol{u}$ in terms of H-units ($= CU/TU$). Consequently, the dot product

$$\boldsymbol{\mu} \cdot \boldsymbol{u} \qquad (H\text{-units})$$

is a legal operation that produces a scalar in exactly the same units as H. This implies we can add this quantity to H to produce a new function

$$\overline{H}(\boldsymbol{\mu}, \boldsymbol{\lambda}, \boldsymbol{x}, \boldsymbol{u}, t) := H(\boldsymbol{\lambda}, \boldsymbol{x}, \boldsymbol{u}, t) + \boldsymbol{\mu}^T \boldsymbol{u}$$

called the ***Lagrangian of the Hamiltonian***. Now observe that

$$\partial_{\boldsymbol{u}} \overline{H} = \partial_{\boldsymbol{u}} H + \boldsymbol{\mu}$$

Comparing this equation to Eq. (2.35), we can write

$$\partial_{\boldsymbol{u}} \overline{H} = \mathbf{0}, \qquad \boldsymbol{\mu} \leq \mathbf{0} \tag{2.36}$$

when $\boldsymbol{u}^* = \boldsymbol{u}^{lower}$.

Study Problem 2.8

Following the same process that led to Eq. (2.36), show that

1. $\partial_{\boldsymbol{u}} \overline{H} = \mathbf{0}, \quad \boldsymbol{\mu} \geq \mathbf{0} \quad \textit{if} \quad \boldsymbol{u}^* = \boldsymbol{u}^{upper}$.
2. $\partial_{\boldsymbol{u}} \overline{H} = \mathbf{0}, \quad \boldsymbol{\mu} = \mathbf{0} \quad \textit{if} \quad \boldsymbol{u}^{lower} \leq \boldsymbol{u}^* \leq \boldsymbol{u}^{upper}$.

Collecting the results of Study Problem 2.8 and Eq. (2.36), the three conditions generated by Study Problem 2.7 on page 117 can be formalized as

- the stationarity condition

$$\frac{\partial \overline{H}}{\partial \boldsymbol{u}} = \mathbf{0} \tag{2.37}$$

- and the complementarity condition

$$\mu_i \begin{cases} \leq 0 & \textit{if} \qquad u_i = u_i^{lower} \\ = 0 & \textit{if} \quad u_i^{lower} < u_i < u_i^{upper} \\ \geq 0 & \textit{if} \qquad u_i = u_i^{upper} \end{cases} \tag{2.38}$$

This set of two conditions is a special case of more general conditions known as the ***Karush-Kuhn-Tucker (KKT) conditions***. Equations (2.37) and (2.38) are the necessary conditions for the special "box-constrained" HMC problem:

$$\text{(box-constrained HMC)} \quad \begin{cases} \underset{\boldsymbol{u}}{\text{Minimize}} & H(\boldsymbol{\lambda}, \boldsymbol{x}, \boldsymbol{u}, t) \\ \text{Subject to} & \boldsymbol{u}^{lower} \leq \boldsymbol{u} \leq \boldsymbol{u}^{upper} \end{cases}$$

2.5.2 KKT Conditions for Problem HMC

Consider the case when $\boldsymbol{u} \mapsto H(\boldsymbol{\lambda}, \boldsymbol{x}, \boldsymbol{u}, t)$ is differentiable and the control space $\mathbb{U}$ is given by functional inequalities

$$\boldsymbol{h}^L \leq \boldsymbol{h}(\boldsymbol{u}) \leq \boldsymbol{h}^U \tag{2.39}$$

where $\boldsymbol{h}^L$ and $\boldsymbol{h}^U$ are the lower and upper bounds on $\boldsymbol{h}(\boldsymbol{u})$, respectively. That is,

$$\mathbb{U} = \left\{ \boldsymbol{u} \in \mathbb{R}^{N_u} : \quad \boldsymbol{h}^L \leq \boldsymbol{h}(\boldsymbol{u}) \leq \boldsymbol{h}^U \right\} \tag{2.40}$$

Then, Problem HMC is a ***nonlinear programming (NLP) problem***:

$$(HMC = NLP) \quad \begin{cases} \underset{\boldsymbol{u}}{\text{Minimize}} & H(\boldsymbol{\lambda}, \boldsymbol{x}, \boldsymbol{u}, t) \\ \text{Subject to} & \boldsymbol{h}^L \leq \boldsymbol{h}(\boldsymbol{u}) \leq \boldsymbol{h}^U \end{cases} \tag{2.41}$$

Notation Alert: Recall that (see page xxiii) we use uppercase letters for cost functions and the corresponding lowercases for constraints. In remaining true to this style, we use the letter $\boldsymbol{h}$ for naming functions and its corresponding bounds for constraints that belong to minimizing H. This is also exactly the same reason why we chose the symbol $\boldsymbol{h}$ for the path function in Problem P in Section 1.4.2 on page 46. This notational convention also implies that $\overline{H}$ goes with the pair $(H, \boldsymbol{h})$.

According to Pontryagin's Principle, the Hamiltonian must be minimized *at each instant of time t*. To do this, we apply the KKT conditions to Problem $HMC = NLP$; that is, we define the ***Lagrangian of the Hamiltonian***

$$\boxed{\overline{H}(\boldsymbol{\mu}, \boldsymbol{\lambda}, \boldsymbol{x}, \boldsymbol{u}, t) := H(\boldsymbol{\lambda}, \boldsymbol{x}, \boldsymbol{u}, t) + \boldsymbol{\mu}^T \boldsymbol{h}(\boldsymbol{u})} \tag{2.42}$$

where $t \mapsto \boldsymbol{\mu} \in \mathbb{R}^{N_h}$ is a time-dependent KKT multiplier function associated with the functional constraint given by Eq. (2.39); that is $\boldsymbol{\mu}$ is a ***path covector*** defined by

$$\boldsymbol{\mu} := \begin{bmatrix} \mu_1 \\ \\ \mu_2 \\ \vdots \\ \mu_{N_h} \end{bmatrix} \quad \begin{matrix} \frac{CU/TU}{h_1\text{-units}} \\ \\ \frac{CU/TU}{h_2\text{-units}} \\ \vdots \\ \frac{CU/TU}{h_{N_h}\text{-units}} \end{matrix} \tag{2.43}$$

From the KKT conditions, the Lagrangian of the Hamiltonian $\overline{H}$ must be stationary with respect to the control $\boldsymbol{u}$:

$$\frac{\partial \overline{H}}{\partial \boldsymbol{u}} = \frac{\partial H}{\partial \boldsymbol{u}} + \left(\frac{\partial \boldsymbol{h}}{\partial \boldsymbol{u}}\right)^T \boldsymbol{\mu} = \boldsymbol{0} \tag{2.44}$$

In addition, *at each instant of time* t the multiplier-constraint pair $(\boldsymbol{\mu}, \boldsymbol{h}(\boldsymbol{u}))$ must satisfy the ***complementarity condition***:

$$\mu_i \begin{cases} \leq 0 & \textit{if} \quad h_i(\boldsymbol{u}) = h_i^L \\ = 0 & \textit{if} \quad h_i^L < h_i(\boldsymbol{u}) < h_i^U \\ \geq 0 & \textit{if} \quad h_i(\boldsymbol{u}) = h_i^U \\ \textit{unrestricted} & \textit{if} \quad h_i^L = h_i^U \end{cases} \tag{2.45}$$

These complementarity conditions determine the ***switching structure*** of the optimal control as they can also be written as

$$\boxed{\begin{array}{ccc} h_i(\boldsymbol{u}(t)) = h_i^L & \textit{if} & \mu_i(t) \leq 0 \\ h_i^L < h_i(\boldsymbol{u}(t)) < h_i^U & \textit{if} & \mu_i(t) = 0 \\ h_i(\boldsymbol{u}(t)) = h_i^U & \textit{if} & \mu_i(t) \geq 0 \\ h_i^L = h_i^U & \textit{if} & \mu_i(t) \ \textit{unrestricted} \end{array}} \tag{2.46}$$

where the time-dependence of the relevant variables is explicitly noted.

2.5.3 Time-Varying Control Space[||]

In many practical problems, the control space $\mathbb{U}$ is not "static" but varies in time as well as with variations in the state $\boldsymbol{x}$. When $\mathbb{U}$ is time-varying, we denote it as $\mathbb{U}(t)$. Pontryagin's Principle continues to hold in all these cases with embarrassing simplicity. To illustrate this point, let

$$\mathbb{U}(t) := \left\{ \boldsymbol{u} \in \mathbb{R}^{N_u} : \quad \boldsymbol{h}^L(t) \leq \boldsymbol{h}(\boldsymbol{u}, t) \leq \boldsymbol{h}^U(t) \right\} \tag{2.47}$$

That is, the inequalities that define the control space depend explicitly on time: Compare Eq. (2.47) with (2.40). Since Problem HMC requires that the Hamiltonian be minimized at each instant of time, the KKT conditions are the same as before except that the conditions must hold explicitly at each instant of time t; thus, we have:

$$\mu_i \begin{cases} \leq 0 & \text{if} \quad h_i(\boldsymbol{u}, t) = h_i^L(t) \\ = 0 & \text{if} \quad h_i^L(t) < h_i(\boldsymbol{u}, t) < h_i^U(t) \\ \geq 0 & \text{if} \quad h_i(\boldsymbol{u}, t) = h_i^U(t) \\ \textit{unrestricted} & \text{if} \quad h_i^L(t) = h_i^U(t) \end{cases} \tag{2.48}$$

2.5.4 Solving Problem HMC

When $\mathbb{U}$ is a continuous set (see Fig. 2.2 on page 87), Problem HMC is an NLP. If $\mathbb{U}$ is discrete, Problem HMC is a discrete/integer programming problem. Hence, in general, Problem HMC is a nonlinear mixed-variable (i.e., continuous-discrete) programming problem. Barring an extremely limited number of special cases, there are no closed-form solutions to NLPs or integer programming problems. *Not surprisingly, in many practical applications, Problem HMC cannot be solved "by hand" or "analytically."* Fortunately, this is not a major deterrent for generating usable solutions to practical optimal control problems. To understand this critical point, it is necessary to appreciate the agonies, nuances, caveats and joys of producing usable solutions.

[||] This section may be skipped without loss in continuity.

The Agonies of Analytical Solutions

An analytical or "closed form" solution is generally defined as a quantity that can be written in terms of elementary operations (e.g., $+, -\ldots$) over elementary functions: sines, cosines, exponentials, etc. A solution is generally considered to be "approximate" or, worse, "numerical" (apparently, of lower quality) if it is expressed in terms of a large number (such as a hundred or more) of elementary operations over elementary functions; for example, a truncated finite sum of an infinite series. The problem with these (mis)perceptions is that the computations of elementary functions are eventually approximate and maybe even done "surreptitiously" by a truncated series expansion (for example, consider how a computer computes $\sin(12.3), e^{4.56}, \pi$ etc.). In other words, when the claims of an analytical solution are dissected, it turns out it is an unfortunate myth that must be rejected with extreme prejudice. There is no grand universal mathematical theorem which guarantees that all solutions to all mathematical problems are expressible in terms of a handful of elementary functions. Within the context of mathematically accurate expectations, Problem HMC is indeed solvable in a vast number of situations.

A solution to Problem HMC can be stated compactly using the arg min notation

$$\boldsymbol{u}^* = \arg\min_{\boldsymbol{u}\in\mathbb{U}} H(\boldsymbol{\lambda}, \boldsymbol{x}, \boldsymbol{u}, t) \tag{2.49}$$

which is shorthand for the statement that the argument (i.e., $\boldsymbol{u}$) of the minimum of H with respect to $\boldsymbol{u}$ is equal to $\boldsymbol{u}^*$. This seemingly high-brow notation is useful and clever at the same time. It is useful in the sense that it avoids the use of the generic symbol $\boldsymbol{g}$, used previously (see Eq. (2.21) on page 105) and replaces it by a more evocative notation given by Eq. (2.49). See also the *Tech Talk* box later on page 196. Equation (2.49) is quite clever because, through the argmin notation, every optimization problem looks solved in an exact analytical form!

The Phantom of Global Optimization

A *necessary* condition for solving Problem B is a globally optimal solution to Problem HMC. Consider the situation illustrated in Fig. 2.15. The globally optimal solution is point d. The necessary conditions for this globally optimal solution are Eqs. (2.33) and (2.34). They are also the necessary conditions for

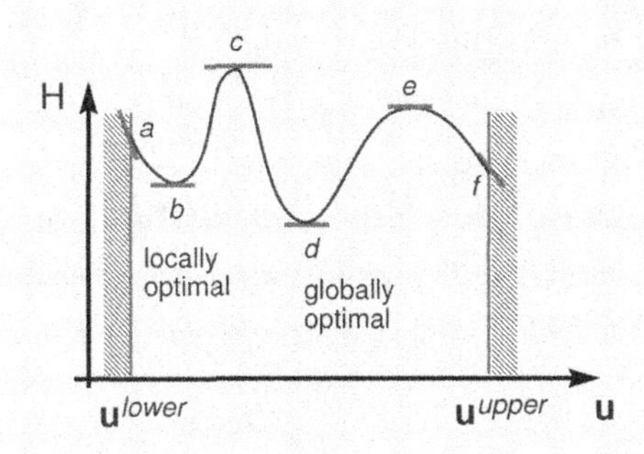

Figure 2.15: The necessary conditions for local and global optimality are the same.

local optimality; hence they are satisfied at point b, as well. In other words, short of enumeration, there is no way to detect a globally optimal solution.

Even if a globally optimal solution to Problem HMC is found, it is only part of the necessary conditions for Problem B. That is, a globally optimal solution to Problem HMC does not even guarantee a locally optimal solution to Problem B.

When $\mathbb{U}$ is parameterized by functional inequalities, the necessary conditions for Problem HMC are given by the KKT conditions. The KKT conditions do not solve Problem HMC. Consequently, replacing Problem HMC with its KKT conditions in the collection of necessary conditions for Problem B produces a gap — and possibly a large one — between the true necessary conditions and its replacement. *The analysis of this gap must be part of the analysis of a candidate optimal solution.*

The Notion of Reigning Optimal Solutions

The preceding discussions appear to suggest that it is extremely hard to produce (globally) optimal solutions to Problem B. It indeed is **if** the requirement is a mathematical proof of global optimality; however, there is a perceptible divide between what is provably optimal and what can be easily generated through a software package like DIDO. Given a Problem B, a software like DIDO generates a system trajectory ($t \mapsto (\boldsymbol{x}, \boldsymbol{u})$) and the associated covector functions ($t \mapsto \boldsymbol{\lambda}$ and $\boldsymbol{\nu}$). That is, DIDO takes Problem B as an input, in nearly the same format as defined in Section 2.1, and outputs a ***candidate***

solution to Problem B^λ; see Fig. 2.16. This candidate solution can then be tested for optimality by applying Pontryagin's Principle "by hand." This is the "analysis space" in Fig. 2.16. *Recognizing the nuances discussed in the preceding paragraphs is critical to carrying out these tests.* These tests help a more careful analysis of many practical problems which, in turn, support the generation of better inputs for DIDO, and possibly alternative problem formulations as discussed in Chapter 1. This is part of the *problem of problems* introduced in Chapter 1.

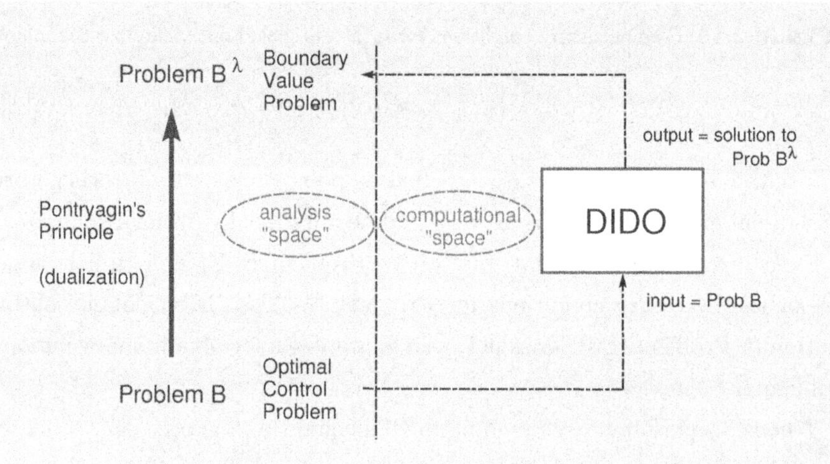

Figure 2.16: Once a paper-pencil analysis of an optimal control problem has been performed through an application of Pontryagin's Principle, a candidate solution generated by DIDO can be tested for optimality. DIDO takes Problem B as an input and outputs a possible solution to Problem B^λ. Compare this to Fig. 2.12.

In the end, a practical test for optimality is not the claim of a globally optimal solution; rather, it is whether a new solution to an old problem is better than the old solution (and by how much) or whether the problem posed is itself new so that any (feasible) solution is desirable to the alternative (of no solution). Short of a mathematical proof to the contrary, a reigning optimal solution can stake the claim of global optimality to a given problem until a better solution is found!

2.5.5 A Tale of Two, Maybe Three Hamiltonians

A solution to Problem HMC generates the function:

$$(\boldsymbol{\lambda}, \boldsymbol{x}, t) \xrightarrow{g} \boldsymbol{u}^* = \arg\min_{\boldsymbol{u}\in\mathbb{U}} H(\boldsymbol{\lambda}, \boldsymbol{x}, \boldsymbol{u}, t) \tag{2.50}$$

When $\boldsymbol{u}^*$ is substituted for $\boldsymbol{u}$ in $H(\boldsymbol{\lambda}, \boldsymbol{x}, \boldsymbol{u}, t)$, it yields the ***minimized Hamiltonian***:

$$\boxed{\mathcal{H}(\boldsymbol{\lambda}, \boldsymbol{x}, t) := \min_{\boldsymbol{u}\in\mathbb{U}} H(\boldsymbol{\lambda}, \boldsymbol{x}, \boldsymbol{u}, t) \quad \equiv H(\boldsymbol{\lambda}, \boldsymbol{x}, \boldsymbol{u}, t)\big|_{u=g(\lambda,x,t)}} \tag{2.51}$$

Illustrating the Concept: The Minimized Hamiltonian

The minimized Hamiltonian for Brac:1 is obtained by substituting Eq. (2.20) in (2.8):

$$\begin{aligned}
\mathcal{H}(\boldsymbol{\lambda}, \boldsymbol{x}) \quad := \quad & \lambda_x v \sin\left(\tan^{-1}\left(\frac{\lambda_x v}{\lambda_y v + \lambda_v g}\right)\right) \\
& +\lambda_y v \cos\left(\tan^{-1}\left(\frac{\lambda_x v}{\lambda_y v + \lambda_v g}\right)\right) \\
& +\lambda_v g \cos\left(\tan^{-1}\left(\frac{\lambda_x v}{\lambda_y v + \lambda_v g}\right)\right)
\end{aligned}$$

*Clearly, $\mathcal{H}$ is **nonlinear in $\boldsymbol{\lambda}$** (while H is always linear).*

At this point, we do not know if the left-hand side of Eq. (2.51) is still a Hamiltonian, that is, if it satisfies Eq. (2.17). It turns out that it indeed does if we assume the function $\boldsymbol{g}$ in Eq. (2.50) is differentiable and $\boldsymbol{u}^*$ is interior to $\mathbb{U}$.

Study Problem 2.9

Show that $\mathcal{H}$ is a Hamiltonian; that is, under appropriate assumptions, it satisfies the equations

$$\begin{aligned}
\dot{\boldsymbol{x}} &= \frac{\partial \mathcal{H}}{\partial \boldsymbol{\lambda}} \\
-\dot{\boldsymbol{\lambda}} &= \frac{\partial \mathcal{H}}{\partial \boldsymbol{x}}
\end{aligned} \tag{2.52}$$

Catalog all the assumptions needed for this proof.

The function, $\mathcal{H}$, is called the ***lower Hamiltonian***. The ***upper Hamiltonian*** is obtained if H were to be maximized instead of minimized. In their original work, Pontryagin et al maximized the Hamiltonian to be consistent with the archetypal or ***Hamilton's Hamiltonian***. This original Hamiltonian is obtained in the classical calculus of variations through a process known as the Legendre transform. Thanks to the simplifications and insights offered by Pontryagin's Principle, the only difference between the two Hamiltonians is a sign change; see also Section 1.4.4 on page 55. In this book, we use the "minimum version" of Pontryagin's Principle.

In general, $\mathcal{H}$ is nonlinear in $\boldsymbol{\lambda}$ and non-differentiable; however, it evolves according to a very simple equation known as the ***Hamiltonian Evolution Equation***:

$$\boxed{\frac{d\mathcal{H}}{dt} = \frac{\partial H}{\partial t}} \tag{2.53}$$

Illustrating the Concept: The Hamiltonian Evolution Equation

From Eq. (2.8) it is clear that the (control) Hamiltonian for Problem Brac:1 does not depend explicitly with time; hence, we have:

$$\frac{\partial H}{\partial t} = 0$$

From the Hamiltonian evolution equation, this implies:

$$\mathcal{H}(\boldsymbol{\lambda}(t), \boldsymbol{x}(t)) = constant \quad (w.r.t.\ time)$$

More explicitly, we can write:

$$\lambda_x(t)v(t)\sin\theta^*[@t] + \lambda_y(t)v(t)\cos\theta^*[@t] + \lambda_v(t)g\cos\theta^*[@t] = constant \tag{2.54}$$

where $\theta^*[@t]$ is the shorthand notation for

$$\theta^*[@t] \equiv \theta^*(\boldsymbol{x}(t), \boldsymbol{\lambda}(t)) = \tan^{-1}\left(\frac{\lambda_x(t)v(t)}{\lambda_y(t)v(t) + \lambda_v(t)g}\right)$$

The Hamiltonian evolution equation is an *integral of motion*: It has significant practical value in solving both academic and industrial optimal control problems.

Study Problem 2.10

In Problem B, introduce an $(N_x + 1)$th state variable

$$\dot{x}_{N_x+1} = 1 \tag{2.55}$$

with initial condition, $x_{N_x+1}(t_0) = t_0$. Then, the data in the modified problem become time-invariant (i.e., not an explicit function of time) with $\boldsymbol{x} \in \mathbb{R}^{N_x+1}$.

1. *Show that $\lambda_{N_x+1}(t) = -\mathcal{H}(\boldsymbol{\lambda}(t), \boldsymbol{x}(t), t)$ where $\mathcal{H}$ is the lower Hamiltonian to the original Problem B.*
2. *In the original Problem B, it is sufficient for $\boldsymbol{f}$ and F to be merely measurable with respect to t. In view of this, criticize the suggestion that time-invariant and time-varying problem data are equivalent under the introduction of Eq. (2.55).*

2.6 A Cheat Sheet for Pontryagin's Principle

The totality of necessary conditions for Problem B can be summarized in the form of a major result:

Theorem 2.1 (Pontryagin's Principle) *Given an optimal solution to Problem B, there exists an absolutely continuous covector function $\boldsymbol{\lambda}(\cdot)$ and a covector $\boldsymbol{\nu}$ that satisfy*

- *the three Hamiltonian conditions:*
  1. *Hamiltonian minimization condition,*
  2. *Hamiltonian value condition,*
  3. *Hamiltonian evolution equation*
- *the adjoint equations, and*
- *the transversality condition.*

Remark 2.1.1 *Pontryagin's Principle has not solved Problem B: it simply states the necessary conditions that a candidate optimal solution must satisfy.*

Moreover, one of these necessary conditions, that is, the Hamiltonian Minimization Condition

$$(HMC) \quad \begin{cases} \textsf{Minimize}_{\boldsymbol{u}} & H(\boldsymbol{\lambda}, \boldsymbol{x}, \boldsymbol{u}, t) \\ \textsf{Subject to} & \boldsymbol{u} \in \mathbb{U} \end{cases} \tag{2.56}$$

is posed as a problem in itself, whose solution can be symbolically stated as:

$$(\boldsymbol{\lambda}, \boldsymbol{x}, t) \mapsto \boldsymbol{u}^* = \arg \min_{\boldsymbol{u} \in \mathbb{U}} H(\boldsymbol{\lambda}, \boldsymbol{x}, \boldsymbol{u}, t) \tag{2.57}$$

Once the function $\boldsymbol{u}^*(\cdot)$ is obtained in the form of Eq.(2.57) (i.e., by solving Problem HMC), then an extremal may be obtained by solving the Hamiltonian system of $2N_x$ ordinary differential equations

$$\dot{\boldsymbol{x}} = \frac{\partial H}{\partial \boldsymbol{\lambda}} \tag{2.58a}$$

$$-\dot{\boldsymbol{\lambda}} = \frac{\partial H}{\partial \boldsymbol{x}} \tag{2.58b}$$

with boundary conditions

$$\boldsymbol{x}(t_0) = \boldsymbol{x}^0 \tag{2.59a}$$

$$\boldsymbol{e}(\boldsymbol{x}(t_f), t_f) = \boldsymbol{0} \tag{2.59b}$$

$$\boldsymbol{\lambda}(t_f) = \frac{\partial \overline{E}}{\partial \boldsymbol{x}_f} \tag{2.59c}$$

that must be satisfied at the start of the clock at the given time

$$t_0 = t^0 \tag{2.60}$$

and at the "optimal" stopping time

$$\mathcal{H}[@t_f] = -\frac{\partial \overline{E}}{\partial t_f} \tag{2.61}$$

where $\mathcal{H}[@t_f]$ is shorthand for for the value of $\mathcal{H}$ at t_f:

$$\mathcal{H}[@t_f] \equiv \mathcal{H}(\boldsymbol{\lambda}(t_f), \boldsymbol{x}(t_f), t_f)$$

Remark 2.1.2 *The Hamiltonian evolution equation*

$$\frac{d\mathcal{H}}{dt} = \frac{\partial H}{\partial t} \tag{2.62}$$

is an integral of motion. It serves an extremely important role in the verification and validation of the computed solution, either by means of analytical equations for simple problems or in providing an equation to check the optimality of a numerical solution.

At first, all these equations might seem a little overwhelming, particularly to a beginner. After some practice (i.e., after studying Chapters 3 and 4), they will seem relatively straightforward and even logical. In the meantime, a beginning student might find it useful to have some quick ways of remembering all the necessary conditions. To this end, we offer the following mnemonics:

- ***3HAT***: This stands for the three Hamiltonian conditions, the adjoint equation and the transversality condition summarized in Theorem 2.1.
- ***HAMVET***: Pontryagin's HAMVET (with apologies to Shakespeare) comes from the following systematic steps that must be carried out for any given problem to develop its necessary conditions for optimality:

  (a) Construct the **H**amiltonian. H
  (b) Develop the **A**djoint equations. A
  (c) **Mi**nimize the Hamiltonian. M
  (d) Evaluate the Hamiltonian **V**alue condition. V
  (e) Integrate the Hamiltonian **E**volution equation. E
  (f) Formulate the **T**ransversality conditions. T

- ***Engineers Value MATH***: This is simply anagramming *HAMVET* to *EV MATH*. Interestingly, it turns out that *MATH* forms the core of Pontryagin's Principle.

Remark 2.1.3 The control Hamiltonian given in Eq. (2.7) is in its "normal form" and not quite complete; the "complete" Hamiltonian for Problem B is actually given by

$$H(\boldsymbol{\lambda}, \boldsymbol{x}, \boldsymbol{u}, t) = \nu_0 F(\boldsymbol{x}, \boldsymbol{u}, t) + \boldsymbol{\lambda}^T \mathbf{f}(\boldsymbol{x}, \boldsymbol{u}, t) \tag{2.63}$$

where ν_0 is called the ***cost multiplier***. Additional necessary conditions include the ***constancy and nonnegativity*** of the cost-multiplier function: $t \mapsto \nu_0 = constant \geq 0$. This condition implies two cases: either $\nu_0 = 0$ or $\nu_0 \neq 0$.

1. $\boldsymbol{\nu_0 = 0}$
   If $\nu_0 = 0$, then the Hamiltonian and all the equations derived from it are independent of the cost! This situation is possible but is deemed "abnormal."

2. $\boldsymbol{\nu_0 \neq 0}$
   If $\nu_0 \neq 0$, then it simply scales the Hamiltonian by a positive constant; hence, we can divide H by ν_0 and analyze the "scaled" Hamiltonian. This scaling is called the ***Normality Condition*** and is equivalent to arbitrarily setting $\nu_0 = 1$ in Eq. (2.63).

Clearly, we must have nontrivial multipliers, $(\nu_0, \boldsymbol{\lambda}) \neq (0, \mathbf{0})$; that is, we must not have all covectors equal to zero. This ***Nontriviality Condition*** can also be stated as

$$\nu_0 + \|\boldsymbol{\lambda}(\cdot)\|_{L^\infty} > 0 \tag{2.64}$$

Pontryagin's Principle as stated in Theorem 2.1 is for the normal case. It can be restated for all cases by writing the Hamiltonian in the form given by Eq.(2.63) and including the additional nuances of constancy, nonnegativity and nontriviality as part of the necessary conditions.

Geek Speak: A lot of technical details are missing in our statement of Pontryagin's Principle as codified in Theorem 2.1. For instance, the standard assumptions for the function space for the state and control trajectories are $\boldsymbol{x}(\cdot) \in W^{1,1}$ and $\boldsymbol{u}(\cdot) \in L^\infty$. Because we assumed $\boldsymbol{f}$ to be Lipschitz in $\boldsymbol{x}$ (actually, C^1 in $\boldsymbol{x}$), we can write $(\boldsymbol{x}(\cdot), \boldsymbol{u}(\cdot)) \in W^{1,\infty} \times L^\infty$.

Study Problem 2.11

Construct the complete set of differential and boundary conditions obtained for any one of the other alternative formulations of the Brachistochrone problem; e.g., Problem Brac : 2 developed in Chapter 1. Does this alternative formulation generate any new information? Discuss.

(Hint: Yes! This is why different mathematical transcriptions of the same word problem are extremely useful in better understanding the structure of optimal solutions.)

Study Problem 2.12

Using the extremal controls obtained in Study Problem 2.11, write down the expressions for $\mathcal{H}$, the lower Hamiltonian, for each of these example problems. Show that $\mathcal{H}$ is not necessarily linear in $\boldsymbol{\lambda}$ or even a differentiable function. Is $\dot{\boldsymbol{x}} = \partial_\lambda \mathcal{H}$? Is $\dot{\boldsymbol{\lambda}} = -\partial_x \mathcal{H}$? Is it possible for $\mathcal{H}$ to be multivalued?

2.7 Pontryagin's Principle for Problem P

Problem P is defined in Section 1.4.2 on page 47, and is repeated here for quick reference:

$$\left.\begin{array}{ll} \mathbb{X}_{search} = \mathbb{R}^{N_x} & \mathbb{U}_{search} = \mathbb{R}^{N_u} \\ \boldsymbol{x} = (x_1, \ldots, x_{N_x}) & \boldsymbol{u} = (u_1, \ldots, u_{N_u}) \end{array}\right\} \text{(preamble)}$$

$$\overbrace{(P)}^{\text{problem}} \left\{\begin{array}{ll} \text{Minimize } J[\boldsymbol{x}(\cdot), \boldsymbol{u}(\cdot), t_0, t_f] := & \\ \qquad E(\boldsymbol{x}_0, \boldsymbol{x}_f, t_0, t_f) + \int_{t_0}^{t_f} F(\boldsymbol{x}(t), \boldsymbol{u}(t), t) & \text{(cost)} \\ \text{Subject to} \qquad \dot{\boldsymbol{x}} = \boldsymbol{f}(\boldsymbol{x}(t), \boldsymbol{u}(t), t) & \text{(dynamics)} \\ \qquad \boldsymbol{e}^L \leq \boldsymbol{e}(\boldsymbol{x}_0, \boldsymbol{x}_f, t_0, t_f) \leq \boldsymbol{e}^U & \text{(events)} \\ \qquad \boldsymbol{h}^L \leq \boldsymbol{h}(\boldsymbol{x}(t), \boldsymbol{u}(t), t) \leq \boldsymbol{h}^U & \text{(path)} \end{array}\right.$$

The pair $(F, \boldsymbol{f})$ is the same in both problems P and B; hence, the definition of the control or Pontryagin's Hamiltonian remains unchanged:

$$H(\boldsymbol{\lambda}, \boldsymbol{x}, \boldsymbol{u}, t) := F(\boldsymbol{x}, \boldsymbol{u}, t) + \boldsymbol{\lambda}^T \boldsymbol{f}(\boldsymbol{x}, \boldsymbol{u}, t)$$

The endpoint Lagrangian depends only on the pair $(E, \boldsymbol{e})$; hence, $\overline{E}$ is modified only in terms of its functional dependencies:

$$\overline{E}(\boldsymbol{\nu}, \boldsymbol{x}_0, \boldsymbol{x}_f, t_0, t_f) := E(\boldsymbol{x}_0, \boldsymbol{x}_f, t_0, t_f) + \boldsymbol{\nu}^T \boldsymbol{e}(\boldsymbol{x}_0, \boldsymbol{x}_f, t_0, t_f)$$

Because a complementarity condition goes hand-in-hand with inequality-type constraints, we define $\boldsymbol{\nu} \dagger \boldsymbol{e}$ as shorthand for the conditions

$$\nu_i \begin{cases} \leq 0 & if \quad e_i(\boldsymbol{x}_0, \boldsymbol{x}_f, t_0, t_f) = e_i^L \\ = 0 & if \quad e_i^L < e_i(\boldsymbol{x}_0, \boldsymbol{x}_f, t_0, t_f) < e_i^U \\ \geq 0 & if \quad e_i(\boldsymbol{x}_0, \boldsymbol{x}_f, t_0, t_f) = e_i^U \\ unrestricted & if \quad e_i^L = e_i^U \end{cases} \tag{2.65}$$

with $\boldsymbol{\mu} \dagger \boldsymbol{h}$ defined similarly.

The control space in Problem P is state-dependent and parameterized by functional inequalities:

$$\mathbb{U}(\boldsymbol{x}, t) := \left\{ \boldsymbol{u} \in \mathbb{R}^{N_u} : \ \boldsymbol{h}^L \leq \boldsymbol{h}(\boldsymbol{x}, \boldsymbol{u}, t) \leq \boldsymbol{h}^U \right\}$$

The symbol $\mathbb{U}(\boldsymbol{x}, t)$ means that $\mathbb{U}$ is not a constant (as in Problem B; see Fig. 1.26 on page 43) but that it depends jointly on $\boldsymbol{x}$ and t. Thus $\mathbb{U}(\boldsymbol{x}, t)$ is also a map from $(\boldsymbol{x}, t)$ to a set or a ***set-valued map***.* Despite this infusion of practically inspired mathematical complication, Problem HMC has the same functional form as before:

$$(HMC) \begin{cases} \underset{\boldsymbol{u}}{\text{Minimize}} & H(\boldsymbol{\lambda}, \boldsymbol{x}, \boldsymbol{u}, t) \\ \text{Subject to} & \boldsymbol{h}^L \leq \boldsymbol{h}(\boldsymbol{x}, \boldsymbol{u}, t) \leq \boldsymbol{h}^U \end{cases} \tag{2.66}$$

*This simple observation illustrates why set-valued analysis has direct practical applications.

Thus, the *Lagrangian of the Hamiltonian* is given by

$$\overline{H}(\boldsymbol{\mu}, \boldsymbol{\lambda}, \boldsymbol{x}, \boldsymbol{u}, t) := H(\boldsymbol{\lambda}, \boldsymbol{x}, \boldsymbol{u}, t) + \boldsymbol{\mu}^T \boldsymbol{h}(\boldsymbol{x}, \boldsymbol{u}, t)$$

and the KKT conditions for Problem HMC can be written compactly as

$$\partial_{\boldsymbol{u}} \overline{H} = \mathbf{0} \quad \text{and} \quad \boldsymbol{\mu} \dagger \boldsymbol{h}$$

where $\boldsymbol{\mu} \dagger \boldsymbol{h}$, as noted previously, is defined similar to Eq. (2.65).

The lower Hamiltonian is also functionally the same as before

$$\mathcal{H}(\boldsymbol{\lambda}, \boldsymbol{x}, t) := \min_{\boldsymbol{u} \in \mathbb{U}(\boldsymbol{x}, t)} H(\boldsymbol{\lambda}, \boldsymbol{x}, \boldsymbol{u}, t)$$

with the caveat that $\mathbb{U}$ depends on $\boldsymbol{x}$ and t.

One of the major modifications to Pontryagin's original principle is that the adjoint equations, while similar to that of Problem B, are now based on $\overline{H}$ and not H:

$$-\dot{\boldsymbol{\lambda}} = \frac{\partial \overline{H}}{\partial \boldsymbol{x}}$$

From the complementarity condition $\boldsymbol{\mu} \dagger \boldsymbol{h}$, it follows that the adjoint covector has the same co-dynamics as Problem B when the state trajectory is "inside" the path constraint (if necessary, refer to Fig. 1.28 on page 46). Obviously, when the state trajectory rides the path constraint, the adjoint equation has additional terms. See Study Problem 2.14 at the end of this section on page 136.

The terminal transversality condition remains the same as before (functionally)

$$\boldsymbol{\lambda}(t_f) = \frac{\partial \overline{E}}{\partial \boldsymbol{x}_f}$$

while we "gain" an ***initial transversality condition***:

$$\boldsymbol{\lambda}(t_0) = -\frac{\partial \overline{E}}{\partial \boldsymbol{x}_0}$$

Likewise, we have ***two Hamiltonian value conditions***

$$\mathcal{H}[@t_0] = \frac{\partial \overline{E}}{\partial t_0} \quad \text{and} \quad \mathcal{H}[@t_f] = -\frac{\partial \overline{E}}{\partial t_f}$$

that correspond to optimal start and stop times. Thus, the *HAMVET* mnemonic and other ones discussed on page 129 still apply, albeit with a few different details.

The collection of all unknowns resulting from Pontryagin's Principle for Problem P are as follows:

1. The system trajectory, $t \mapsto (\boldsymbol{x}, \boldsymbol{u}) \in \mathbb{R}^{N_x} \times \mathbb{R}^{N_u}$;
2. The adjoint covector funtion, $t \mapsto \boldsymbol{\lambda} \in \mathbb{R}^{N_x}$;
3. The path covector function, $t \mapsto \boldsymbol{\mu} \in \mathbb{R}^{N_h}$;
4. The endpoint covector, $\boldsymbol{\nu} \in \mathbb{R}^{N_e}$; and
5. The initial and final times, $t_0 \in \mathbb{R}$ and $t_f \in \mathbb{R}$.

These unknowns must satisfy a collection of differential and algebraic constraints, with the latter structured in terms of conditional inequalities. Collecting all these equations and inequalities we can constitute Problem P^λ as follows:

$$
(P^\lambda)\left\{
\begin{array}{rcll}
\dot{\boldsymbol{x}}(t) - \partial_\lambda \overline{H}(\boldsymbol{\mu}(t), \boldsymbol{\lambda}(t), \boldsymbol{x}(t), \boldsymbol{u}(t), t) & = & \mathbf{0} & \text{(state eqns)} \\
\dot{\boldsymbol{\lambda}}(t) + \partial_x \overline{H}(\boldsymbol{\mu}(t), \boldsymbol{\lambda}(t), \boldsymbol{x}(t), \boldsymbol{u}(t), t) & = & \mathbf{0} & \text{(costate eqns)} \\
\boldsymbol{h}^L \leq \boldsymbol{h}(\boldsymbol{x}(t), \boldsymbol{u}(t), t) & \leq & \boldsymbol{h}^U & \text{(path condition)} \\
\partial_u \overline{H}(\boldsymbol{\mu}(t), \boldsymbol{\lambda}(t), \boldsymbol{x}(t), \boldsymbol{u}(t), t) & = & \mathbf{0} & \text{(Hamiltonian} \\
 & \boldsymbol{\mu} \dagger \boldsymbol{h} & & \text{Minimization)} \\
\boldsymbol{e}^L \leq \boldsymbol{e}(\boldsymbol{x}_0, \boldsymbol{x}_f, t_0, t_f) & \leq & \boldsymbol{e}^U & \text{(endpoint eqns)} \\
\boldsymbol{\lambda}(t_0) + \partial_{x_0} \overline{E}(\boldsymbol{\nu}, \boldsymbol{x}_0, \boldsymbol{x}_f, t_0, t_f) & = & \mathbf{0} & \text{(initial and} \\
\boldsymbol{\lambda}(t_f) - \partial_{x_f} \overline{E}(\boldsymbol{\nu}, \boldsymbol{x}_0, \boldsymbol{x}_f, t_0, t_f) & = & \mathbf{0} & \text{final transversality} \\
 & \boldsymbol{\nu} \dagger \boldsymbol{e} & & \text{conditions)} \\
\mathcal{H}[@t_0] - \partial_{t_0} \overline{E}(\boldsymbol{\nu}, \boldsymbol{x}_0, \boldsymbol{x}_f, t_0, t_f) & = & 0 & \text{(Hamiltonian} \\
\mathcal{H}[@t_f] + \partial_{t_f} \overline{E}(\boldsymbol{\nu}, \boldsymbol{x}_0, \boldsymbol{x}_f, t_0, t_f) & = & 0 & \text{value conditons)}
\end{array}
\right.
$$

Problem P^λ is not a "simple" BVP (compare with Problem B^λ) because the inequalities and complementarity conditions generate switches, jumps, phases and so on in one or more variables at one or more (unknown) interior points. It will be apparent later (see Section 2.9 on page 150) that solving even a simple BVP is cursed with extreme sensitivity because of the symplectic structure (see page 114) of the Hamiltonian system (state-costate pair). This is why the simpler path to finding a solution to Problem P^λ is through the covector mapping principle discussed later in Section 2.9.2 on page 157.

Study Problem 2.13

Show that the "additional" necessary conditions for Problem B are given by

1. $\boldsymbol{\lambda}(t_0) = -\boldsymbol{\nu}_{x_0}$
2. $\mathcal{H}[@t_0] = \nu_{t_0}$

where $(\boldsymbol{\nu}_{x_0}, \nu_{t_0})$ *is the covector pair associated with the initial conditions* $\boldsymbol{x}_0 = \boldsymbol{x}^0$ *and* $t_0 = t^0$*, respectively.*

A Caveat or Two About Problem P^λ

In Pontryagin et al[75], the development and formulation of Problem P^λ is different from our presentation: It does not contain the mixed state-control constraints, and the formulation of the necessary conditions is based on differentiating $\boldsymbol{h}$. The addition of a path constraint (to Problem B), first as a pure state constraint (i.e., $\boldsymbol{h}$ is a function of $\boldsymbol{x}$ only) and later as a mixed state-control constraint (as in Problem P) has generated a substantial amount of literature since Pontryagin's ground-breaking work. See [24] and the references contained therein. New technical difficulties arise in extending the necessary conditions for Problem P. As a result, there are a number of different "versions" of "Pontryagin's" Principle. So, the question becomes which one should we choose?

We have chosen the version presented on page 134 because it aligns with the covector mapping principle implemented in DIDO. Refer to the discussions associated with Fig. 2.28 on page 164. According to Dmitruk[26, 27], this version of Pontryagin's Principle was developed by Milyutin and Dubovitskii and published in Russian over the period 1963–1981. Starting with his doctoral thesis[20], Clarke[21, 22, 24] and others (see [99] and the references contained therein) have shown that *nonsmooth analysis* is the "simpler" and unified framework for a general theory of optimal control. From this starting point, Problem P^λ is also derived in [22, 24] with additional abstractions.

We have ignored a large number of technical assumptions that go along with Problem P^λ, the most dominant of these being in the form of constraint qualifications. In other words, problem formulation, once again, takes center stage! Because Problem P is widely used for practical applications, it is important to acquire a working knowledge of Problem P^λ. These elements are described next; however, to better understand these details, we strongly recommend an attempt to solve Study Problem 2.14.

Study Problem 2.14

Recall that the control space $\mathbb{U}$ in Problem Brac:3 defined on page 28 was state dependent. Pontryagin's Principle for Problem B is inapplicable to this problem formulation. Using the results from this section, develop the necessary conditions for Problem Brac:3. Do these conditions generate any new insights?

Computational Tip: The obstacle avoidance problem and other RTOC implementations require moving state constraints in the problem formulation. See, for example, page 71 and [17, 40, 59]. As a result, it is every easy to inadvertently formulate problems that violate many of the assumptions needed in the statement of Problem P^λ. In such situations, great care must be exercised in how the problem is formulated, particularly in a DIDO environment, because the coded problem may either produce incorrect results, or may generate correct results but only after a long computational time[84]. Reformulating a problem is critical to producing correct solutions. See, for example, Section 4.1.4 on page 253.

A Deeper Dive on The Impact of Path Constraints

The costate and path covector trajectories $t \mapsto (\boldsymbol{\lambda}, \boldsymbol{\mu})$ have characteristically different shapes in the presence of active state and mixed constraints. This can be quickly observed by expanding the costate equations for Problem P as

$$\begin{aligned} -\dot{\boldsymbol{\lambda}} &= \partial_{\boldsymbol{x}} \overline{H} \\ &= \partial_{\boldsymbol{x}} H + \left(\frac{\partial \boldsymbol{h}}{\partial \boldsymbol{x}}\right)^T \boldsymbol{\mu} \qquad (with\ \boldsymbol{\mu} \dagger \boldsymbol{h}) \end{aligned} \tag{2.67}$$

If $\mu_i = 0$, then the corresponding costate trajectory has "unconstrained" co-dynamics, but over regions where $\mu_i \neq 0$, the related adjoint equation acquires additional terms that reshape the (costate) trajectory.

Illustrating the Concept: A Constrained Brachistochrone Problem

Consider Brac:1 with an added path constraint:

$$y \leq ax + b \tag{2.68}$$

With $h(\boldsymbol{x}) := y - ax$, $h^L = -\infty$, $h^U = b$, *we have*

$$-\dot{\lambda}_x(t) = -a\mu(t) \tag{2.69a}$$

$$-\dot{\lambda}_y(t) = \mu(t) \tag{2.69b}$$

$$-\dot{\lambda}_v(t) = \lambda_x \sin\theta + \lambda_y \cos\theta \tag{2.69c}$$

where $\mu(t)$ *is complementary to* $h(\boldsymbol{x}(t))$ *according to*

$$\mu(t) \begin{cases} = 0 & if \quad y(t) - ax(t) < b \\ > 0 & if \quad y(t) - ax(t) = b \end{cases} \tag{2.70}$$

Compare this with Eq. (2.16) on page 101. Equations (2.69) and (2.70) are illustrated in Fig. 2.17.

Study Problem 2.15

Consider Brac:1 with with the addition of Eq. (2.68).

1. *Prove that* λ_x *must be non-decreasing while* λ_y *must be non-increasing as apparent from Fig. 2.17.*
2. *Prove that* $t \mapsto \lambda_v$ *is affine over the constraint boundary* $y = ax + b$.
3. *Is the trajectory* $t \mapsto \mu$ *continuous?*

4. Are the costate trajectories $t \mapsto (\lambda_x, \lambda_y, \lambda_v)$ continuous? Continuously differentiable?

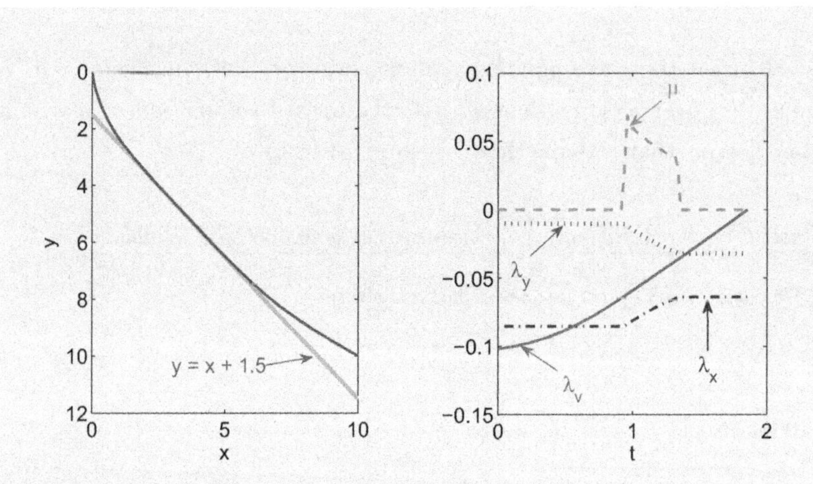

Figure 2.17: Solution to a sample constrained Brachistochrone problem illustrating the effects of the path constraint $y \le x+1.5$. These plots were generated by DIDO using the data: $g = 9.8$, $x^f = 10.0$, $y^f = 10.0$. See also Section 3.2 on page 177.

A quick examination of Problem P^λ reveals that there are no dynamics associated with the covector $\boldsymbol{\mu}$; that is, there is no $\dot{\boldsymbol{\mu}}$ anywhere; hence, $\boldsymbol{\mu}$ is inertia-less. This implies that the trajectory $t \mapsto \boldsymbol{\mu}$ may have discontinuities — a point quite apparent from Fig. 2.17. Because we assumed $\boldsymbol{h}$ was differentiable with respect to $\boldsymbol{x}$, a discontinuous path-covector trajectory $t \mapsto \boldsymbol{\mu}$ has the same effect on the costate trajectory as a discontinuous control. See Eq. (2.67) and observe that $\boldsymbol{\mu}$ is like a co-control on the co-dynamics.

Now, recall that Pontryagin's Principle for Problem B states the existence of an *absolutely continuous* costate trajectory $t \mapsto \boldsymbol{\lambda}$. Absolute continuity is simply a mathematician's way of saying "differentiable ... but not the way you think." See $\lambda_x(t)$ and $\lambda_y(t)$ in Fig. 2.17, particularly over the two points where $\boldsymbol{\mu}(t)$ jumps. The discussions would end here if we were assured that $t \mapsto \boldsymbol{\mu}$ behaves "no worse" than a control trajectory; that is, piecewise continuous (technically,

L^∞). Unfortunately, this is not the case:

$t \mapsto \boldsymbol{\mu}$ may have impulses.

An impulsive $\boldsymbol{\mu}$ implies (see Eq. (2.67)) that the costate trajectory may have jumps; hence, $t \mapsto \boldsymbol{\lambda}$ is no longer assured to be absolutely continuous. This phenomenon is not some mathematical minutia that can be pushed under the rug; it is quite real and quite practical, as illustrated by the following physical problem due to Breakwell[16].

Illustrating the Concept: Impulsive Path Covector Trajectory

The Breakwell problem is a simple double integrator $\ddot{x} = u$ with a path constraint $x \le \ell$ and quadratic cost $\int_0^1 u^2(t)dt$; see Section 4.4.3 on page 270 for details. As outlined in §4.4.3, we have

$$-\dot{\lambda}_x(t) = \mu(t) \begin{cases} = 0 \;\; if & x(t) \;\; < \ell \\ = 0 \;\; if & x(t) \;\; = \ell \end{cases} \tag{2.71}$$

If we hastily conclude that $\lambda_x(t)$ must be a constant, then we get no solution for the boundary condition $x(0) = x(1) = 0,\ \dot{x}(0) = -\dot{x}(1) = 1$, an erroneous result that is quite apparent from elementary analysis. This "paradox" is easily solvable if we "allow" $\mu(t)$ to have precise and ***finite impulses*** *such that*

$$\int_{t_e-\epsilon}^{t_e+\epsilon} \mu(t)\, dt = \eta \qquad (> 0 \;\; as \;\; 0 < \epsilon \to 0) \tag{2.72}$$

where t_e is an entry or exit time; that is, t_e is the time when $x(t)$ enters or leaves the boundary $x(t) = \ell$; see Fig. 2.18. With this expansion of our mathematical horizons, Eq. (2.71) now tells us that $t \mapsto \lambda_x$ is ***piecewise constant*** *with allowable jumps at t_e. As a result of the sign of $\mu > 0$ (or $\eta > 0$) these jumps must be downward for $\lambda_x(t)$.*

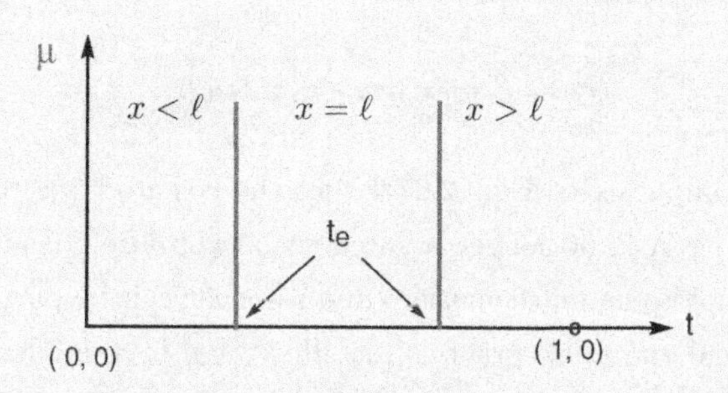

Figure 2.18: Schematic of the covector trajectory $t \mapsto \mu$ for the Breakwell Problem[16]. See also Section 4.4.3 on page 270 for additional details on this problem.

A quick approach for detecting what may happen with finite impulsive elements in $\boldsymbol{\mu}(t)$ is to integrate Eq. (2.67) over an epsilon neighborhood of a candidate jump point

$$\int_{t_e-\epsilon}^{t_e+\epsilon} -\dot{\boldsymbol{\lambda}}\, dt = \int_{t_e-\epsilon}^{t_e+\epsilon} \left[\partial_{\boldsymbol{x}} H + \left(\frac{\partial \boldsymbol{h}}{\partial \boldsymbol{x}}\right)^T \boldsymbol{\mu} \right] dt$$

and note that we can write (with some minor abuse of ϵ):

$$\boldsymbol{\lambda}(t_e^-) - \boldsymbol{\lambda}(t_e^+) = \left(\frac{\partial \boldsymbol{h}}{\partial \boldsymbol{x}}\right)^T_{\boldsymbol{x}=\boldsymbol{x}_e} \underbrace{\int_{t_e-\epsilon}^{t_e+\epsilon} \boldsymbol{\mu}(t)\, dt}_{\boldsymbol{\eta}} \tag{2.73a}$$

$$= \left(\frac{\partial \boldsymbol{h}}{\partial \boldsymbol{x}}\right)^T_{\boldsymbol{x}=\boldsymbol{x}_e} \boldsymbol{\eta} \qquad (\textit{with } \boldsymbol{\eta} \dagger \boldsymbol{h}[@t_e]) \tag{2.73b}$$

Equation (2.73b) can be used as follows: If $\boldsymbol{h}$ does not depend on a particular state variable, then its corresponding costate is continuous. See for example Eq. (2.69c) on page 137. This is a sufficient condition; it is not necessary. That is, simply because $\partial_x \boldsymbol{h} \neq 0$ does not mean there will be jumps. See, for example, Eqs. (2.69a) and (2.69b) and Fig. 2.17. This is because $t \mapsto \boldsymbol{\mu}$ may not have finite impulses.

Equation (2.73b) is discussed in some textbooks and papers as a ***jump condition*** but without the insight of the connection between $\boldsymbol{\eta}$ and $\boldsymbol{\mu}$. The complementarity of $\boldsymbol{\nu}$ with $\boldsymbol{h}$ at t_e (denoted by $\boldsymbol{\eta} \dagger \boldsymbol{h}[@t_e]$) is inherited from $\boldsymbol{\mu}$ because the integration is over a positive measure.

Tech Talk: Because of the need to incorporate atomic measures, Eq. (2.73a) can be written more "cleanly" as a Lebesgue-Stieltjes integral:

$$\int \partial_x \boldsymbol{h}[@t] \ \boldsymbol{\mu}(t)\, dt = \int \partial_x \boldsymbol{h}[@t] \ d\tilde{\boldsymbol{\mu}}(t)$$

Such an approach is commonly found in the mathematical literature with extensive use of Radon measures. Another approach is to write $\boldsymbol{\mu}(t)$ as the sum of two functions

$$\boldsymbol{\mu} = \boldsymbol{\mu}_{L^\infty} + \boldsymbol{\mu}_\delta$$

where $\boldsymbol{\mu}_\delta$ is of atomic measure. We have used this approach implicitly in characterizing $\boldsymbol{\mu}_\delta$ as a collection of finite impulses.

The Hamiltonian Evolution Equation for Problem P^λ

Recall that the Hamiltonian evolution equation was an integral of motion for Problem B. For Problem P, this evolution equation is given analogously by

$$\frac{d\mathcal{H}}{dt} = \frac{\partial \overline{H}}{\partial t} \tag{2.74}$$

Similar to Problem B, Eq. (2.74) serves an extremely important role in the *verification and validation* of a computed solution. Similar to Eq. (2.67), we can write Eq. (2.74) as

$$\frac{d\mathcal{H}}{dt} = \partial_t H + \left(\frac{\partial \boldsymbol{h}}{\partial t}\right)^T \boldsymbol{\mu} \qquad (\textit{with } \boldsymbol{\mu} \dagger \boldsymbol{h})$$

Because $t \mapsto \boldsymbol{\mu}$ may have finite impulses, $\mathcal{H}$ may jump similarly to $\boldsymbol{\lambda}(t)$. This jump condition may be derived similarly to the procedure used for the costate trajectory and written as

$$\mathcal{H}[@t_e^+] - \mathcal{H}[@t_e^-] = \left(\frac{\partial \boldsymbol{h}}{\partial t}\right)^T_{t=t_e} \boldsymbol{\eta} \qquad (\textit{with } \boldsymbol{\eta} \dagger \boldsymbol{h}[@t_e]) \tag{2.75}$$

Illustrating the Concept: A Constrained Brachistochrone Problem

Because $t \mapsto \mu$ had no impulses in the constrained Brachistochrone problem (see Fig. 2.17 on page 138) we have $\eta = 0$; hence, the lower Hamiltonian is continuous. Furthermore, because $\partial_t H = 0$, we must have

$$\mathcal{H}(\boldsymbol{\lambda}(t), \boldsymbol{x}(t)) = -1 \qquad (\forall\, t)$$

That $\mathcal{H}[@t]$ must not jump and be a constant of value -1 even in the presence of the state constraint is a powerful statement for a V & V of a solution. This aspect of gaining confidence in a computed solution is illustrated in Fig. 2.19.

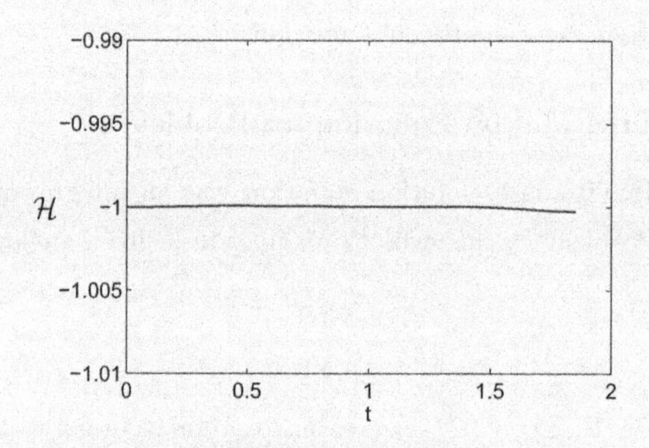

Figure 2.19: The lower Hamiltonian $t \mapsto \mathcal{H}(\boldsymbol{\lambda}(t), \boldsymbol{x}(t))$ generated by DIDO for the sample constrained Brachistochrone problem discussed on page 138.

How to Use Problem P^λ: Some Dos and Don'ts

The caveats and warnings that go along with Problem P^λ are a further indication of the maxim that it is best to use the necessary conditions as necessary conditions, that is, as checks on the optimality of a candidate solution instead of a sanction for a technique to solve problems. Doing so would generate a "hard" problem *by choice.*

At the other end of the spectrum is the use of a software package like DIDO.

Because DIDO makes problem-solving "easy," it is essential to exercise great restraint in solving problems without analysis; that is, generating solutions without analyzing (not solving!) Problem P^λ. An analysis of Problem P (and hence P^λ) includes a cognizance of the structure of $t \mapsto \boldsymbol{\mu}$. Recall that the Breakwell problem warned us that $t \mapsto \boldsymbol{\mu}$ may have finite impulses, while the constrained Brachistochrone problem only gave us discontinuities in $\boldsymbol{\mu}(t)$. It is critical to understand Problem P^λ in advance of coding (see Section 4.1.4 on page 253) or modifying a working code before declaring a problem to be "hard."

In a practical and computational setting, a useful rule of thumb for the detection of jumps is the notion of the ***order of a state constraint***. If $\boldsymbol{h}$ depends on $\boldsymbol{x}$ only, and possibly t, then its time derivative is given by

$$\begin{aligned}\frac{d\boldsymbol{h}(\boldsymbol{x},t)}{dt} &= \partial_{\boldsymbol{x}}\boldsymbol{h}(\boldsymbol{x},t)^T\dot{\boldsymbol{x}} + \partial_t\boldsymbol{h}(\boldsymbol{x},t)\\ &= \partial_{\boldsymbol{x}}\boldsymbol{h}(\boldsymbol{x},t)^T\boldsymbol{f}(\boldsymbol{x},\boldsymbol{u},t) + \partial_t\boldsymbol{h}(\boldsymbol{x},t)\end{aligned} \tag{2.76}$$

If Eq. (2.76) depends explicitly on $\boldsymbol{u}$, then $\boldsymbol{h}$ is said to be of order one. For example, Eq. (2.68) is of order one because

$$\begin{aligned}\dot{h}(\boldsymbol{x}) &= \dot{y} - a\dot{x}\\ &= v\cos\theta - av\sin\theta \qquad (\theta \text{ is control})\end{aligned}$$

When a constraint is of order one, it is typical for $t \mapsto \boldsymbol{\mu}$ to not have impulses; see Fig. 2.17.

It is quite possible that Eq. (2.76) may contain no control terms. For example, in the Breakwell problem we have

$$\begin{aligned}\dot{h}(\boldsymbol{x}) &= \dot{x}\\ &= v\end{aligned}$$

In such a situation, we can differentiate $\boldsymbol{h}$ again. If $\boldsymbol{u}$ appears explicitly for the second derivative, then $\boldsymbol{h}$ is called a path constraint of order two. Performing this exercise for the Breakwell problem we have

$$\begin{aligned}\ddot{h}(\boldsymbol{x}) &= \dot{v}\\ &= u\end{aligned}$$

Hence, the constraint $h(\boldsymbol{x}) = x \le \ell$ is of second order. It is typical for second and higher-order constraints to have impulsive path covector trajectories; see Fig. 2.18.

Note from these discussions that as the problem becomes more practical, Pontryagin's Principle provides many more conditions for analysis.

Computational Tip. In one of its "dual" operating modes, DIDO uses a Galerkin approximation for the covector trajectories. This mode "smooths out" impulses and discontinuities in dual spaces to achieve computational efficiency. For higher *dual accuracy*, other operating modes may be used but at the price of reduced computational speed. Because accuracy of the control trajectory is the last word in optimal control, an "efficient" computational strategy in the presence of path constraints is to use the Galerkin option for the dual operating mode. Under this option, the controls converge more rapidly than the dual variables; hence, the covector outputs of DIDO must be used as approximations while performing V & V. This mathematical insight on the mechanics of pseudospectral optimal control theory is due to Q. Gong.

2.8 Avoiding Common Errors # 3

Pontryagin's Principle is a statement of the *necessary* conditions for Problem B. It asserts the *existence* of certain covector functions. The origin of many common errors can be traced to overlooking these fundamentals.

1. Necessary, Not Sufficient

A solution to an optimal control problem (B) is also a solution to the boundary value problem (B^λ). The converse is not necessarily true:

$$\text{Solution to Problem } B \quad \begin{matrix} \Rightarrow \\ \not\Leftarrow \end{matrix} \quad \text{Solution to Problem } B^\lambda$$

See Section 4.1.3 on page 252 for a demonstration of the insufficiency of Pontryagin's Principle based on a simple problem. It is very tempting to use the backward implication and claim optimality, for instance, by solving the BVP. Fortunately, the forward implication is much easier to use; see Section 2.9.

2. Bane of Non-unique Existence

Pontryagin's Principle simply asserts the existence of covector functions; it does not imply their uniqueness. The non-uniqueness of covectors is not part of some fine print. It occurs frequently and often goes unrecognized as shown by the following problem (inspired by the unpublished results of M. Karpenko and Q. Gong). See also Section 4.9 on page 306.

Study Problem 2.16

The kinematics of a rigid body can be expressed in terms of Euler parameters or "quaternions" as

$$\dot{\boldsymbol{q}} = \frac{1}{2}\left(q_4\, \boldsymbol{\omega} + \boldsymbol{q} \times \boldsymbol{\omega}\right)$$
$$\dot{q}_4 = -\frac{1}{2}\boldsymbol{\omega} \cdot \boldsymbol{q}$$

where $\boldsymbol{q} \in \mathbb{R}^3$ *is the Euler vector and* q_4 *is the scalar (component of the quaternion). Assuming no running cost, show that the adjoint covector functions obey the same dynamics as the quaternions; i.e.,*

$$\dot{\boldsymbol{\lambda}}_q = \frac{1}{2}\left(\lambda_4\, \boldsymbol{\omega} + \boldsymbol{\lambda}_q \times \boldsymbol{\omega}\right)$$
$$\dot{\lambda}_4 = -\frac{1}{2}\boldsymbol{\omega} \cdot \boldsymbol{\lambda}_q$$

Consequently, $\boldsymbol{\lambda}_q \cdot \boldsymbol{\lambda}_q + \lambda_4^2 =$ *constant. A useful calculus for this proof is the rule*

$$\frac{\partial}{\partial \vec{x}}\left(\vec{a} \cdot \left(\vec{b} \times \vec{x}\right)\right) = \vec{a} \times \vec{b}$$

Discuss the conditions that generate an infinite number of costate trajectories. (Hint: Consider the transversality conditions). *See also Section 4.9 on page 306.*

3. Analyze, Not Solve

An application of Pontryagin's Principle to Problem B generates a different, presumably simpler problem (B^λ); see Fig. 2.12 on page 112. Embedded inside

Problem B^λ is another problem (HMC). At first, it seems reasonable to assume that the inner problem (HMC) needs to be solved to analyze the outer problem (B^λ). This is not necessarily true because Problem HMC can be analyzed via its KKT conditions.

The KKT conditions are necessary conditions; they do not solve Problem HMC. Fortunately, there is no need to "solve" Problem HMC. Its inclusion in Theorem 2.1 facilitates an analysis of the the inner and outer problems as a single unit. Such an analysis is facilitated through an application of DIDO as illustrated in Fig. 2.16 on page 124.

4. Sufficient, But Unnecessary

Based on the misconception that it is necessary to produce an analytic solution to Problem HMC, many beginners and some practitioners fall into the trap of formulating the wrong problem (e.g., "augmenting" unnecessary cost functions or performing ruthless linearizations). Recall (see page 122) that problem solving does not imply analytic solutions. As we gradually move away from the abacus to a computer, it becomes increasingly clear that is better to solve the correct problem "approximately" than solve the wrong problem "exactly." This is why it is more important to formulate the correct problem than perform unnecessary simplifications based on the presumed supremacy of analytic solutions. Analytic solutions may indeed enter in the formulation of the *problem of problems* as noted in Chapter 1, but in their proper context, for instance, as part of the development of the right problem, or as a means to verify and validate (V&V) a computer code.

5. Chasing Phantoms

Pontryagin's Principle is a formulation of the necessary conditions for optimality; consequently, it is an analysis tool, not a problem-solving tool. It can be correctly used as a necessary condition, that is, to test the optimality of a candidate solution regardless of how it is obtained. In Chapter 3, we take the liberty of using it as a problem-solving tool because it can indeed be used as such on a very small class of academic-strength problems. In industrial applications, Pontryagin's Principle is used at its fundamental level: as a necessary condition. That is, if a candidate solution fails Pontryagin's test, it is not an optimal solution, but an optimal solution must satisfy the necessary conditions.

Solutions that satisfy Pontryagin's Principle are called ***extremals***. By definition, the extremals are feasible but they are not necessarily optimal. Not even locally optimal. These facts are not a deterrent to generating extremely good and usable solutions to many problems provided sufficient care and caveats are used. In many instances, an extremal solution might be fully satisfactory if its cost is sufficiently low. For example, if the cost of an extremal solution is lower than the best prevailing solution, it can be crowned the reigning optimal solution until a better solution is found. See remarks on page 123 in Section 2.5.4.

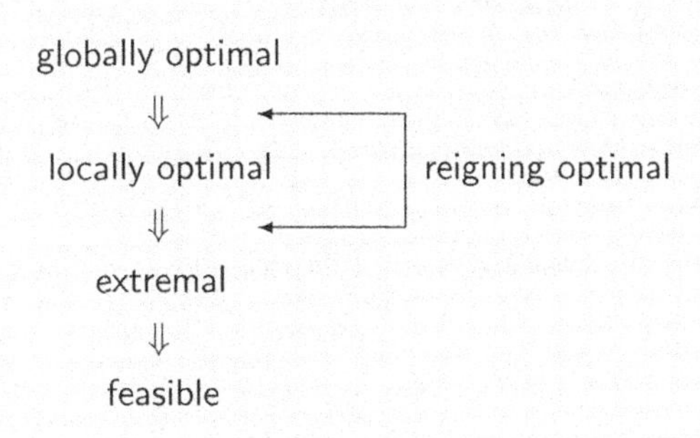

Figure 2.20: Any extremal that is better than a prevailing solution is a reigning optimal solution. A mathematical proof of local optimality for an extremal solution requires higher-order conditions (not discussed in this book).

Sometimes, the claim or demand of global optimality is uninformed; however, there are many application problems where it is possible to estimate a lower bound for $J[\boldsymbol{x}(\cdot), \boldsymbol{u}(\cdot), t_f]$ and analyze the optimality of an extremal solution. That is, suppose we can find an a priori number, $\mathcal{J}^{lower}$, such that

$$J[\boldsymbol{x}(\cdot), \boldsymbol{u}(\cdot), t_f] \geq \mathcal{J}^{lower}$$

for all feasible decision variables, $(\boldsymbol{x}(\cdot), \boldsymbol{u}(\cdot), t_f)$. Now let $(\boldsymbol{x}^{\#}(\cdot), \boldsymbol{u}^{\#}(\cdot), t_f^{\#})$ be an extremal solution (e.g., one that is successfully computed by DIDO). Then, based on the difference,

$$J[\boldsymbol{x}^{\#}(\cdot), \boldsymbol{u}^{\#}(\cdot), t_f^{\#}] - \mathcal{J}^{lower} \qquad (\geq 0)$$

any number of practical conclusions can be made:

1. Is the difference small enough† that it is not economical to find the globally optimal solution?

2. Is the difference so large that there might be an error in the estimation of $\mathcal{J}^{lower}$?

3. Is the difference so large that it is worth exploring finding another extremal solution closer to $\mathcal{J}^{lower}$?

4. Is there a mistake in the problem formulation itself? See Chapter 1, particularly Fig. 1.50 on page 84.

As an illustrative example of these considerations, let us examine a problem from astronautics. In space maneuvering, nearly all problems are fundamentally driven by propellant usage because of the extremely high dollar-cost of launch. Thus, when the cost functional is propellant, it is easy to write

$$\mathcal{J}^{lower} = 0$$

In Bedrossian's celebrated zero propellant maneuvers (ZPMs) implemented onboard the International Space Station[5], the extremal solutions satisfy

$$J[\boldsymbol{x}^{\#}(\cdot), \boldsymbol{u}^{\#}(\cdot), t_f^{\#}] = 0$$

Hence, the triple $\left(\boldsymbol{x}^{\#}(\cdot), \boldsymbol{u}^{\#}(\cdot), t_f^{\#}\right)$ is indeed globally optimal. See Fig. 2.21.

In many situations, (provable) global optimality is not necessarily the most important criterion. A user may be fully satisfied with a feasible solution if its cost is lower than some targeted value $\mathcal{J}^{target}$. Then, any non-zero difference

$$\mathcal{J}^{target} - J[\boldsymbol{x}^{\#}(\cdot), \boldsymbol{u}^{\#}(\cdot), t_f^{\#}] \qquad (\geq 0)$$

is considered a bonus, and any extremal solution that bests the other is the reigning optimal solution. For instance, in 2012 (August), the "optimal" propellant maneuver[7] replaced the ZPM on the Space Station even though it expended propellant because the time to complete the maneuver was less than one orbital period, a "soft" requirement that was highly valued by NASA.

†In an industrial environment, it is common to use phrases like "within single digits" for small or "double-digit improvements in performance" for large.

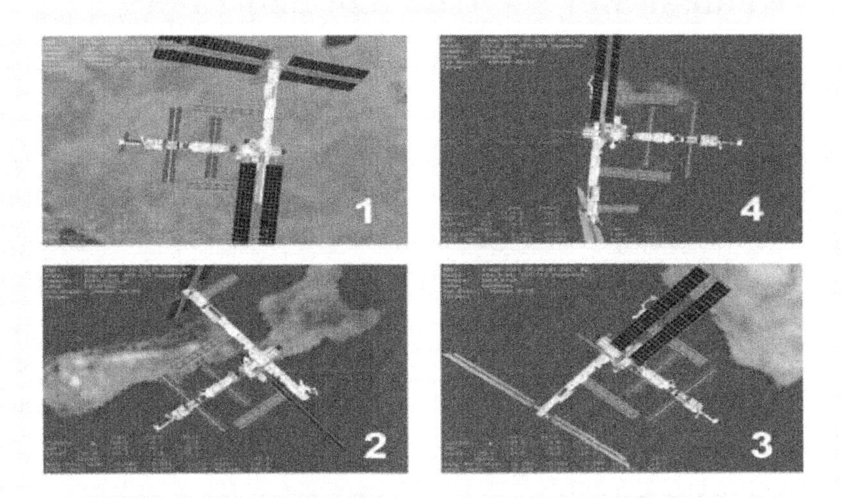

Figure 2.21: "Snapshots" of the first (November 2006) *Zero Propellant Maneuver* implemented onboard the International Space Station. The extremal solutions (computed via DIDO) are indeed globally optimal.

Thus, the practice of optimal control is not necessarily the quest for globally or even locally optimal solutions, rather, it is a quest for usable solutions that are candidates for the reigning optimal. Pontryagin's Principle helps eliminate spurious solutions through mathematical tests.

5. Feasible = Optimal

There are times when some users/customers imply that they are more interested in finding feasible solutions than optimal solutions. In this case, all feasible solutions are globally optimal! In DIDO, this can be done by simply turning off the cost function.

2.9 Kalman and the Curse of Sensitivity

In 1964, R. Kalman presented a paper in Yorktown Heights, N.Y., at an IBM symposium on control[46].‡ Based on the challenges he and others faced in computing optimal controls, Kalman argued that the problem was "intrinsically difficult"; hence, he suggested, there must be some underlying fundamental mathematical principles that could form the basis for a "theory of difficulty." Roughly speaking, Kalman contended that any computational method based on numerically integrating (simulating) differential equations was doomed, and that the only escape from this eventual disaster was some "newfangled algebra" that could "algebraize" the dynamics. Based on the somewhat contentious discussions that ensued at the end of his presentation (as recorded in [46]) and the subsequent lack of citations to his paper, it is apparent that Kalman's ideas were not quite well received. Inspired by Bellman's *curse of dimensionality*, we refer to Kalman's concept of intrinsic difficulty as the *curse of sensitivity*. The escape clause on the curse of sensitivity (i.e., the newfangled algebra that Kalman predicted) is the basis for *pseudospectral (PS) optimal control theory*[87]. The curse and its escape are illustrated and introduced in this section. The details are dispersed over several papers cited in [87].

2.9.1 An Introduction to the Curse of Sensitivity

Consider the following problem:

$$\left\{\begin{array}{lrcl} & x \in \mathbb{R} \quad u \in \mathbb{R} & & \\ \text{Minimize} & J[x(\cdot), u(\cdot)] & = & \dfrac{1}{2}\displaystyle\int_{t_0}^{t_f} u^2(t)\, dt \\ \text{Subject to} & \dot{x} & = & ax + bu \\ & x_0 & = & x^0 \\ & t_0 & = & 0 \\ & x_f & = & x^f \\ & t_f & = & t^f \end{array}\right. \tag{2.77}$$

‡This symposium was attended by a veritable Who's Who in applied mathematics and control: One of the presenters at this conference was none other than Pontryagin himself!

where a and b are given real numbers. An application of Pontryagin's Principle generates (show this!) the following pair of state-costate differential equations:

$$\begin{aligned} \dot{x} &= ax - b^2\lambda && \text{(primal)} \\ \dot{\lambda} &= -a\lambda && \text{(dual)} \end{aligned} \tag{2.78}$$

It can be shown (do it!) that the exact solution to the state trajectory is given by

$$x(t) = x^0 e^{at} + \frac{b^2\lambda_0}{2a}\left(e^{-at} - e^{at}\right) \tag{2.79}$$

where $\lambda_0 \equiv \lambda(0)$ is the unknown quantity. Evaluating Eq. (2.79) at $t = t_f$, and rearranging the terms, we get:

$$\lambda_0 = \frac{2a}{b^2}\left(\frac{x_f - x^0 e^{at_f}}{e^{-at_f} - e^{at_f}}\right) \tag{2.80}$$

Substituting the boundary conditions, $t_f = t^f$ and $x(t_f) = x^f$ in Eq. (2.80), we can compute the exact value of $\lambda(0)$ as

$$\lambda^0 := \frac{2a}{b^2}\left(\frac{x^f - x^0 e^{at^f}}{e^{-at^f} - e^{at^f}}\right) \tag{2.81}$$

If λ^0 is substituted for λ_0 in Eq. (2.79), we will have obtained the exact solution for the state trajectory, $t \mapsto x$; see Fig. 2.22.

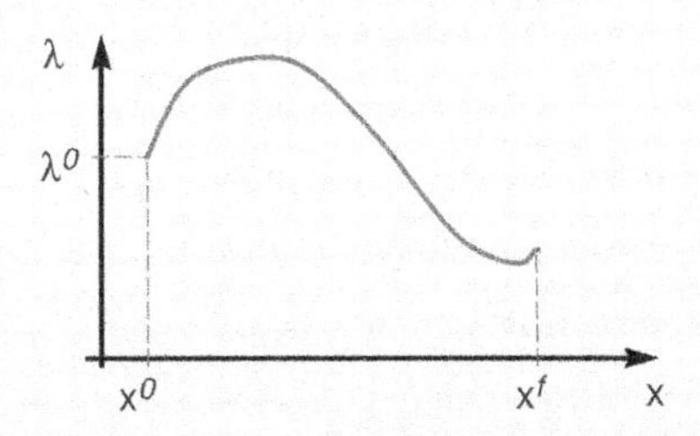

Figure 2.22: Using the data (x^f, t^f, x^0), the exact value of the missing initial condition (λ^0) is computable from Eq. (2.81). When this value is used as the initial condition for the primal-dual system, it achieves the final time condition $x(t^f) = x^f$ (precisely).

Study Problem 2.17

Show that:

1. *The Pontryagin extremal control to the problem defined by Eq. (2.77) is given by*

$$u = -\lambda b$$

2. *The lower Hamiltonian to Problem 2.77 is given by*

$$\mathcal{H}(\lambda, x) := ax\lambda - \frac{b^2\lambda^2}{2}$$

3. *The partials $\partial_\lambda \mathcal{H}$ and $\partial_x \mathcal{H}$ produce the right-hand sides of the primal-dual system given by Eq. (2.78). Will such a relationship hold for all problems? (Hint: Examine this for Brac:1.)*
4. *A particular solution to the primal system (defined in Eq. (2.78)) is given by*

$$x_{particular}(t) = \frac{b^2\lambda_0}{2a} e^{-at}$$

In a general problem, the key formula, $(x^f, t^f, x^0) \mapsto \lambda^0$, as given by Eq. (2.81) for Problem 2.77, is not known. This is because of the absence of closed-form solutions to generic nonlinear differential equations; see also page 122 for a discussion of analytical solutions. Despite this "setback," it is easy to generate the map $\lambda_0 \mapsto x_f$ through numerical integration (e.g., a standard Runge-Kutta procedure; see Fig. 2.23). As a means to understand the consequences of solving BVPs through a sequence of initial value problems (IVPs), let us assume that we had to solve Eq. (2.78) iteratively. This BVP is given by

$$\begin{aligned} \dot{x} &= ax - b^2\lambda & t_0 &= 0, & x(t_0) &= x^0 \\ \dot{\lambda} &= -a\lambda & t_f &= t^f, & x(t_f) &= x^f \end{aligned} \qquad (2.82)$$

The intuitive, and prima facie logical-sounding ***shooting algorithm***, would proceed as follows: Guess a value for λ_0, say λ^{Guess}. Solve the *initial value*

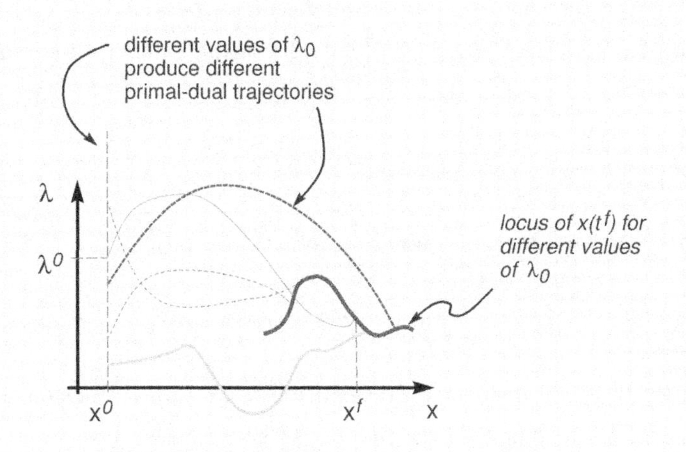

Figure 2.23: When λ_0 is not equal to its exact value (λ^0), it generates $x(t^f)$, whose locus can be determined through numerical propagation.

problem

$$\begin{aligned} \dot{x} &= ax - b^2\lambda & t_0 &= 0, \quad x(t_0) = x^0, \quad \lambda(t_0) = \lambda^{Guess} \\ \dot{\lambda} &= -a\lambda & t_f &= t^f \end{aligned} \tag{2.83}$$

all the way up to the given value of $t_f = t^f$. This process produces a number $x(t^f)$ that we can use to compare to its specified value x^f; see Fig. 2.23. Assuming we did not correctly guess the exact value of λ_0 (i.e., the value given by Eq. (2.81)) we will have

$$x(t^f) \neq x^f$$

The shooting algorithm is based on the idea that we can use some measure of the difference, $(x(t^f) - x^f)$, to generate a new guess for λ_0 and repeat the process until $x(t^f) = x^f$. Because the process is iterative,[§] it is unlikely that we will obtain $x(t^f) = x^f$ exactly; hence, we simply require that

$$x(t_f) \approx x^f \tag{2.84}$$

[§]This iterative process is a nonlinear programming problem; e.g., the objective may be to minimize error norms subject to tolerance constraints. Consequently, it is erroneous to imply that a shooting method is not based on nonlinear programming techniques as is sometimes indicated in the literature.

within some pre-specified tolerance. Furthermore, because $x(t^f)$ is itself computed through numerical integration, its value will also be approximate; hence, we can write Eq. (2.84) a little more precisely as

$$x^N(t^f) \approx x^f \tag{2.85}$$

where $x^N(t^f)$ is itself an approximation to $x(t^f)$ such that

$$\lim_{N \to \infty} x^N(t^f) \to x(t^f)$$

The number N is an abstract representation of an approximation process: It may be construed as denoting the number of integration steps (i.e., the inverse of the step size) or similar other notion of discretization. This concept is codified in Fig. 2.24 where $B^{\lambda N}$ represents the computational problem that is actually solved. Because we expect convergence as $N \to \infty$, an argument can be made that the better the accuracy of the numerical integration, the higher the accuracy of the solution.

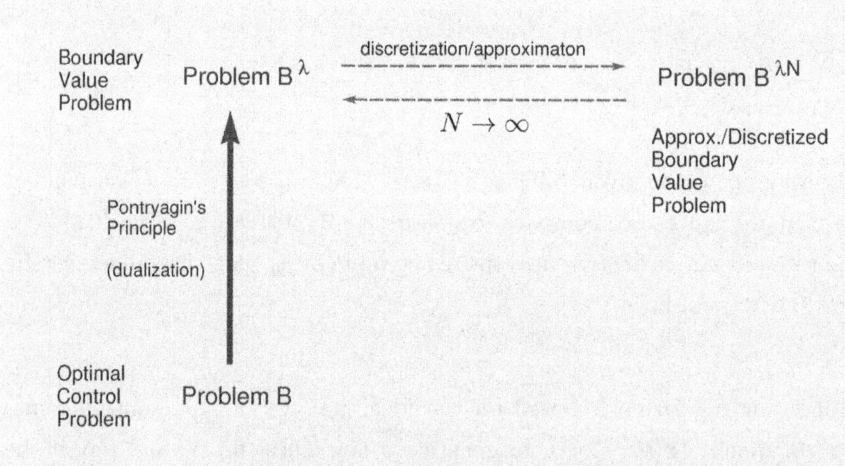

Figure 2.24: A BVP cannot be solved exactly. Problem $B^{\lambda N}$ is a representation of the problem that is actually solved. Compare with Fig. 2.12 on page 112.

This procedure is so intuitive and simple that it smells like it ought to work seamlessly. *It doesn't*! Here's why: *Assume we have perfect integration*; then, subtracting Eq. (2.81) from (2.80) and rearranging the terms, we get

$$x_f - x^f = \frac{b^2}{2a}\left(e^{-at_f} - e^{at_f}\right)\left(\lambda_0 - \lambda^0\right) \tag{2.86}$$

This equation shows that for different values of $\lambda_0 \neq \lambda^0$, we get different values of $x_f \neq x^f$; hence, for different values of $\lambda^{Guess}(= \lambda_0)$, the computed value of $x(t_f) = x_f$ is not equal to the specified number x^f. The difference between x_f and x^f varies as

$$\mathcal{S} := \left(e^{-at_f} - e^{at_f}\right)$$

The absolute value of $\mathcal{S}$ is exponentially large and is independent of the sign of a. The greater the quantity

$$|a|\, t_f \tag{2.87}$$

the larger the number $\left|x_f - x^f\right|$, *even when* $\lambda_0 \equiv \lambda^{Guess}$ *is near* λ^0. In other words, the value of $x(t_f)$ is quite sensitive to the value of λ_0. See Fig. 2.25. This sensitivity relationship can also be inferred from Eq. (2.79) by evaluating

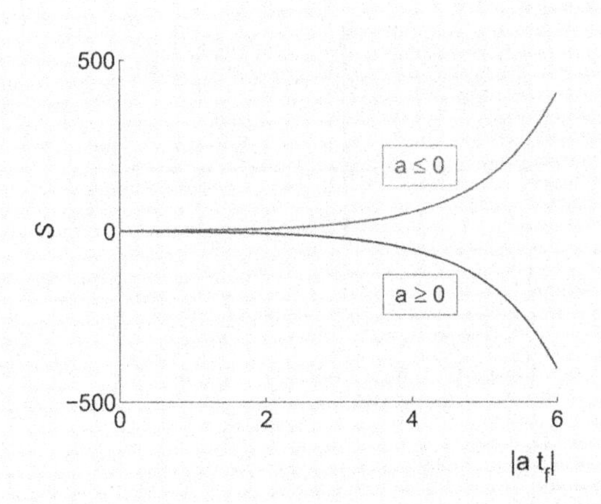

Figure 2.25: A graphic that highlights the curse of sensitivity.

$x(t)$ at $t = t_f$ and taking the partial derivative

$$\frac{\partial x_f}{\partial \lambda_0} = \frac{b^2}{2a}\left(e^{-at_f} - e^{at_f}\right)$$

This leads us to the following observations:

1. Small changes in λ_0 produce exponentially large changes in x_f.

2. The sensitivity of the mapping $\lambda_0 \mapsto x_f$ is independent of the sign of a; hence, it is a myth that "stable" systems produce stable BVPs.

The last statement about stability comes from the observation that the quantity a is an eigenvalue of the dynamical system. The eigenvalues of the primal-dual system given by Eq. (2.78) are a and $-a$; see Fig. 2.26. This fact follows quite readily from the eigenvalues of the system matrix,

$$\begin{bmatrix} a & -b^2 \\ 0 & -a \end{bmatrix}$$

Hence, from Eq. (2.87) we conclude that the product of the absolute value of the eigenvalue of the system and the horizon play a central role in quantifying the sensitivity of the system: the greater this product the higher the sensitivty.

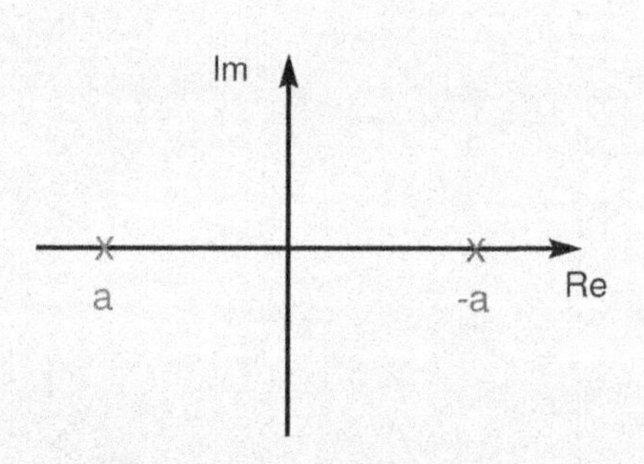

Figure 2.26: Eigenvalues of the primal-dual system for the problem given by Eq. (2.77) on page 150.

Study Problem 2.18

1. *If the dynamical system in an optimal control problem has linear time-varying coefficients, discuss its impact on the curse of sensitivity. Is it possible to generate a formula similar to Eq. (2.87) for such a system?*

2. *If the dynamical system in an optimal control problem is nonlinear (e.g., given by a generic equation,* $\dot{x} = f(x, u, t)$*), what is the meaning of the eigenvalue of the system? Can a connection be made to the Lipschitz constant? See [92].*

3. *Does an infinite horizon problem have infinite sensitivity?*

Study Problem 2.19

Explain how the non-uniqueness of the costates (refer back to Section 2.8) impacts a procedure for solving the BVP (Problem B^λ*).*

Study Problem 2.20

Critique the argument that geometric or symplectic integration can provide an escape from the curse of sensitivity.

2.9.2 Escaping the Curse of Sensitivity

The shortcomings of the shooting method have been observed since the early 1960s. In 1962, Morrison et al[72] wrote,

> [As] happens altogether too often, the differential equations are so unstable that they "blow up" before the initial value problem can be completely integrated. This can occur even in the face of extremely accurate guesses for the initial values.

It is now apparent that this observation is a direct consequence of Eq. (2.87). The eigenvalues of a dynamical system are intrinsic; hence, we cannot change the quantity a in formula Eq. (2.87). The only variable is t_f. Based on this observation, Morrison et al[72] proposed a ***multiple shooting method***. The concept involves dividing the time interval $[t_0, t_f] := [0, t^f]$ into M smaller intervals so that the $|a|\, t^f$ is locally small:

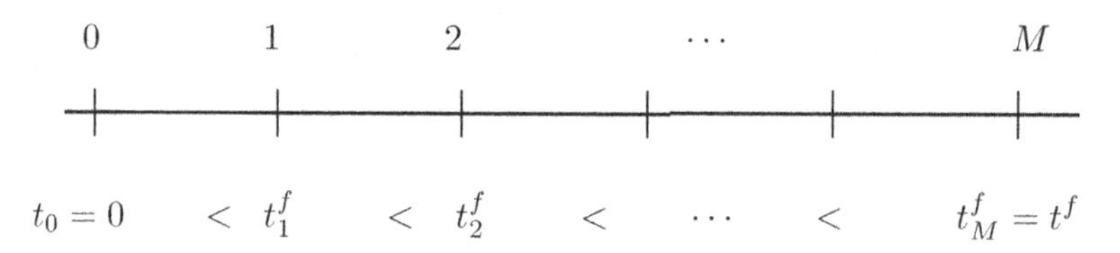

Because the local initial and final-time conditions are unknown, the concept requires an enforcement of continuity conditions:

$$\begin{array}{llllll} x(t_0) = 0 & \cdots & x(t_2^{f-}) = x(t_2^{f+}) & \cdots & \cdots & x(t_M^f) = x^f \\ \lambda(t_0) = \lambda_0^{guess} & \cdots & \lambda(t_2^{f-}) = \lambda(t_2^{f+}) & \cdots & \cdots & \lambda(t_M^f) = N/A \end{array}$$

These conditions form the basis for concatenating local shooting elements where sensitivity is reduced as a consequence of shorter horizons.

Observe that each shooting element in a multiple shooting method involves discretization (via integration):

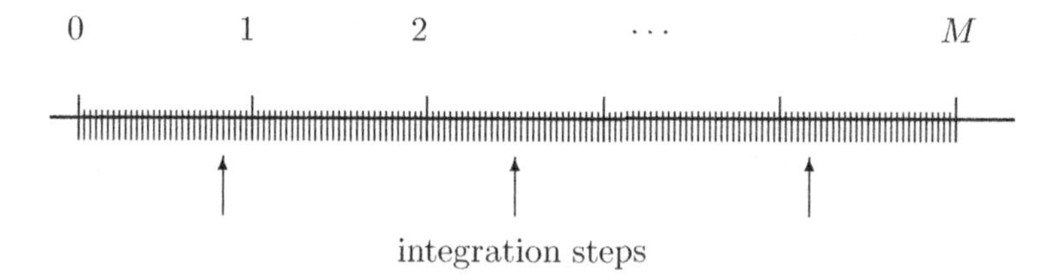

Thus, the multiple shooting element may be re-framed as a procedure that involves a "micro-discretization" at the element level coupled with a "macro-discretization" of the elements themselves. Therefore, it stands to reason that if shooting were done at the micro-discretization level, the result would be the greatest reduction in sensitivity coupled with simplifications resulting from the elimination of the macro-discretizations. This concept is the basis of ***collocation methods***.

In 1987, Hargraves and Paris[36] suggested that instead of discretizing and solving Problem B^λ, it might far easier and computationally more efficient to discretize and solve Problem B directly. They based their argument of computational efficiency from the simple observation that while Problem B^λ involves discretizing $2N_x$ equations (in $\boldsymbol{x}$ and $\boldsymbol{\lambda}$ variables), Problem B involves only half as many: A discretization in N_x variables only (in just $\boldsymbol{x}$ variables). Their concept came to be known as ***direct collocation methods*** in contrast to the prior art of ***indirect collocation methods***; see Fig. 2.27.

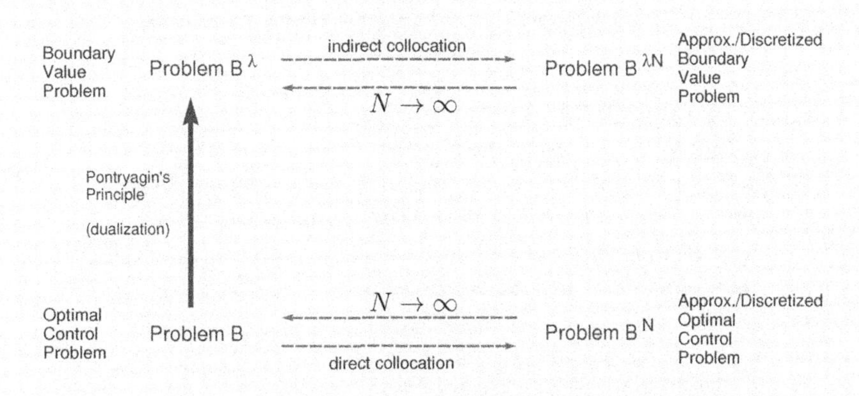

Figure 2.27: Visualizing the difference between direct and indirect collocation methods. Compare this figure with Fig. 2.24.

Study Problem 2.21

1. *By comparing Figures 2.24 and 2.27 explain the meaning of a* ***direct shooting method****.*
2. *Does direct shooting alleviate the curse of sensitivity?*
3. *Pick any three optimal control software packages from the industry and categorize them (i.e., direct or indirect) based on the method used.*

Collocation methods formed the basis of software packages developed in the 1990s. The main reason why these concepts were not developed and exploited in the 1960s is because of the absence of the requisite computer memory available at that time. To appreciate this limitation, observe that a problem with N_x variables and N points requires the manipulation of $N_x N$ variables that may involve matrices of $N_x N \times N_x N$ dimensions. Ignoring sparsity, a simple problem with $N_x = 6$ and $N = 1000$ requires about $6^2 \times (1000)^2 \times 8$ Bytes or about 275 MB of memory — a number that seemed inconceivable in the 1960s. If we limit N to 100 or less, then, the memory requirements reduce to less than 3 MB, a more realistic number for a high-end computer of that era. Taking step sizes in Runge-Kutta methods with $N \leq 100$ generally produces low accuracy; hence, these methods got branded (incorrectly) as low-accuracy methods.

Study Problem 2.22

Estimate the computer memory required to solve an optimal control problem by a multiple shooting method. Based on this estimate, discuss the scale (i.e., numerical value of N_x) of problems that can be solved in a present-day computer under the absence of the curse of sensitivity.

Using the number $N_x^2 \times N^2$ as the basis of computational memory requirements it is apparent that a curse-free method is limited by the scale of the problem (N_x) for a prescribed accuracy (determined inversely by N). A low value of N implies low accuracy. Although computational capacities in the 1990s were substantially larger than those of the 1960s, collocation methods were viewed with suspicion because the practice of these methods necessitated a choice of N that was frequently low for reasonable accuracy. That is, the problems that were actually being solved were too far to the right in Fig. 2.27. As a result, the reigning idea in the 1990s was to use the low-accuracy solution of collocation methods as a seed for a shooting method on the presumption that a good guess would alleviate the curse of sensitivity (i.e., almost ignoring Eq. (2.87)). Direct collocation methods were viewed with even greater suspicion because there was no connection between the computed solution and Pontryagin's Principle. The wide gap between theory and practice opened the door for an increasingly large collection of ad hoc methods that further fueled the folklore that optimal control problems were just hard. A turning point occurred around the year 2000 through the introduction of ***pseudospectral (PS) optimal control theory***[87].

The theory of PS optimal control is based on functional analysis. It is a "third" concept that is separate from either Pontryagin's or Bellman's. The approach is theoretically founded on the classical ***Stone–Weierstrass theorem*** and practically implemented by a "sufficiently high" order polynomial representation of a function in much the same way as other elementary functions (like sines, cosines, exponentials, etc.) are implemented through polynomial representations. By cutting the "middle man" out (i.e., polynomials masquerading as sines, cosines, etc.) PS optimal control theory directly seeks ***designer functions*** that solve Problem B. As a result, it fundamentally intertwines theory with computation. This is why it is sometimes confused as a pure computational method, albeit it did start out as such.

PS optimal control is founded on two fundamental notions:

- The idea of using a solution-centric framework that uses the differential and other constraints to shape the solution instead of "solving" the equations; and,
- The exploitation of the connection between the covectors of Problems B and B^N — a concept known as the ***Covector Mapping Principle (CMP)***.

In a spectral method, the state trajectory $\boldsymbol{x}(\cdot)$ is expressed as an infinite series expansion

$$\boxed{\boldsymbol{x}(t) = \sum_{m=0}^{\infty} \boldsymbol{a}_m P_m(t)} \tag{2.88}$$

where $P_m(t)$ is a polynomial in t of degree m. The justification for expressing $\boldsymbol{x}(\cdot)$ in terms of an infinite degree polynomial stems from the fact that the state trajectory is absolutely continuous over the finite horizon; hence the Stone–Weierstrass theorem guarantees the existence of Eq. (2.88). Note also that elementary functions (sines, cosines, logs, exponentials, etc.) are also often expressed or computed through polynomial expansions. Hence, we may view elementary functions as shorthand notations for a special set of polynomials. From this perspective, Eq. (2.88) seeks to express the solution to an optimal control problem through a ***"designer set" of polynomials*** by cutting out the "middle man" (of elementary functions).

Note that Eq. (2.88) has the "look and feel" of a Fourier series expansion. This is not an accident! Equation (2.88) is indeed a ***generalized Fourier expansion*** with "amplitude" $\boldsymbol{a}_m$ and "frequency" $P_m(t)$. Typical choices for the frequency basis functions ($P_m(t)$) are the "big two" orthogonal polynomials: ***Legendre*** and ***Chebyshev***.

A key principle in a PS approach is that the coefficients $\boldsymbol{a}_m$, called the spectral coefficients, are computed indirectly by transforming Eq. (2.88) to an equivalent form

$$\boldsymbol{x}(t) = \sum_{j=0}^{\infty} \boldsymbol{b}_j \phi_j(t) \tag{2.89}$$

where t_j, $j = 0, 1, 2, \ldots$ are discrete points in time called ***nodes*** and $\phi_j(t)$ is a Lagrange interpolating polynomial that satisfies the Kronecker relationship:

$$\phi_j(t_k) = \delta_{jk} \tag{2.90}$$

Satisfaction of the Kronecker relationship implies that

$$\boldsymbol{x}(t_k) = \sum_{j=0}^{\infty} \boldsymbol{b}_j \phi_j(t_k) = \boldsymbol{b}_k \tag{2.91}$$

That is, the coefficient $\boldsymbol{b}_j$ in Eq. (2.89) is the value of $\boldsymbol{x}(t)$ at $t = t_j$. It is this ***sampling property***, which is absent in Eq. (2.88), that makes a PS approach distinct from the direct use of Eq. (2.88). As a result, we can now rewrite Eq. (2.89) using a more evocative notation:

$$\boxed{\boldsymbol{x}(t) = \sum_{j=0}^{\infty} \boldsymbol{x}_j \phi_j(t)} \tag{2.92}$$

Equation (2.92) is called a ***nodal representation*** of $\boldsymbol{x}(\cdot)$ to distinguish it from its equivalent ***modal representation*** given by Eq. (2.88).

Using the same arguments, we write

$$\boldsymbol{\lambda}(t) = \sum_{j=0}^{\infty} \boldsymbol{\lambda}_j \phi_j(t) \tag{2.93}$$

Lastly, although it may seem logical to write the control function in the same way, that is, $\boldsymbol{u}(t) = \sum_{j=0}^{\infty} \boldsymbol{u}_j \phi_j(t)$, it turns out that this is not necessary and is potentially very limiting in applicability. The optimal control is expressed as

$$\boxed{\boldsymbol{u}(t) = \sum_{j=0}^{\infty} \boldsymbol{u}_j \psi_j(t)} \tag{2.94}$$

where $\psi_j(t)$ is a special interpolating function (not necessarily Lagrange) that makes the pair $t \mapsto (\boldsymbol{x}, \boldsymbol{u})$ dynamically feasible. This aspect of ψ_j has inspired an alternative PS method known as the ***Bellman pseudospectral method***[84, 86]. The Bellman PS method facilitates real-time optimal control (RTOC) through the use of pseudospectral theory.

Using these elementary ideas, in addition to some newfangled algebra as Kalman predicted, it can be shown that PS optimal controls can be generated to very high accuracy (e.g., 10^{-6}) with fairly low N (e.g., $N \leq 100$). For instance, the accuracy achievable with a Runge-Kutta method for $N_{RK} \sim 1000$ can typically be obtained with an order of magnitude fewer points ($N_{PS} \sim 100$) using a spectral method. (See Section 3.2.4 on page 187 for a quantitative discussion with regard to the Brachistochrone problem.) Thus, for instance, a six-dimensional problem (i.e., $N_x = 6$) can be solved to high accuracy with less than 3 MB of computer memory — a feat that could have been achieved in the 1960s (see page 160).

One of the most powerful features of PS optimal control theory is that coefficients of the finite expansion adjoint covector function

$$\boldsymbol{\lambda}^N(t) = \sum_{j=0}^{N} \boldsymbol{\lambda}_j^N \phi_j(t) \tag{2.95}$$

and other covectors that arise in optimal control transform linearly to the covectors of the finite Problem B^N. This implies that Pontryagin's Principle and computation are fundamentally connected in much the same way as Bellman's Principle is connected to the adjoint covectors through the gradient of the value function.This connectivity is called the ***Covector Mapping Principle (CMP)***; see Fig. 2.28. The CMP removes the suspicions (of the 1990s) of computed solutions by intimately connecting the results to Pontryagin's Principle.

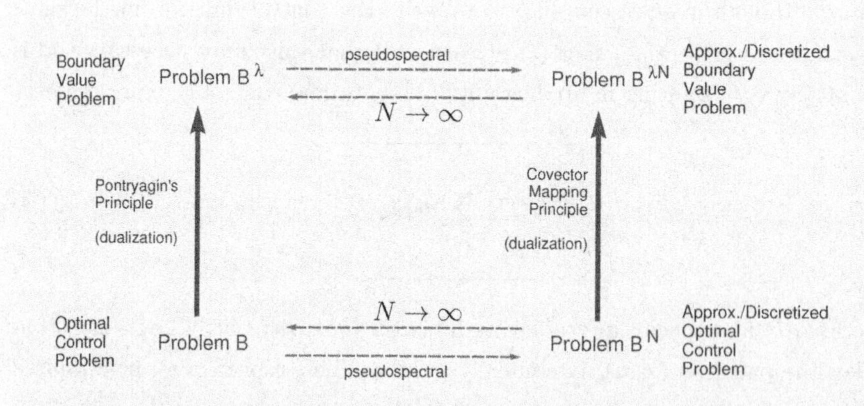

Figure 2.28: The Covector Mapping Principle fills the theoretical and computational gaps apparent in Figs. 2.27, 2.24 and 2.12. See also Fig. 2.16 for additional context. Note the absence of "direct" and "indirect" qualifiers. Further details and finer points associated with this figure are reviewed in [87].

In effect, the CMP completes the quest for filling the gaps that are now apparent in Figs. 2.24 and 2.27.

Note also that the CMP renders the terms "direct" and "indirect" obsolete. The plot in the story of the CMP also reveals that the obvious and "shorter path" of dualizing first and computing afterward is strewn with difficulties while a "longer path" of reversing the operations eliminates the curse of sensitivity. Thus, the CMP facilitates replacing the historical *plug-and-pray* approach with a *plug-and-play* technique by simply ***commuting the key operations*** indicated in Fig. 2.28.

The story of optimal control theory is replete with similar discoveries of commuting operations to produce better ways to both formulate and solve problems[84]. This notion of commuting operations to produce new and useful results can also be contextualized as a central theme in distinguishing Pontryagin's calculus of optimal control from the classical calculus of variations. In the former, we differentiate first and minimize afterward while in the latter we minimize first and differentiate afterward. Recall that in Pontryagin's calculus we need to determine $\partial_x H$, which is very easy to compute. By re-mapping this idea to the classical calculus of variations, we need to find the lower Hamiltonian first

$$\mathcal{H}(\boldsymbol{\lambda}, \boldsymbol{x}, t) := \min_{\boldsymbol{u} \in \mathbb{U}} H(\boldsymbol{\lambda}, \boldsymbol{x}, \boldsymbol{u}, t)$$

and differentiate afterward

$$-\dot{\boldsymbol{\lambda}} = \frac{\partial \mathcal{H}}{\partial \boldsymbol{x}}$$

As evident from the problems and examples discussed in the preceding pages, this procedure ($\partial_x \mathcal{H} = \partial_x (\min_{\boldsymbol{u}} H)$) is considerably painful or impossible to perform as $\mathcal{H}$ is, quite often, not differentiable even when H is very smooth.

Study Problem 2.23

Pick any recent journal or conference paper on optimal control. Critique it using the concepts developed in this chapter, and particularly, this section.

2.10 Pontryagin and the Calculus of Variations¶

Today, Problem B may seem quite a natural way to formulate many problems in disparate fields. Prior to 1955, the language of optimal control did not exist: It was invented by Pontryagin and his students[33, 66]. Obviously, without a language, optimal control problems went unrecognized in many fields. Such problems — the domain of an extremely small group of specialists[66] — had to be formulated using the old (i.e., non-control) language of the "calculus of variations." The calculus of variations — a term coined by Leonhard Euler (1707–1783) — was invented by Joseph-Louis Lagrange (1736–1813) when he was only 19 years old! Prior to Lagrange's invention, Euler (Bernoulli's student) had formulated the first general problem of a "new calculus," which can be formally stated (in modern notation) as

$$\left\{ \begin{array}{lrl} \text{Minimize} & J[\boldsymbol{x}(\cdot), t_f] = & \displaystyle\int_{t_0}^{t_f} L(\boldsymbol{x}(t), \dot{\boldsymbol{x}}(t), t)\, dt \\ \text{Subject to} & \boldsymbol{x}(t_0) = & \boldsymbol{x}^0 \\ & \boldsymbol{e}(\boldsymbol{x}_f, t_f) = & \boldsymbol{0} \end{array} \right. \tag{2.96}$$

¶This section may be skipped without loss of continuity, but why skip a good story?

Note that there is no concept of a control function in Problem 2.96, although $\dot{\boldsymbol{x}}(t)$ is treated somewhat independently of $\boldsymbol{x}(t)$. Ignoring the possible confusions arising from the meaning of such problems, an optimal control problem, prior to Pontryagin, had to be "fitted" inside of Problem 2.96. (Try this as an exercise!)

Euler derived the first necessary condition for Problem 2.96 as

$$\frac{d}{dt}\frac{\partial L(\boldsymbol{x}(t), \dot{\boldsymbol{x}}(t), t)}{\partial \dot{\boldsymbol{x}}} - \frac{\partial L(\boldsymbol{x}(t), \dot{\boldsymbol{x}}(t), t)}{\partial \boldsymbol{x}} = \boldsymbol{0} \tag{2.97}$$

He derived this condition through a process of discretization and limit taking[69, 95] — see bottom part of Fig. 2.28 on page 164 — obviously well before the advent of computers (which also reminds us that discretization and limit taking are independent of computation and computers). Lagrange derived the same condition using his invention of "variations." Euler was so impressed by this idea that he abandoned his own and branded the new calculus as the calculus of variations[95]; this is why Eq. (2.97) is called the ***Euler-Lagrange equation***. In the decades and centuries that followed, Problem 2.96 drew the attention of some of the greatest mathematicians. Legendre, Weierstrass, Jacobi and many others derived additional necessary conditions that bear their names. In the early 20th century (1930s), the hub of this activity was at the University of Chicago[66, 74, 95].

In the 1950s, aerospace engineers were facing a number of problems in optimal control (when the term did not even exist) and the closest available tool at that time was the calculus of variations[33]. Even if they were successful in force-fitting their problems inside the language of the calculus of variations, they had to subsequently wade through a fog of arcane necessary conditions and a large number of theorems to pull out something useful. As McShane notes, *mastery of this subject gave answers to questions no one was asking*[66]. In sharp contrast, Pontryagin — a well-known topologist at that time — completely abandoned his field to answer engineering questions that arose in the 1950s[33].

In the Spring of 1955 two (Russian) Air Force Colonels visited Pontryagin at the Steklov Mathematical Institute and posed a problem on time-optimal aircraft maneuvers[33]. Pontryagin recognized that this problem, and many other emerging ones at that time, required a new calculus. This new calculus of optimal control, developed by Boltyanski, Gamkrelidze and Pontryagin[27, 74, 75], not only generalized the old one but unified, simplified and extended it in a way that engineers could quickly apply and use the results to develop insights into the problems they were facing in the 1950s. Among many, one of the fundamental shifts was the introduction of the *state-space model*, $\dot{\boldsymbol{x}} = \boldsymbol{f}(\boldsymbol{x}, \boldsymbol{u}, t)$, as a constraint in a *dynamic optimization problem*. This "simple" perspective changed everything! For instance, Problem (2.96) can be formulated in this "new language" as

$$\left\{ \begin{array}{lrl} \text{Minimize} & J[\boldsymbol{x}(\cdot), \boldsymbol{u}(\cdot), t_f] = & \displaystyle\int_{t_0}^{t_f} L(\boldsymbol{x}(t), \boldsymbol{u}(t), t)\, dt \\ \text{Subject to} & \boldsymbol{x}(t_0) = & \boldsymbol{x}^0 \\ & \dot{\boldsymbol{x}}(t) = & \boldsymbol{u}(t) \\ & \boldsymbol{e}(\boldsymbol{x}_f, t_f) = & \boldsymbol{0} \end{array} \right. \tag{2.98}$$

What is more remarkable about this new perspective is that through the introduction of Pontryagin's Hamiltonian, the awkward Euler-Lagrange equation given by Eq. (2.97) transforms to the very elegant ***dual form*** given by

$$\frac{\partial H}{\partial \boldsymbol{u}} = \boldsymbol{0} \tag{2.99}$$

The price for this elegance is, of course, the introduction of covectors as a fundamental analysis tool. Equation (2.99) is *equivalent* to, but is not the same as, the Euler-Lagrange equation. Equation (2.99) follows from Problem HMC, but under three severe conditions:

1. The control space $\mathbb{U}$ must be continuous (see Fig. 2.2 on page 87),

2. The H-function must be differentiable with respect to $\boldsymbol{u}$, and

3. The optimal control must be interior to $\mathbb{U}$.

In Chapter V of [75], Pontryagin et al derived the Euler-Lagrange equation and all the other necessary conditions of the calculus of variations using their new *simpler* tools while simultaneously showing the limitations of the former. Despite this, many in the mathematical community were reluctant to accept their results as something completely new[1]. On the other hand, engineers wholeheartedly accepted and endorsed the new ideas[1, 66] because they were practical and systematic: There was no longer a need to understand or apply the chaos of limitless theorems of the calculus of variations. Not to be beaten, the traditionalists reorganized the calculus of variations using Pontryagin's new ideas and "showed" that the Hamiltonian minimization condition was "merely" a generalization of Weierstrass' condition. Using similar rewrites, it can be shown that

$$\frac{\partial^2 H}{\partial \boldsymbol{u}^2} \geq 0 \tag{2.100}$$

is a restatement of the ***Legendre-Clebsch condition***. Hence, the argument went, it was "obvious" from equations (2.99) and (2.100) that the Hamiltonian must be minimized. This gave credibility to those educators who chose to inflict the classical calculus of variations upon unsuspecting students. A genuine new calculus blending the old with the new had to wait until Clarke's breakthrough of nonsmooth calculus[20, 23].

In 1989, McShane warned[66] that mathematicians had not learned from the history of the 1950s. New "baroque theorems," he criticized, were being generated of "increasing intricacy [that is] of interest to a steadily shrinking collection of experts in the subject." Even worse, theory and computation got completely divorced in the 1990s, leading to such notions as "direct" methods (apparently not based on theory) and "indirect" methods (apparently based on Pontryagin's Principle). See Section 2.9.2 on page 157 and [77] on the origins of this "divorce," and Section 3.7 on page 240 for a counter example. A turning point occurred around the year 2000 when the Covector Mapping Principle was introduced[87]. In one fell swoop, Pontryagin's theory, computation and problem-solving became fundamentally intertwined; see Fig. 1.50 on page 84. In 2006, *SIAM News* heralded the dawn of a new era: Pseudospectral optimal control theory had debuted flight.

2.11 Endnotes

Pontryagin's Principle is named in honor of Lev S. Pontryagin (1908–1988), who, along with his associates at the Steklov Institute of Mathematics, Moscow, was the first (1956) to articulate all major elements of this fundamental concept[1]. Barring a footnote on page 45 in [75] where they refer to *covariant* vectors, Pontryagin et al never used the word *covector* in their entire book; however, they were absolutely clear everywhere — through their tensor notation — that their "auxiliary variables" (i.e., covectors in this book) were indeed covariant. The word covector also means a *covariant vector*.

After Pontryagin, many applied textbooks introduced these covectors as Lagrange multipliers — a name change that does little to unlock their mystery.[||] In the mathematics literature, Pontryagin's auxiliary variables were indeed identified as covectors but in their abstract form as linear functionals. The mystery and suspicion of Lagrange multipliers, their apparent lack of physical meaning funneled through the curse of sensitivity fueled the folklore that optimal control problems were hard. They indeed were! But everything changed around the year 2000 when the escape clause on the curse of sensitivity was thoroughly exploited. See [84] for historical and technical details.

The concept of covectors, as presented in Sections 2.2 and 2.3, grew as an outgrowth of the scaling and balancing procedure used in DIDO when the software was first created in 2001. Subsequent experience with DIDO helped codify a re-interpretation of covectors as presented in this chapter. Interestingly, the physical interpretation of covectors presented in this chapter is more closely aligned with Pontryagin's original geometric view of the calculus of optimal control. The presentation in Section 2.2, particularly the bottom of Fig. 2.5, was inspired by the combination of Pontryagin's geometric insight on duality and the classic interpretation of covectors in physics as presented by Misner, Thorne and Wheeler[68].

Our treatment of covectors is in sharp contrast to their tricky introduction as Lagrange multipliers presented in many textbooks on applied optimal control where there is a frequent lamentation that the multipliers have no physical meaning. Having dispensed with this myth, the measuring and scaling procedure offered by the covectors can be used very effectively to solve many optimal control problems in DIDO. This implies that a practical understanding of Pontryagin's Principle is more important than ever before. Simply running a code does little in understanding Pontryagin's Principle or the problem being solved. That is, to solve a "hard" optimal control problem "easily," it is critical to apply Pontryagin's Principle in a manner that exploits the problem-specific covectors using the "new rules" and insights presented in this chapter. Simple examples illustrating part of this methodical procedure are presented in Chapter 3 with additional problems and insights in Chapter 4. For application problems, such as some of those presented in Chapter 4, Pontryagin's Principle must be applied

[||] Lagrange multipliers were actually invented by Euler as part of his quest to solve Queen Dido's problem[23, 96]: See Section 4.6 on page 282 for a discussion on this problem and [96] for a well-researched historical account.

as indicated in Fig. 2.16, page 124. This is why a pragmatic understanding of Pontryagin's Principle is essential to producing good DIDO application codes. In other words, *theory is not divorced from computation!* More importantly, we now have an intimate connection between theory, computation and problem formulation itself!! Solutions from an initial problem formulation feed new problem formulations. This is part of the *problem of problems* concept (described in Chapter 1) where the DIDO solutions from a given problem are analyzed and used to reformulate the problem repeatedly until the correct problem is formulated; see the work flow outlined in Fig. 1.50 on page 84. Welcome to a journey of discovery!

Chapter 3

Example Problems

The solution is the problem.
To question is to answer.

Pontryagin's Principle provides not only open-loop solutions but also ***feedback solutions*** to many of the example problems discussed in this chapter. *It is a widely held misconception that feedback solutions are not possible by way of Pontryagin's Principle.*

In addition to being illustrative of Pontryagin's Principle, the following examples also reveal some key ideas on formulating and solving problems in ***optimal feedback control***.

3.1 Solving the Brachistochrone Problem

From Study Problem 2.6 (page 113), it appears that obtaining a closed-form solution to the Brachistochrone problem from Pontryagin's Principle may be quite daunting. It is. But the "trick" in arriving at an analytical solution is to fully exploit the *integrals of motion* through possibly ad hoc techniques. One such approach for the Brac:1 problem formulation is as follows.

Combining the Hamiltonian value condition, Eq. (2.31), with the Hamiltonian evolution equation, (2.54), we get

$$\lambda_x(t)v(t)\sin\theta(t) + \lambda_y(t)v(t)\cos\theta(t) + \lambda_v(t)g\cos\theta(t) \quad = \quad -1 \tag{3.1}$$

For simplicity, we drop the $*$ superscripts while keeping in mind that all equations correspond to a candidate optimal solution.

Recall from page 104 that the Hamiltonian minimization condition generates the equation (see Eq. (2.18))

$$\lambda_x(t)v(t)\cos\theta(t) - \lambda_y(t)v(t)\sin\theta(t) - \lambda_v(t)g\sin\theta(t) \quad = \quad 0 \tag{3.2}$$

Equations (3.1) and (3.2) form two integrals of motion. By eliminating $\lambda_v(t)$ from these equations we get

$$\lambda_x(t)v(t) + \sin\theta(t) = 0 \tag{3.3}$$

From the adjoint equation, (2.16), and the transversality condition, Eq. (2.28), we have a third integral of motion:

$$\lambda_x(t) = \nu_1 \tag{3.4}$$

Substituting Eq. (3.4) in (3.3) we get

$$\nu_1 v(t) + \sin\theta(t) = 0 \tag{3.5}$$

Study Problem 3.1

Equation (3.5) can be written as

$$\theta = \sin^{-1}(-\nu_1 v)$$

Is this a feedback control solution? (Hint: Examine if ν_1 is a constant with respect to the initial states.)

Because equation (3.5) is true for every t, differentiating both sides of it generates

$$\nu_1 \dot{v}(t) + \dot{\theta}(t) \cos\theta(t) = 0$$

Substituting the equation of motion $\dot{v}(t) = g\cos\theta(t)$ in the above equation we get

$$\dot{\theta}(t) = -\nu_1 g \tag{3.6}$$

That is, $\dot{\theta}(t)$ is a constant in time; hence, we can write

$$\theta(t) = -\nu_1 g t + \cancelto{0}{\text{constant}} \tag{3.7}$$

The constant is zero as a result of Eq. (3.3), and the initial condition $v(0) = 0$. Note that Eq. (3.7) implies that the unit of ν_1 and hence $\lambda_x(t)$ is time per length or the inverse of the units of velocity.

Substituting Eq. (3.7) in Eq. (3.3) generates a closed-form result for velocity:

$$v(t) = \frac{\sin(\nu_1 g t)}{\nu_1} \tag{3.8}$$

Substituting equations (3.8) and (3.7) in the kinematics equations we get

$$\begin{aligned} \dot{x}(t) &= -\frac{\sin^2(\nu_1 g t)}{\nu_1} &&= \frac{\cos(2\nu_1 g t) - 1}{2\nu_1} \\ \dot{y}(t) &= \frac{\sin(\nu_1 g t)}{\nu_1}\cos(-\nu_1 g t) &&= \frac{\sin(2\nu_1 g t)}{2\nu_1} \end{aligned} \tag{3.9}$$

where the second equalities in Eq. (3.9) follow from half-angle formulas. Integrating these equations and using the initial conditions $x(0) = 0 = y(0)$, we get

$$x(t) = \frac{g}{\omega^2}(\omega t - \sin\omega t) \tag{3.10a}$$

$$y(t) = \frac{g}{\omega^2}(1 - \cos\omega t) \tag{3.10b}$$

where ω is defined as

$$\omega := -2\nu_1 g \tag{3.11}$$

Study Problem 3.2

1. *Fill in the details of the missing steps (calculus and trigonometric) that were taken in arriving at Eq. (3.10).*
2. *What happens if ω in Eq. (3.11) was defined with a plus sign?*
3. *Show that if a disc of radius r rolls without slipping, then the trace of the point on the edge of the disc has Cartesian coordinates given by*

$$x = r(\phi - \sin\phi) \tag{3.12a}$$

$$y = r(1 - \cos\phi) \tag{3.12b}$$

where ϕ is the angle of rotation. The curve obtained by graphing $x \mapsto y$ is called a cycloid. See Fig. 3.1.

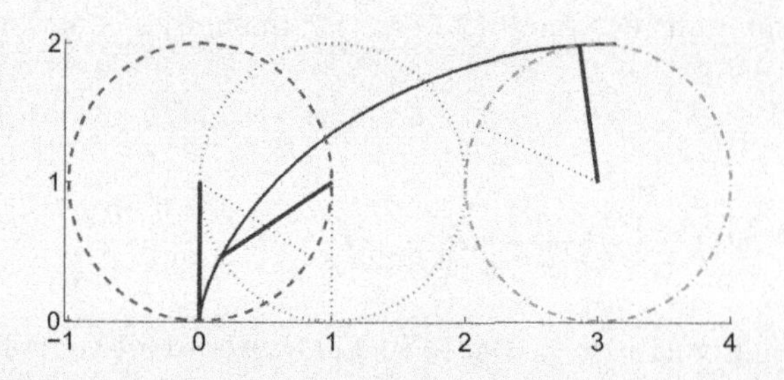

Figure 3.1: "Snapshots" of the rolling disc that generates a cycloid.

Note that while Eq. (3.10) has the look and feel of a solution, it is not! It simply states that the solution to the Brachistochrone problem is the same as the equation for a cycloid given by Eq. (3.12).

In Eq. (3.10) we do not know the value of ω. Moreover, the value of t_f is also unknown. These two unknowns are obtained by setting the final values of the state trajectory to be equal to the target conditions:

$$\frac{g}{\omega^2}(\omega t_f - \sin \omega t_f) = x^f \tag{3.13a}$$

$$\frac{g}{\omega^2}(1 - \cos \omega t_f) = y^f \tag{3.13b}$$

In other words, to generate a complete solution to the Brachistochrone problem, we need to solve for the zeroes (roots) of the pair of nonlinear transcendental equations in the two unknowns, ω and t_f:

$$\mathbf{c}(\omega, t_f) := \begin{bmatrix} c_1(\omega, t_f) \\ \\ c_2(\omega, t_f) \end{bmatrix} := \begin{bmatrix} \frac{g}{\omega^2}(\omega t_f - \sin \omega t_f) - x^f \\ \\ \frac{g}{\omega^2}(1 - \cos \omega t_f) - y^f \end{bmatrix} = \mathbf{0} \tag{3.14}$$

Finding the zeroes of a scalar function of a scalar variable is easy (e.g., a graphical technique can be used). Finding the zeroes of a vector function of vector variables is not so-easy; in fact, this is a central problem in *nonlinear programming*. Thus, we have "reduced" the Brachistochrone problem to a nonlinear programming problem!

Study Problem 3.3

1. *Critique the claim that the Brachistochrone problem has an analytical solution.*
2. *Solve Eq. (3.14) for the following data*

$$g = 9.8, \quad x^f = 10.0, \quad y^f = 10.0 \tag{3.15}$$

   using the fixed-point iteration described on page 249.
3. *Solve Eq. (3.14) using* any method *for the data given by Eq. (3.15). Evaluate the merits of the chosen method against sensitivity to initial guess (if any).*
4. *For the data given by Eq. (3.15) assess the accuracy of the following solution*

$$\omega = 1.3079, \quad t_f = 1.8442 \tag{3.16}$$

   by substituting the values in Eq. (3.14).
5. *Prove that the solution to Eq. (3.14) generates a globally optimal solution to the Brachistochrone problem.*

A simple and interesting solution is obtained if we set $y(t_f)$ to be free instead of fixing it at the value y^f. Then, from the transversality condition, we get $\lambda_y(t_f) = 0$. Substituting this in Eq. (3.2) implies that $\theta(t_f) = \pm\pi/2$; hence, we have

$$\omega t_f = \pm\pi$$

Substituting this in Eq. (3.13a) we get

$$\omega = \sqrt{\pm\pi g/x^f} \quad \Rightarrow \quad t_f = \sqrt{\pm\pi x^f/g} \tag{3.17}$$

Obviously the $\pm$ sign in Eq. (3.17) must be chosen depending on the sign of x^f. The value of y_f for this case is obtained by substituting Eq. (3.17) in Eq. (3.13b); this yields

$$y^f = \frac{2\,x^f}{\pm\pi}$$

where the $\pm$ sign is chosen to ensure that $y^f > 0$.

Study Problem 3.4

1. *Determine the exact value of t_f in Eq. (3.17) for the case when $x^f \equiv 10.0$ exactly.*
2. *Find the exact values of ω and t_f for the data given by Eq. (3.15). (Hint: See page 122 in Chapter 2.)*

Study Problem 3.5

1. *In the derivation of the solution to the Brachistochrone problem, several assumptions were made implicitly. List all these assumptions and discuss their ramifications.*
2. *Instead of taking all initial conditions to be zero, re-derive the results for the Brachistochrone problem for arbitrary initial conditions (t_0, x_0, y_0, v_0). Using these results discuss how you would obtain a feedback solution.*

3.2 The Brachistochrone Problem via DIDO

The Brachistochrone problem is used in this book as a running proposition to illustrate the various themes involved in solving a present-day optimal control problem. As illustrated in Fig. 3.2, solving an optimal control problem via

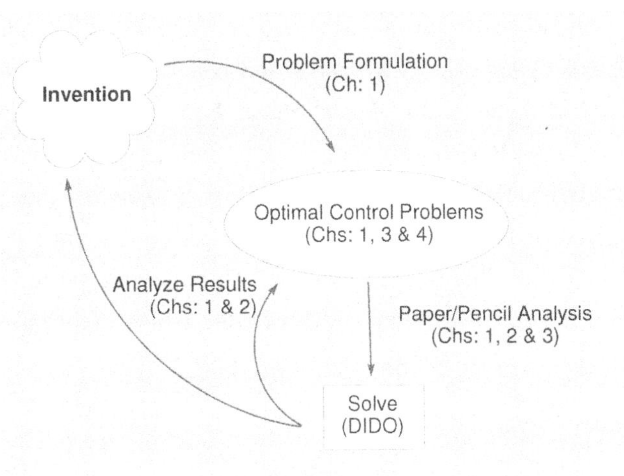

Figure 3.2: A typical work-flow in solving a present-day optimal control problem. Same as Fig. 1.50 repeated here for quick reference.

DIDO is the *last step* of the *first iteration.* That is, DIDO is the last of the four major steps involved in the first iteration of solving an optimal control problem:

- **1. Invention Step:**

  Define a "word problem." For the Brachistochrone problem, this was done by Johann Bernoulli in 1696 in the June issue of the journal *Acta Eruditorum.* A 21st-century version of a similar gauntlet was thrown by Bedrossian in his *attitude guidance* challenge for the $100B International Space Station. This step of defining a word problem is largely discipline-specific. Knowledge of the discipline is a necessary condition.

- **2. Transcription Step:**

  Transcribe the word problem into several (initial) mathematical problem formulations. A significant part of Chapter 1 is devoted to this aspect of optimal control using the thread of Bernoulli's challenge.

- **3. Analysis Via Pontryagin's Principle:**

  Apply Pontryagin's Principle. An application of Pontryagin's Principle to the problems developed in Step 2 provides many useful tips on the correctness and completeness of the mathematical formulation. For the Brac:1 problem, this aspect was illustrated in detail in Chapter 2 as well as in Section 3.1. The Study Problems covered the rest of the Brachistochrone problem formulations. Additional examples are discussed in the remainder of this Chapter. This step — the main theme of this book — also provides a large number of necessary conditions that are checkable in the analysis portion of Step 4.

- **4. Analysis Via DIDO:**

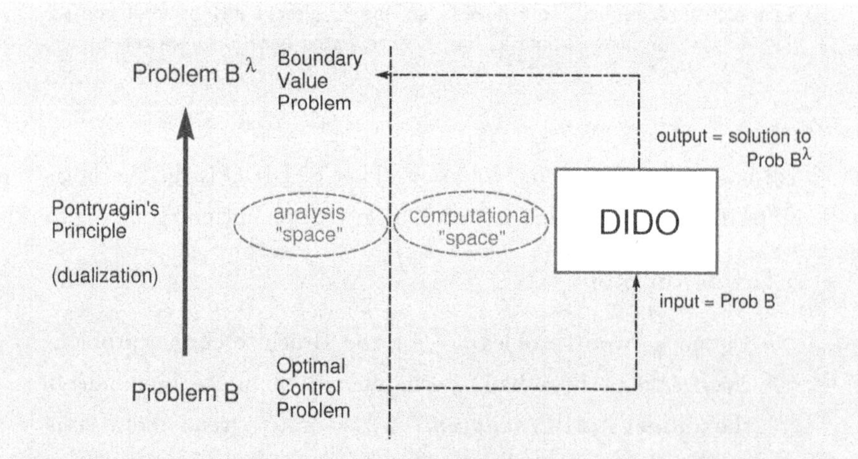

Figure 3.3: DIDO takes Problem B as an input and outputs a candidate solution to Problem B^{λ}. Same as Fig. 2.16, repeated here for quick reference. A free version of DIDO can be downloaded from http://www.ElissarGlobal.com.

Solve the problem using DIDO. As indicated in Fig. 3.3, DIDO takes the entire problem as an input and generates a *candidate* extremal solution as an output:

$$[\ \text{extremal}\ ] = \text{dido}\,(\text{problem})$$

The task of the analyst is to analyze the candidate solution (generated by DIDO) using the results from Step 3 and decide if there needs to be a subsequent iteration (there almost always is!) and where this iteration should commence. See Fig. 3.2 on page 177 and note the circular work flow.

A Pseudocode for Brac:1

BEGIN problem definition

$$\begin{aligned} \textit{problem.}\underline{\textit{cost}} &= \text{'Brac1Cost'} \\ \textit{problem.}\underline{\textit{dynamics}} &= \text{'Brac1Dynamics'} \\ \textit{problem.}\underline{\textit{events}} &= \text{'Brac1Events'} \end{aligned}$$

END problem definition

CALL DIDO

$$[\textit{cost, primal, dual}] = \textit{dido}\,(\textit{problem})$$

% This pseudocode is fairly close to the actual format of a DIDO code.

$$\overbrace{(Brac:1)}^{\text{problem}} \left\{ \begin{array}{lrl} \text{Minimize} & J[\boldsymbol{x}(\cdot), \boldsymbol{u}(\cdot), t_f] := t_f & \} \ \text{(cost)} \\ \text{Subject to} & \dot{x} = v \sin\theta & \\ & \dot{y} = v \cos\theta & \text{(dynamics)} \\ & \dot{v} = g \cos\theta & \\ & t_0 = 0 & \\ & (x_0, y_0, v_0) = (0, 0, 0) & \text{(events)} \\ & (x_f - x^f, y_f - y^f) = (0, 0) & \end{array} \right.$$

From the pseudocode of the Brac:1 problem formulation it is apparent that it closely matches the mathematical problem formulation presented on page 12 and repeated on page 179 for quick reference. Transforming the pseudocode to a DIDO code, Brac:1 can be solved fairly quickly. A plot of a candidate ***primal*** solution is shown in Fig. 3.4.

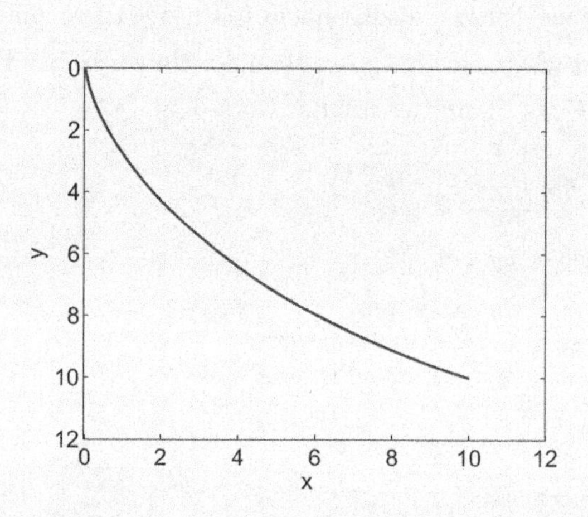

Figure 3.4: A candidate DIDO solution to the Brachistochrone problem, Brac : 1, for the data given by Eq. (3.15).

Once such a candidate solution is obtained, the next set of tasks is the most important ... starting with how do we know if the solution is correct? Answering this question is the essence of a verification and validation (V&V) procedure of *any* computed solution (generated from *any* software).

3.2.1 Basic Verification and Validation (V&V)

Pontryagin's Principle is a necessary condition; hence, it is a "negative" test on optimality. That is, satisfaction of Pontryagin's Principle does not ensure optimality but it can certainly eliminate spurious solutions. Using the *HAMVET* mnemonic (see page 129) a basic V&V can be constructed as follows:

1. The Adjoint Equations

A plot of the costates generated by DIDO is shown in Fig. 3.5. Clearly λ_x and λ_y are constants in time: precisely as predicted by Pontryagin's Principle (see

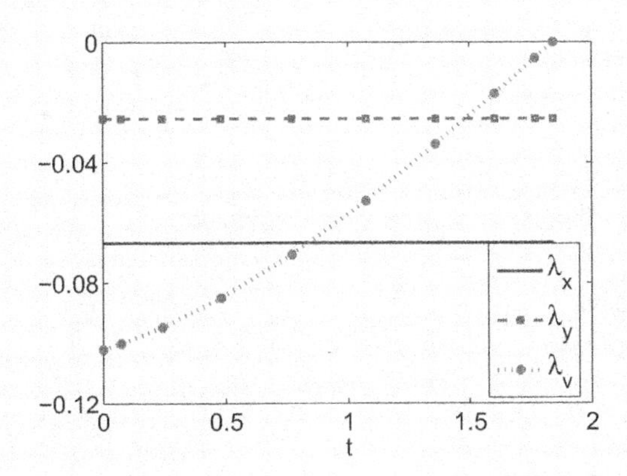

Figure 3.5: Costate trajectories generated by DIDO for Problem Brac : 1, and the data given by Eq. (3.15).

Eq. (2.16) on page 101). This provides good initial evidence that DIDO has likely produced an extremal.

At the present time, there is not much we can do in using λ_v for V&V. In many problems, not all costates are checkable; however, the costates provide valuable information in scaling and balancing the equations for efficient and reliable computation[81].

2. The Hamiltonian Minimization Condition

The Hamiltonian minimization condition provides joint conditions on the controls, states and costates. This is why in many problems it can be used as a powerful V&V technique. In the case of Brac:1, we have the condition

$$\partial_{\boldsymbol{u}} H(\boldsymbol{\lambda}, \boldsymbol{x}, \boldsymbol{u}) := \lambda_x v \cos\theta - \lambda_y v \sin\theta - \lambda_v g \sin\theta = 0 \qquad (\forall\, t) \tag{3.18}$$

This condition was derived on page 104; see Eq. (2.18). A plot of $\partial_\theta H[@t]$ evaluated according to Eq. (3.18) is shown in Fig. 3.6. It is apparent from this figure that one can claim a higher confidence with regard to the extremality of the solution generated by DIDO.

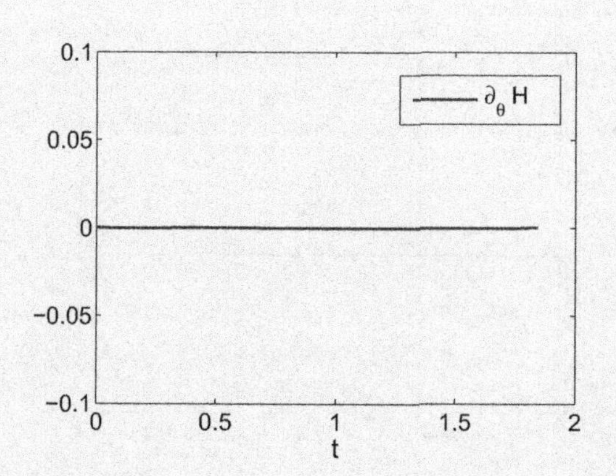

Figure 3.6: An evaluation of the satisfaction of the Hamiltonian minimization condition for Problem Brac : 1 using the results generated by DIDO.

3. The Hamiltonian Value Condition

The Hamiltonian value condition provides the "optimal" stopping time equation. For minimum-time problems, this is given by

$$\mathcal{H}[@t_f] = -1 \tag{3.19}$$

The value of the Hamiltonian at the final time produced by DIDO is

$$\mathcal{H}_{DIDO}[@t_f] = -1.00002$$

In addition to supporting an increasingly higher confidence in the DIDO results, this small deviation from -1 also suggests that the value of the final time produced by DIDO ($t_f = 1.8442$) is likely the minimum time.

4. The Hamiltonian Evolution Equation

The Hamiltonian evolution equation is an integral of motion. Because the (Pontryagin) Hamiltonian for Brac:1 does not depend explicitly on time, we have

$$\frac{\partial H}{\partial t} = 0$$

This implies (see Eq. (2.53) on page 126)

$$\frac{d\mathcal{H}}{dt} = 0 \qquad \Rightarrow \qquad \mathcal{H}[@t] = \text{constant (with respect to time)} \tag{3.20}$$

A plot of the time history of the lower Hamiltonian generated by DIDO is shown in Fig. 3.7. In addition to its reasonable numerical constancy, it is apparent from

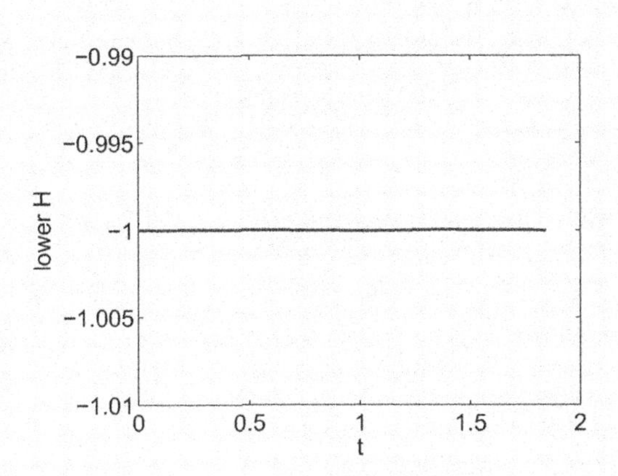

Figure 3.7: Time history of the minimized Hamiltonian generated by DIDO.

the figure that $\mathcal{H}_{DIDO}[@t] = -1.000$. That $\mathcal{H}[@t] = -1$ follows from the joint satisfaction of equations (3.19) and (3.20).

5. The Transversality Conditions

The transversality condition requires that $\lambda_v(t_f) = 0$. From Fig. 3.5, it is apparent that λ_v at the final time is indeed zero. The actual value of $\lambda_v(t_f)$ generated by DIDO is -6.4054×10^{-12}.

While the transversality condition on λ_v provided a specific numerical value, the same is not true of λ_x and λ_y. Nonetheless, it is possible to check these conditions, as well.

The values of ν_1 and ν_2 generated by DIDO are

$$\nu_1 = -0.0667 \qquad \nu_2 = -0.0255 \tag{3.21}$$

These numbers match fairly precisely the value of $\lambda_x(t_f)$ and $\lambda_y(t_f)$; in fact, they are identically equal to the values of ν_1 and ν_2, respectively, well beyond 10 significant digits. See also Fig. 3.5.

Computational Tip: The precision in the satisfaction of the various necessary conditions for Brac:1 must not be construed as universal. Although DIDO is expected to provide fairly accurate results for many problems, a large number of factors enters in the generation of "exact" solutions. One of the most dominant of these factors is the proper *scaling and balancing* of the equations; see Section 1.1.4. See also Section 3.2.4 later on page 187.

Study Problem 3.6

Develop the necessary conditions for the various formulations of the Brachistochrone problem discussed in Chapter 1. Which of these formulations provides the most useful information in verification and validation of a solution?

3.2.2 Problem-Specific V&V

The V&V discussed in the preceding subsection follow a check of the basic necessary conditions (encapsulated in the *HAMVET* mnemonic). In many problems, it is possible to perform additional checks based on a paper-and-pencil analysis. We illustrate this point for the Brac:1 problem formulation.

1. Check On θ

From Eq. (3.7), the optimal control trajectory $t \mapsto \theta$ must be a linear function of time. The DIDO control shown in Fig. 3.8 is obviously linear.

2. Check On Energy Integral

Recall that mechanical energy Brachistochrone problem is conserved. Refer back to Study Problem 1.3. This implies that we must have

$$\frac{v^2(t)}{2} - g\,y(t) = 0 \qquad (\forall\ t)$$

A plot of this energy function over time generated by DIDO is shown in Fig. 3.9. This integral of motion is not a check on the optimality of the trajectory; rather,

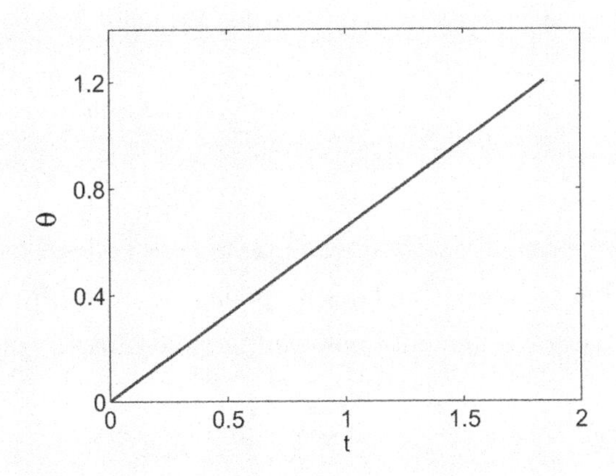

Figure 3.8: Control trajectory for Brac:1 generated by DIDO.

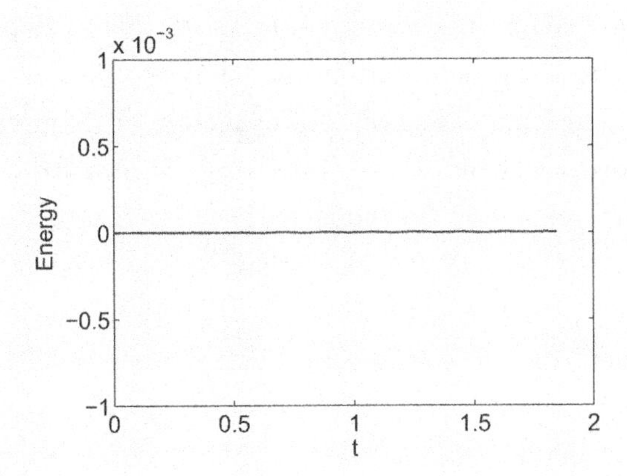

Figure 3.9: Energy integral generated by DIDO for Brac:1.

it is a part of the validation of the equations of motion. See also the **Tech Talk** comment on page 28 for an additional perspective on this concept.

3.2.3 Some Finer Points

Recall that Problem P contains more dual information than B; see Section 2.7 on page 131. Because DIDO solves Problem $P \supset$ Problem B, it is often instructive to check these additional necessary conditions. From Study Problem 2.13 (see page 135) the additional necessary conditions for Problem B are:

$$\boldsymbol{\lambda}(t_0) = -\boldsymbol{\nu}_{x_0} \qquad \textit{initial transversality condition}$$
$$H[@t_0] = \nu_{t_0} \qquad \textit{initial Hamiltonian value condition}$$

From a theoretical perspective, these conditions are "useless" as they equate one unknown number to another unknown number; however, from a computational perspective, they can be quite powerful in validating the optimality of a candidate solution.

For Brac:1, DIDO generates the following dual information:

$$\boldsymbol{\nu}_{x_0} = (0.0667,\ 0.0255,\ 0.1020) \qquad \nu_{t_0} = -1.0000$$
$$\boldsymbol{\lambda}(t_0) = (-.0667, -0.0255, -0.1020) \qquad H[@t_0] = -1.0000$$

See also Fig. 3.5, and compare the values of $\boldsymbol{\nu}_{x_0}$ with that of Eq. (3.21).

Part of a standard V&V procedure in the flight application of DIDO (and pseudospectral theory) is an *independent* propagation of the initial conditions through an interpolation of the control trajectory. Such a requirement is imposed by many flight operators (including those at NASA) because the propagated result is agnostic to DIDO or any other means of generating a control solution. In addition, such results are "trusted" because the propagation is based on legacy methods and software that may have a track record of successful flights. Furthermore, key operational constraints can be independently checked as part of a pre-flight validation.

In following this standard procedure, Figure 3.4 shown on page 180 was, in fact, generated by propagating the initial conditions through a trusted Runge-Kutta (4/5) routine by linearly interpolating the control trajectory over 10 nodes:

$$\dot{\boldsymbol{x}} = \boldsymbol{f}(\boldsymbol{x}, \boldsymbol{u}^{\#}(t), t), \quad \boldsymbol{x}(t_0) = \boldsymbol{x}_0, \quad \boldsymbol{u}^{\#}(t) = \sum_{j=0}^{N_{DIDO}} \boldsymbol{u}_j \psi_j(t) \tag{3.22}$$

This trusted *dynamically feasible* state trajectory is shown in Fig. 3.10 along with DIDO's 10-node solution.

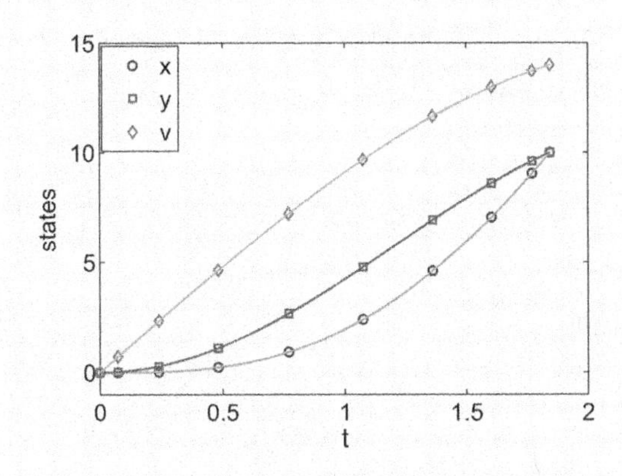

Figure 3.10: A 10-node DIDO solution for Brac:1 overlaid with a "trusted" propagation (RK4/5) of the initial condition (solid lines) performed according to Eq. (3.22).

3.2.4 Pseudospectral Theory vs Exact Solutions

For all practical purposes, the 10-node solution generated by DIDO for Brac:1 shown in Fig. 3.10 is accurate and "flyable." For all practical purposes, it is also an "optimal" solution in the sense that it satisfies all the necessary conditions of optimality. A question of theoretical importance with deep practical implications is the relationship between the computed solution and the usually unknown exact solution. The following theorem[50, 84, 87] provides this connection:

Theorem 3.1 (Kang-Ross-Gong) *Let* $\left\{t \mapsto (x^{*N}, u^{*N})\right\}_{N=N_0}^{\infty}$ *be a sequence of optimal solutions generated by Problem* B^N. *Then, there exists a subsequence* $\{t \mapsto (x^{*n}, u^{*n})\}_{n=n_0}^{\infty}$ *and an optimal solution* $t \mapsto (x^*, u^*)$ *of Problem B such that the following limits converge uniformly:*

$$\begin{aligned} \lim_{n\to\infty} x^{*n}(t) &= x^*(t) \\ \lim_{n\to\infty} u^{*n}(t) &= u^*(t) \\ \lim_{n\to\infty} J^n[x^*(\pi^n), u^*(\pi^n)] &= J[x^*(\cdot), u^*(\cdot)] \end{aligned} \tag{3.23}$$

Remark 3.1.1 *Theorem 3.1 is completely independent of Pontryagin's Principle. It relies on the Stone–Weierstrass Theorem and the Arzelà-Ascoli Theorem; see [87] for a perspective and [50, 84] for details.*

A practical approach to using Theorem 3.1 is to capitalize on the convergence of the partial sums. This point is illustrated in Fig. 3.11 where the DIDO solution to Brac:1 is shown for 10 and 30 nodes. It is apparent that the 10-node solution

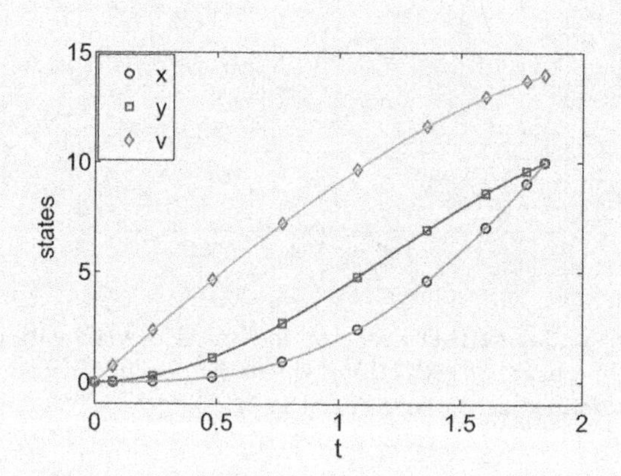

Figure 3.11: A practical usage of Theorem 3.1: A 10-node DIDO solution for Brac:1 overlaid with a 30-node solution (solid lines). Compare with Fig. 3.10.

has practically converged. Note also that this figure is substantially similar to Fig. 3.10 although it is independent of a Runge-Kutta V & V.

Remark 3.1.2 *Theorem 3.1 can also be rigorously applied by verifying the assumptions that go with it; see [50]. Although these assumptions are quite checkable, and the process would be independent of Pontryagin's Principle, it would require the user to learn a new set of tools. An alternative and simpler approach to using this theorem is to invoke the Covector Mapping Principle (see Fig. 2.28 on page 164) and apply Pontryagin's Principle. This is precisely the approach used in the preceding subsections via DIDO, which merely automates this process as per Fig. 3.3.*

Theorem 3.1 seems to suggest that we need an infinite-order polynomial to get the exact solution, while Fig. 3.11 and those preceding it seem to imply that fairly low-order polynomials generate accurate solutions. In other words, it is simply a "waste of resources" to use a very large N because a proper use of pseudospectral theory is based on the notion that carefully designed low-order polynomials can generate highly accurate solutions.* To appreciate this point a little further, we begin by reexamining the exact solution for $x(t)$ given by Eq. (3.10a) and repeated below for quick reference:

$$x(t) = \frac{gt}{\omega} - \frac{g}{\omega^2}\sin\omega t$$

The first term on the right-hand side of this equation is easy to compute by simple multiplication and division, but how do we compute $\sin(\omega t)$? Obviously, we punch in the value of $\phi = \omega t$ on the "sin" button on a calculator app and read off the result. But how does the app do it? Or, how does any computer compute $\sin(\phi)$?

In many computing machines, the computation of *elementary functions* (sin, cos, arctan, ln, exp, square-root, etc.) is done through some combination of table lookups and algorithms: For instance, the CORDIC algorithm[100] has been the workhorse of many microprocessors since its invention in 1959. These computations are not exact! For the purposes of this discussion, the "chip-level" representation of sine and other elementary functions can be thought of in terms of an equivalent "digital" computation of the values of these functions by a truncation of its Taylor series

$$x_{chip}(t) = \sum_{n=0}^{N_{chip}} \frac{x^{(n)}(t_a)}{n!}(t - t_a)^n$$

where $x^{(n)}$ is the n-th derivative of x evaluated at t_a. In other words, elementary functions are essentially different polynomials under different labels! In DIDO, the computation of the state trajectory is done by a truncation of a different infinite series (see page 161)

$$x_{DIDO}(t) = \sum_{m=0}^{N_{DIDO}} a_m P_m(t) \tag{3.24}$$

where P_m is either a ***Legendre*** or ***Chebyshev*** polynomial. The fact that a fairly small value of N_{DIDO} may be needed to generate high accuracy is due to the exponential convergence rate of Legendre and Chebyshev polynomials for analytic functions. Thus, DIDO holds the potential to compute chip-level

*Not all pseudospectral methods converge. See Section 4.4.4 on page 274.

"exact" solutions because all exact solutions are eventually approximate. As an example, the 10-node representation of Eq. (3.24) is given by

$$^{10}x_{DIDO}(t) = 0.0027\, t^9 - 0.0163\, t^8 + 0.0411\, t^7 - 0.0273\, t^6 - 0.1801\, t^5 + 0.0059\, t^4 + 2.1349\, t^3 - 0.0002\, t^2 \quad (3.25)$$

In contrast, the "exact" solution is given by

$$x_{exact}(t) = \frac{gt}{\omega} - \frac{g}{\omega^2}\sin\omega t \approx 7.4929\, t - 5.7290\, \sin(1.3079\, t) \quad (3.26)$$

The second equality in the exact solution is approximate (irony implied!) because we do not know the exact value of ω; hence, we have taken it to be the approximate value from Eq. (3.16) on page 175. The exact solution undergoes a second layer of approximation through its chip-level computation. A plot of Eq. (3.26), as computed by an Intel® Core^{TM} i7 processor, and the 9th-order polynomial solution generated by DIDO is shown in Fig. 3.12.

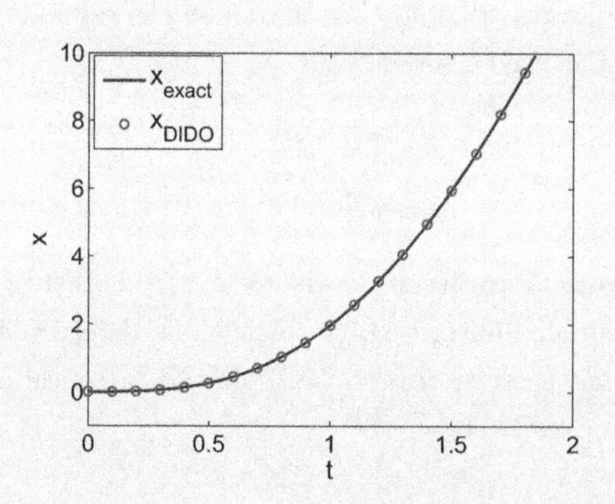

Figure 3.12: Comparison of a micro-processor computed "exact" solution (Eq. (3.26)) and its DIDO-polynomial (Eq. (3.25)).

PS solutions are founded on generating exact solutions by cutting out the "middle-man representation" of an exact solution that may be given in terms of sums, products, powers, convolutions etc. of elementary functions. Thus, a PS solution generates a problem-specific polynomial for the state trajectory. This is, in essence, the meaning of a ***designer function***. The *Stone-Weierstrass theorem* provides the license to represent these designer functions in terms of polynomials, while the exponential convergence rate of Legendre and Chebyshev polynomials provides the foundations to generate high-precision solutions. Additional discussions pertaining to the convergence of PS solutions to exact solutions are presented in [78] and [87].

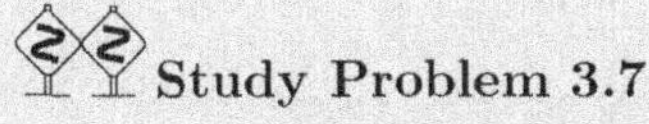

Study Problem 3.7
Develop the Taylor (Maclaurin) series expansion for Eq. (3.26) and compare it to the DIDO-generated polynomial representation given by Eq. (3.25). Explain why the DIDO polynomial is not a Taylor series.

3.3 A Double-Integrator Quadratic Problem

A double integrator ($\ddot{x} = u$) is a ubiquitous dynamical system in electromechanical systems. Its origins are Newton's laws of motion; see Fig. 3.13 for a small sample of applications.

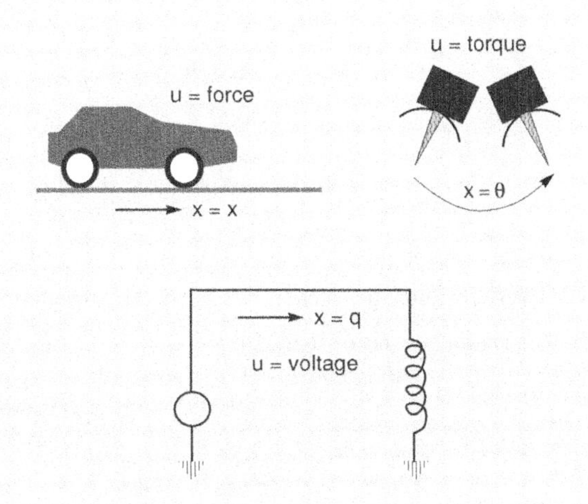

Figure 3.13: A double integrator can be used as a simplified model for a very large number of electromechanical systems.

Consider the problem of finding the optimal controller that transfers a double integrator from its initial rest position, $(x, \dot{x}) = (0,0)$, to another rest position, $(x, \dot{x}) = (1,0)$. What do we take as our optimality criterion? In certain applications (like the maneuvering of a spacecraft), the optimality criterion may be well defined (see section 3.5 for time-optimal maneuvers). In many other applications, *defining the optimality criteria in mathematical terms is indeed a major task.* A popular optimality criterion among control engineers is the minimization of a quadratic cost functional. In electrical systems, if the control is the voltage, then minimizing the integral of the square of the voltage is equivalent to minimizing energy. This motivates the cost functional

$$J[u(\cdot)] = \frac{1}{2}\int_{t_0}^{t_f} u^2(t)\, dt$$

Collecting all the preceding information, we can pose the ***double-integrator quadratic-control (DQC)*** problem in the format of Problem B (see Section 1.4 on page 42) as

$$\boldsymbol{x} := (x, v) \in \mathbb{R}^2 \qquad \boldsymbol{u} := u \in \mathbb{R} \qquad \mathbb{U} = \mathbb{R}$$

$$(\mathsf{DQC}) \begin{cases} \textsf{Minimize} & J[x(\cdot), v(\cdot), u(\cdot), t_f] = \frac{1}{2}\int_{t_0}^{t_f} u^2(t)\, dt \\ \textsf{Subject to} & \dot{x} = v \\ & \dot{v} = u \\ & (x_0, v_0, t_0) = (0,0,0) \\ & (x_f - 1, v_f, t_f - 1) = (0,0,0) \end{cases} \tag{3.27}$$

From the *HAMVET* mnemonic (see page 129), the first step in analyzing any problem is to construct the Hamiltonian.

1. Construct the Hamiltonian:

Because $\boldsymbol{x} \in \mathbb{R}^2$, we need to have $\boldsymbol{\lambda} \in \mathbb{R}^2$. Instead of writing $\boldsymbol{\lambda} = (\lambda_1, \lambda_2)$, we choose deliberate subscripts

$$\boldsymbol{\lambda} := \begin{bmatrix} \lambda_x \\ \lambda_v \end{bmatrix}$$

so that each component of $\boldsymbol{\lambda}$ is tagged to the relevant variable. The Hamiltonian is then given by

$$H(\boldsymbol{\lambda}, \boldsymbol{x}, u) := \overbrace{\frac{u^2}{2}}^{F} + \overbrace{\begin{bmatrix} \lambda_x & \lambda_v \end{bmatrix}}^{\boldsymbol{\lambda}^T} \overbrace{\begin{bmatrix} v \\ u \end{bmatrix}}^{\boldsymbol{f}}$$
$$= \frac{u^2}{2} + \lambda_x v + \lambda_v u$$

2. Develop the Adjoint Equations:

$$-\dot{\lambda}_x := \frac{\partial H}{\partial x} = 0 \qquad \Rightarrow \lambda_x = a, \tag{3.28a}$$
$$-\dot{\lambda}_v := \frac{\partial H}{\partial v} = \lambda_x \qquad \Rightarrow \lambda_v = -at - b \tag{3.28b}$$

where a and b are constants of integration. Thus, in dual space, the costate trajectories are straight lines.

3. Minimize the Hamiltonian (with respect to u only):

The static problem $\min_{u \in \mathbb{R}} H(\boldsymbol{\lambda}, \boldsymbol{x}, u)$ is unconstrained ($\mathbb{U} = \mathbb{R}$); hence, the extremal control is given by

$$\frac{\partial H}{\partial u} = 0 \qquad \Rightarrow u + \lambda_v = 0$$

from which we get the control function, $g_1 : \lambda_v \mapsto u$, as

$$u = g_1(\lambda_v) := -\lambda_v \tag{3.29}$$

Note that the convexity condition is automatically satisfied for *any control*, and in strict form

$$\frac{\partial^2 H}{\partial u^2} = 1 > 0$$

4. Evaluate the Hamiltonian Value Condition:

By inspection, the endpoint function, $\boldsymbol{e}$ is given by

$$\boldsymbol{e}(\boldsymbol{x}_f, t_f) := \begin{bmatrix} x_f - 1 \\ v_f \\ t_f - 1 \end{bmatrix}$$

Hence, we must choose $\boldsymbol{\nu} \in \mathbb{R}^3$ to construct the endpoint Lagrangian:

$$\overline{E}(\boldsymbol{x}_f, t_f) := \overbrace{0}^{E} + \overbrace{\begin{bmatrix} \nu_1 & \nu_2 & \nu_3 \end{bmatrix}}^{\boldsymbol{\nu}^T} \overbrace{\begin{bmatrix} x_f - 1 \\ v_f \\ t_f - 1 \end{bmatrix}}^{e}$$

$$= \nu_1(x_f - 1) + \nu_2 v_f + \nu_3(t_f - 1)$$

The Hamiltonian Value Condition is now given by

$$\mathcal{H}[@t_f] = -\frac{\partial \overline{E}}{\partial t_f} = -\nu_3$$

Clearly, it provides no useful analytical information; however, recall that it provides valuable computational information for V & V: see Section 3.2.1.

5. Analyze the Hamiltonian Evolution Equation:

Because

$$\frac{\partial H}{\partial t} = 0$$

we have

$$\dot{\mathcal{H}} = 0$$

Hence, the minimized (or lower) Hamiltonian is a constant with respect to time. The function $\mathcal{H}$ can be obtained by substituting Eq. (3.29) in the expression for H:

$$\begin{aligned}\mathcal{H}(\lambda_x, \lambda_v, x, v) &:= H(\lambda_x, \lambda_v, x, v, -\lambda_v) \\ &= \frac{\lambda_v^2}{2} + \lambda_x v - \lambda_v^2 \\ &= -\frac{\lambda_v^2}{2} + \lambda_x v\end{aligned}$$

Hence, the function $t \mapsto -\lambda_v(t)^2/2 + \lambda_x(t)\, v(t)$ is a constant.

6. Determine the Transversality Conditions:

The terminal transversality conditions are obtained by simply differentiating the endpoint Lagrangian:

$$\begin{aligned}\lambda_x(t_f) := \frac{\partial \overline{E}}{\partial x_f} &= \nu_1 \\ \lambda_v(t_f) := \frac{\partial \overline{E}}{\partial v_f} &= \nu_2\end{aligned}$$

Obviously, we get no new *analytical* information (although it does provide useful computational information for V & V as discussed in Section 3.2.1).

Study Problem 3.8

Show that $\lambda_x(t) = \nu_1$, and ν_2 satisfies the equation

$$\nu_2 + \nu_1 + b = 0$$

Generating a Feedback Solution via RTOC:

From Eqs.(3.29) and (3.28b), the extremal control can also be written as a function $t \mapsto u$:

$$u = g_2(t; a, b) := at + b \tag{3.30}$$

Tech Talk: Eqs.(3.29) and (3.30) represent the same extremal control but have different functional expressions: Eq.(3.29) is the function $\lambda_v \mapsto u$ while Eq.(3.30) is $t \mapsto u$ (with a and b as paramters). To a beginning student, such distinctions may be distractions in minutia; however, in the end, they are quite important to understanding and implementing practical optimal control. Quite often, blurring these distinctions leads to significant confusion in understanding optimal control theory. For example, suppose we had simply written both equations (3.29) and (3.30) as $u = -\lambda_v$ and $u = at + b$, respectively; then, had we desired $\partial u/\partial t$, we would get 0 if we used Eq. (3.29) and a if we used Eq. (3.30). No confusion arises when we use different symbols for the ***function*** (g_1, g_2, etc.) and the ***value of the function*** (control, u). We frequently abuse notation by writing $u(\lambda_v)$ and $u(t; a, b)$ for Eqs.(3.29) and (3.30), respectively, to prevent running out of symbols or "bookkeeping" numerical subscripts. In such notational abuses, we have to merely keep writing which function $u(\cdot)$ we are referring to. It will be apparent quite shortly (see Eq.(3.35) later on page 198) that the extremal control can be expressed in more ways than these two equations.

Substituting Eq. (3.30) in the state dynamics and integrating the equations (i.e., solving the differential equations) we get

$$
\begin{aligned}
v^*(t; a, b) &= \frac{at^2}{2} + bt + c \\
x^*(t; a, b) &= \frac{at^3}{6} + \frac{bt^2}{2} + ct + d
\end{aligned}
$$

Substituting the boundary conditions generates the following values for the constants:

$$
a = -12 \quad b = 6 \quad c = 0 \quad d = 0 \tag{3.31}
$$

A plot of the extremals (states, costates and control) is shown in Fig. 3.14.

Study Problem 3.9

1. *Perform the necessary computational steps used in arriving at the values of the constants indicated in Eq. (3.31).*
2. *Show that the optimal value of J is 6.*
3. *Redo the above computational steps for $x(0) = 0$, $v(0) = 1$. Develop a code to automate this procedure.*

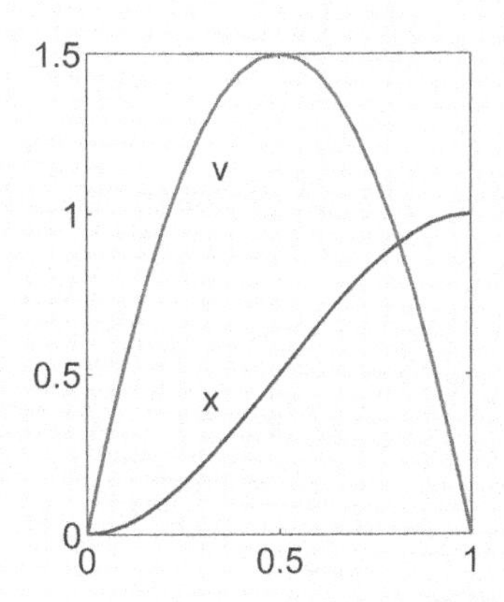

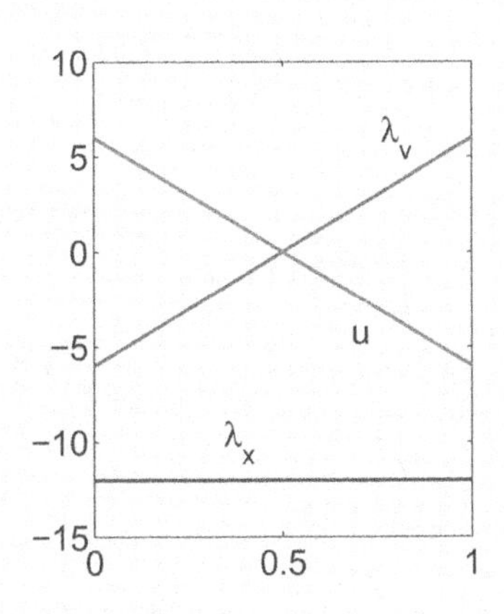

Figure 3.14: Extremal states, costates and control for Problem DQC.

It is clear that if the initial conditions were changed to different numerical values, we could easily produce new values for the constants a, b, c and d. Regardless of the specifics on how these values are produced, we obtain them by solving the system of linear equations:

$$\underbrace{\begin{pmatrix} t_0^3/6 & t_0^2/2 & t_0 & 1 \\ t_0^2/2 & t_0 & 1 & 0 \\ 1/6 & 1/2 & 1 & 1 \\ 1/2 & 1 & 1 & 0 \end{pmatrix}}_{\boldsymbol{T}} \underbrace{\begin{pmatrix} a \\ b \\ c \\ d \end{pmatrix}}_{\boldsymbol{p}} = \underbrace{\begin{pmatrix} x_0 \\ v_0 \\ 1 \\ 0 \end{pmatrix}}_{\boldsymbol{q}} \tag{3.32}$$

We can solve Eq. (3.32) for the 4-vector of parameters $\boldsymbol{p}$ quite easily from

$$\boldsymbol{p} = \boldsymbol{T}^{-1}\boldsymbol{q} \tag{3.33}$$

Recall that *a fundamental rule in linear algebra is <u>not</u> to solve a system of linear equations by computing inverses even when inverses exist*; hence, Eq. (3.33) must be viewed as a conceptual solution and not a recommendation for solving Eq. (3.32) by computing $\boldsymbol{T}^{-1}$. In any case, it is apparent that we can quickly and easily compute the function

$$(t_0, x_0, v_0) \mapsto (a, b)$$

for arbitrary values of initial conditions. A sample solution for $x_0 = 0, v_0 = 1, t_0 = 0$ is shown in Fig. 3.15.

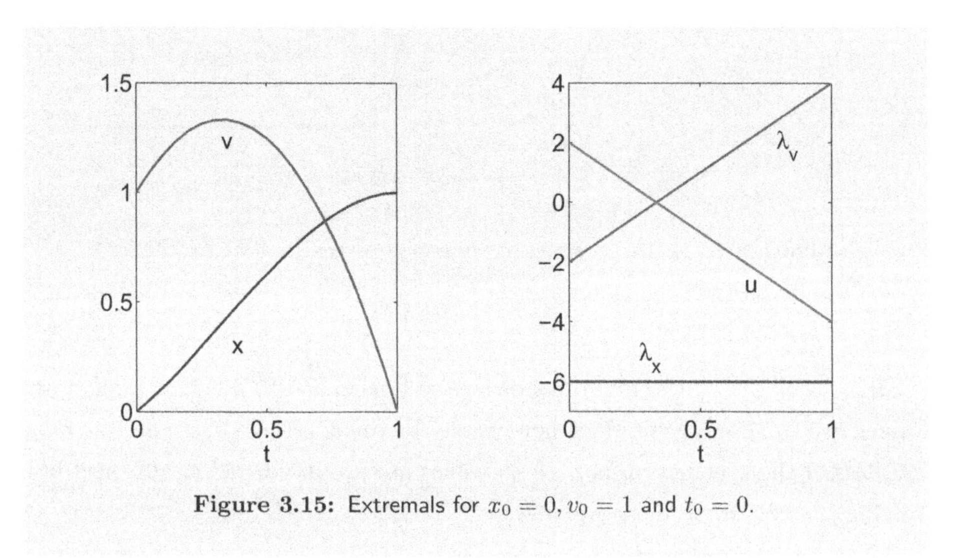

Figure 3.15: Extremals for $x_0 = 0, v_0 = 1$ and $t_0 = 0$.

The elementary observation that we can compute $\boldsymbol{p}$, and hence (a, b), in real-time is surprisingly far-reaching. By replacing arbitrary initial values (x_0, v_0) and initial time t_0 by *current values* for state $\big(x(t) = x, v(t) = v\big)$ and time t, we can rewrite Eq. (3.33) as

$$\boldsymbol{p} = \boldsymbol{T}^{-1}(t)\, \boldsymbol{q}(x, v) \tag{3.34}$$

Because we are largely interested in the first two elements of $\boldsymbol{p}$, we can generate the function $(t, x, v) \mapsto (a, b)$. Denoting this function as $\big(a(t, x, v), b(t, x, v)\big)$, and substituting it in Eq. (3.30), we can find the *optimal*[†] *closed-loop controller*:

$$u = g_3(t, x, v) = a(t, x, v)\, t + b(t, x, v) \tag{3.35}$$

Now, if Eq.(3.30) is implemented but a and b are not updated, then the con-

[†]Strictly speaking, this is an *extremal* closed-loop controller and not necessarily the optimal control since optimality has not been proved.

troller is open-loop. On the other hand, if a and b are *continuously* updated as suggested by Eq.(3.35), then the controller is closed-loop; see Fig. 3.16. This is

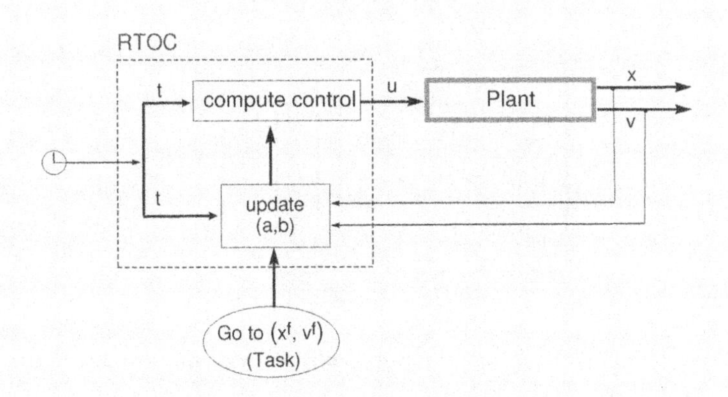

Figure 3.16: When the coefficients a and b of the open-loop control Eq. (3.30) are updated (in real-time) it generates a closed-loop controller. See also Section 1.5.1 on page 64.

the notion of achieving feedback via ***real-time optimal control (RTOC)***.

This exercise illustrates that the only difference between open- and closed-loop controllers is whether or not (state) information at the current time is used to compute/update the control (at the current time). The best way to understand RTOC is to explore it via simulation. This is left as an exercise. See Section 4.3 on page 260.

Tech Talk: There are two frequencies at play in Fig. 3.16. One is the output from the RTOC box and the other is its input. The input updates must be done at or greater than the Lipschitz frequency (see section 1.5 on page 63) while the output can be sampled at any frequency because at each update of a and b, the entire control signal $t \mapsto u$ is generated. If the coefficients a and b are not updated (e.g., the parallel wires at the bottom of Fig. 3.16 are cut) the signal to the plant for the remainder of the time is the open-loop solution. This open-loop signal may continue to be sampled at a frequency greater than or equal to the Nyquist frequency for a zero-order-hold implementation. See [85, 92] for details.

3.3.1 Closed-Loop Versus Closed-Form

The closed-loop implementation of the optimal feedback control implied in Fig. 3.16 is done by solving the following equation

$$\underbrace{\begin{pmatrix} t^3/6 & t^2/2 & t & 1 \\ t^2/2 & t & 1 & 0 \\ 1/6 & 1/2 & 1 & 1 \\ 1/2 & 1 & 1 & 0 \end{pmatrix}}_{\boldsymbol{T}(t)} \underbrace{\begin{pmatrix} a \\ b \\ c \\ d \end{pmatrix}}_{\boldsymbol{p}} = \underbrace{\begin{pmatrix} x \\ v \\ 1 \\ 0 \end{pmatrix}}_{\boldsymbol{q}(x,v)} \tag{3.36}$$

online and without computing $\boldsymbol{T}^{-1}(t)$; that is, by using standard tools from linear algebra. It is possible to determine a and b in *closed-form* by inverting the $\boldsymbol{T}$-matrix anyway; this generates the solution (see Study Problem 3.10)

$$a(t,x,v) = \frac{6\,v}{(1-t)^2} - \frac{12\,(1-x)}{(1-t)^3} \tag{3.37a}$$

$$b(t,x,v) = -\frac{2(t+2)\,v}{(1-t)^2} + \frac{6(t+1)\,(1-x)}{(1-t)^3} \tag{3.37b}$$

Implementing a closed-loop control using this closed-form expression also produces a feedback solution, but without using linear algebra online; in fact, the linear algebra is used off-line. The details on the ramifications of this implementation are left as an exercise; see Section 4.3.1 on page 260.

Study Problem 3.10

1. *Invert the* $\boldsymbol{T}$*-matrix and derive Eq. (3.37). You may need a symbolic toolbox.*
2. *Discuss the situations*

$$\lim_{t\to 1} a(t,x(t),v(t)) \quad \textit{and} \quad \lim_{t\to 1} b(t,x(t),v(t))$$

(You may want to plot these functions to aid your discussions.)

3. *What is the difference, if any, between the closed-loop control obtained through the use of Eq. (3.37) and the one shown in Fig. 3.14?*

It is clear that if $v \neq 0$ or $x \neq 1$, the coefficients a and b will approach infinity as $t \to 1$. From Study Problem 3.10, it follows that this does not happen *along* the optimal solution; however, the closed-loop system may indeed exhibit this phenomenon in the presence of uncertainties; see Section 4.3.1 on page 260 for additional details. When a or b become unbounded, the closed-loop controls also become unbounded. *It is important to note that this problem is not with the closed-loop implementation but with the problem formulation itself!* That is, the problem formulation (Problem DQC) has no explicit requirements on uncertainty management; hence, its performance should not be measured on such an a posteriori specification. Nonetheless, because the feedback principle manages uncertainties, an a posteriori knowledge of the performance of RTOC, or other optimal feedback implementations, can be used to reformulate the deterministic problem to support a better performance of the closed-loop implementation. One particular concept is discussed in Section 4.4.1 on page 266.

Tech Talk: An explicit approach for handling uncertainties is through the concept of ***tychastic optimal control*** introduced in Section 1.6 on page 75. The uncertainty that can be included in a tychastic optimal control problem is an inertia parameter, m, in the double integrator:

$$m\ddot{x} = u$$

The characteristics of the closed-loop system for a uniformly distributed inertia is explored in Section 4.3.1. All of these notions underline the fact that a proper problem formulation is indeed a central issue in the use and exploitation of optimal control.

3.3.2 An Optimal-Control Perspective of PD Controllers

Substituting Eq. (3.37) in Eq. (3.35) and grouping terms, it is straightforward to show that we can write

$$u = -\frac{4}{(1-t)}\, v + \frac{6}{(1-t)^2}\,(1-x) \tag{3.38}$$

The quantity

$$t_{go} := 1 - t$$

is called ***time-to-go***, particularly in missile guidance parlance[106]. Equation (3.38) can also be written as

$$u = k_x(t_{go})\,(1 - x) - k_v(t_{go})\,v \tag{3.39}$$

where the quantities

$$k_x(t_{go}) := \frac{6}{t_{go}^2} \quad \text{and} \quad k_v(t_{go}) := \frac{4}{t_{go}} \tag{3.40}$$

are called ***feedback gains***. These gains are time-varying. Certain implementations of time-varying gains are known as ***gain scheduling***, although gain scheduling is frequently not optimal.

If the gains in Eq. (3.39) were set to some constant values, the controller is called ***proportional-plus-derivative*** or simply a ***PD controller*** because the control is a linear function of "errors" in x and its derivative, $\dot{x} = v$. A block diagram of this implementation is shown in Fig. 3.17.

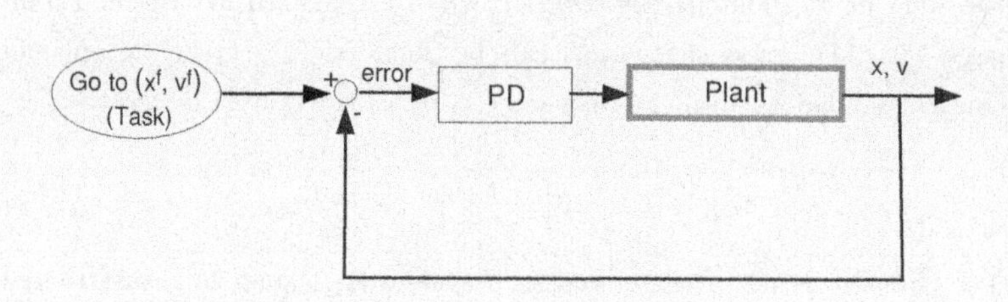

Figure 3.17: A PD controller can be construed as a sub-optimal implementation of the RTOC version shown in Fig. 3.16. The "error" is $x^f - x(t) = 1 - x(t)$ and $v^f - v(t) = -v(t)$.

PD controllers predate optimal control theory. A large part of feedback control theory is based on assuming that the control is given in terms of a linear function of the states and the task is to find an appropriate set of gains. The non-constancy of the gains in Eq. (3.39) implies that, with proper tuning, a PD controller may be constructed as a sub-optimal solution for this problem. This is one reason why a well-tuned PD controller can be extremely efficient as an inner-loop.

Study Problem 3.11

Assume that a PD controller has been sufficiently tuned so that its performance compares well with the RTOC implementation. If the bottom wires in Figs. (3.17) and (3.16) are cut, how do these implementations perform under this failure mode?

Note that the gains $k_x(t_{go})$ and $k_v(t_{go})$ tend to infinity as t_{go} approaches zero. As a result, storage of the gains becomes a problem well before $t_{go} = 0$. To circumvent this problem, the feedback control in the form given by Eq. (3.39) can be implemented by storing the gains up to some time $t_f - t_b = 1 - t_b$ where $t_b > 0$ is a designer-chosen time interval. The choice of this interval is usually based on the maximum value of gain that can be stored onboard. Then, the control can be executed in feedback form all the way up to $t_f - t_b$, followed by an open-loop command from $t_f - t_b$ to t_f. The time interval t_b is called the ***blind time***, yet another phrase borrowed from missile guidance.

Study Problem 3.12

Show that the matrix $\boldsymbol{T}$ in Eq. (3.36) becomes singular when $t = 1$. Near $t = 1$, $\boldsymbol{T}$ is ill-conditioned; hence, implementing feedback controls by way of Eq. (3.36) requires techniques to solve ill-conditioned systems of linear equations. The blind-time concept may also be used, but t_b in such an implementation is typically much smaller than the gain-based solution; see Section 4.3.1 on page 260. Note that $\boldsymbol{T}$ being singular at $t = 1$ is actually required (by the "physics" of the problem).

3.3.3 Applications to Optimal Missile Guidance

As illustrated in Fig. 3.13 on page 191, Problem DQC can be interpreted and applied to many electromechanical systems. One particularly interesting application of this problem is in its derivation of classical missile guidance laws. In a two-dimensional version of this problem, a missile speeds at a constant velocity V_m, to intercept its target. The direction of V_m at $t = t_0$ is called the initial line of sight (ILOS); see Fig. 3.18. If nothing is done, the missile will miss the target by a normalized distance of

$$y(t_0) = -1$$

The guidance objective is to apply a normal acceleration command

$$u = n_c$$

that drives the miss distance to zero

$$y(t_f) = 0$$

Modeling the dynamics of the missile in the normal direction by a double integrator, we can readily apply the results of Problem DQC through a few small

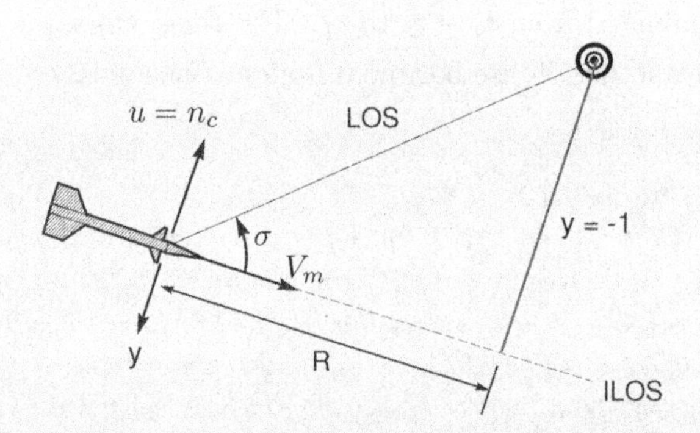

Figure 3.18: Schematic for missile guidance at $t = t_0$.

modifications.

First, note that the boundary conditions, $y(t_0) = -1$ and $y(t_f) = y(1) = 0$, can be mapped to those of Problem DQC by a shift in the coordinate system:

$$x(t) := y(t) + 1 \tag{3.41}$$

Thus, $x(t_0) = 0$ and $x(t_f) = 1$ as required. Next, it is not necessary for $v(t_f)$ to be equal to zero as stipulated in Problem DQC; in fact, we set v_f to take on any value (i.e., free). As a result, we need to modify the boundary condition on

Problem DQC, let us call this new problem missile-DQC.

Applying Pontryagin's Principle to Problem missile-DQC generates the same conditions as Problem DQC, except for the transversality condition pertaining to v_f; this condition is given by

$$\lambda_v(t_f) = 0 \tag{3.42}$$

In other words, we lose the old condition $v_f = 0$ while gaining a new one; hence, the resulting set of new equations can be solved to produce a new solution. Equation (3.42) implies (see Eq. (3.28b)

$$a + b = 0$$

Consequently, we have

$$u = at + b = a(t - 1) \tag{3.43}$$

Carrying out the remainder of the steps leading to Eq. (3.32), we now get a 2×2 system

$$\begin{pmatrix} T_0^3/6 & T_0 \\ T_0^2/2 & 1 \end{pmatrix} \begin{pmatrix} a \\ c \end{pmatrix} = \begin{pmatrix} x_0 - 1 \\ v_0 \end{pmatrix}$$

where $T_0 = t_0 - 1$. Solving for a and substituting it in Eq. (3.43) we get

$$u = 3\left(\frac{v_0 T_0 - (x_0 - 1)}{T_0^3}\right)(t - 1)$$

By setting $t_0 = t$, the current time, we get a closed-loop guidance law:

$$u = -\frac{3}{t_{go}^2}\Big(v\, t_{go} + (x - 1)\Big) \tag{3.44}$$

An implementation of Eq. (3.44) requires knowledge of the missile's state, x and v, relative to the target. Instead of estimating these quantities for feedback, we can map these variables to the sensor output σ called the line of sight (LOS) angle; see Fig. 3.18. For small angles, we can write

$$\sigma \approx \frac{y}{R} = \frac{y}{V_m t_{go}}$$

Differentiating this equation, we get

$$\dot{\sigma} \approx \frac{\dot{y}}{V_m t_{go}} + \frac{y}{V_m t_{go}^2} = \frac{1}{V_m t_{go}^2}\Big(v\, t_{go} + (x-1)\Big)$$

Hence, Eq. (3.44) can be written as

$$u = -3V_m \dot{\sigma}$$

This equation is a special case of ***proportional navigation guidance*** law[106], written more generally as,

$$u = n_c = -NV_m \dot{\sigma}$$

where N, called the ***navigation ratio***, is a designer-chosen gain in the range of 3–5.

The foregoing derivation of the "optimality" of proportional navigation guidance is based on a slightly different approach taken by Bryson and Ho[16].

3.3.4 Everything is Optimal

The preceding subsections highlight the observation that anything can be made "optimal" under the right set of assumptions. Hence, when designers claim that they are not seeking optimal solutions but something "simple" or something "that works," that statement is not entirely true: The real problem is that we do not have a mathematical model of the cost function.

PD controllers and proportional navigation guidance laws were conceived by engineers through intuition and equipment available in the first half of the 20th century. A demonstration of their "optimality" is essentially reverse-engineering their design to determine the optimal control problem.

The previous subsections also show that many "simple" solutions can be framed as yet another app of optimal control theory. In other words, disparate concepts can be brought under one unifying framework.

Sidebar: The power of a unified theory cannot be underestimated. Imagine how difficult dynamics would be without Newton's "unifying" laws of motion: Every motion (car, airplane, planets, etc.) would seem "complicated." Newton's three basic laws help explain everything! Similarly, every color would seem "complex" and unfathomable to mathematize if we did not know that they could be generated by just three basic colors. In much the same way, optimal control and optimization, in general, help explain a lot of "complexity" through the power of three: decision variables, cost function and constraints.

Study Problem 3.13

Problem DQC is a time-fixed problem. Discuss what happens if t_f is free. In particular, show that

$$t_f = -\frac{b}{a} \tag{3.45}$$

and that Eq. (3.32) gets modified to

$$\begin{pmatrix} t_0^3/6 & t_0^2/2 & t_0 & 1 \\ t_0^2/2 & t_0 & 1 & 0 \\ t_f^3/6 & t_f^2/2 & t_f & 1 \\ t_f^2/2 & t_f & 1 & 0 \end{pmatrix} \begin{pmatrix} a \\ b \\ c \\ d \end{pmatrix} = \begin{pmatrix} x_0 \\ v_0 \\ 1 \\ 0 \end{pmatrix} \tag{3.46}$$

1. *Using Eqs. (3.45) and (3.46) design both open- and closed-loop solutions. Compare your open-loop solution with the one displayed in Fig. 3.19 for the same initial conditions used in Fig. 3.15 ($x_0 = 0, v_0 = 1$).*
2. *Does the closed-loop solution solve the infinite-gain problem?*

Study Problem 3.14

Instead of freeing up the final time completely, consider the cost functional

$$J[\boldsymbol{x}(\cdot), \boldsymbol{u}(\cdot), t_f] := \int_0^{t_f} \left(T(t) + \frac{R}{2} u^2(t) \right) dt$$

where $T(t)$ is a tuning function and $R > 0$ is a another knob called a weight. Redo all the steps taken to solve

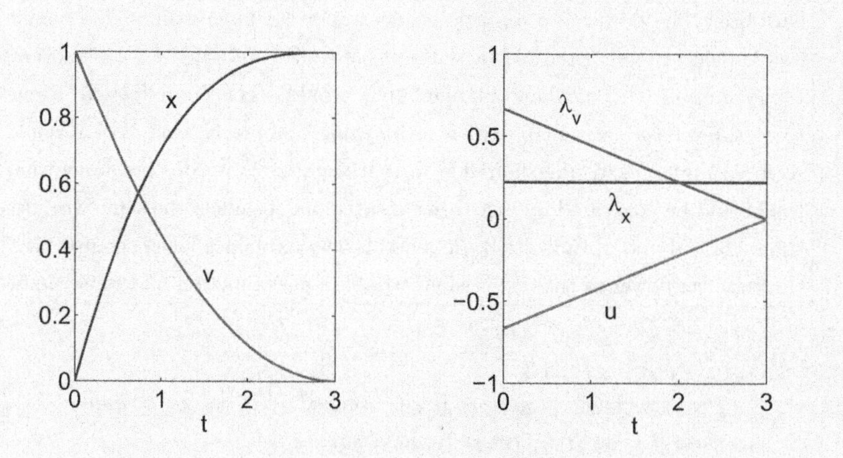

Figure 3.19: Time-free extremals for $x_0 = 0, v_0 = 1$.

Problem DQC for this new problem. Analyze the resulting equations for the following special cases:

1. $T(t) = 1$.
2. $T(t) = e^{-t/\tau}$, *where* τ *is yet another tuning parameter.*

Using these results and those from the entirety of this section, discuss the meaning and relevance of quadratic costs.

3.4 Kalman's Linear-Quadratic Control Problem

Many electromechanical systems and their modes of operation can be well approximated by linear system dynamics. Examples are electric motors used in almost every machine, automobile cruise control, electronic circuitry, and so on. In addition, the inner loops of many nonlinear control systems are designed using linear system theory. Refer back to Section 1.5.1, particularly pages 67–69, for a quick overview on the design of feedback control systems.

In linear system dynamics, the function $\boldsymbol{f}$ is linear in $\boldsymbol{x}$ and $\boldsymbol{u}$

$$\dot{\boldsymbol{x}} = \boldsymbol{A}(t)\boldsymbol{x} + \boldsymbol{B}(t)\boldsymbol{u} \tag{3.47}$$

where $\boldsymbol{A}(t) \in \mathbb{R}^{N_x \times N_x}$ and $\boldsymbol{B}(t) \in \mathbb{R}^{N_x \times N_u}$ are time-varying system matrices. In many systems, $\boldsymbol{A}$ and $\boldsymbol{B}$ are also time-invariant. In the 1950s and '60s, a linear system could be easily simulated using an analog computer: All that was needed was an amplifier, a summer and an integrator; see Fig. 3.20. If $\boldsymbol{u}(t) = \mathbf{0}$

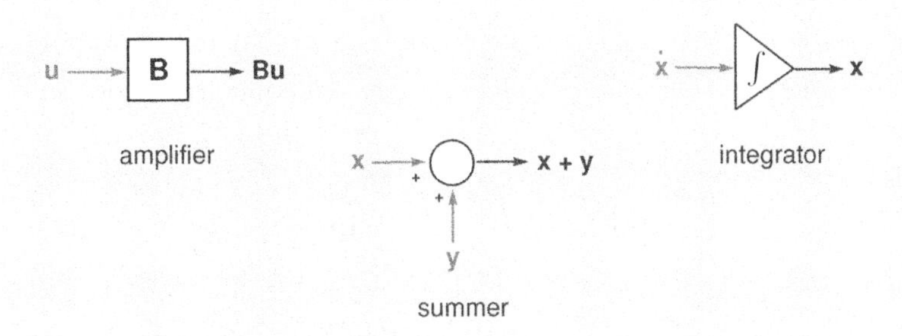

Figure 3.20: Analog computer components to simulate a linear system (above) and their assembly (below) to simulate a linear system.

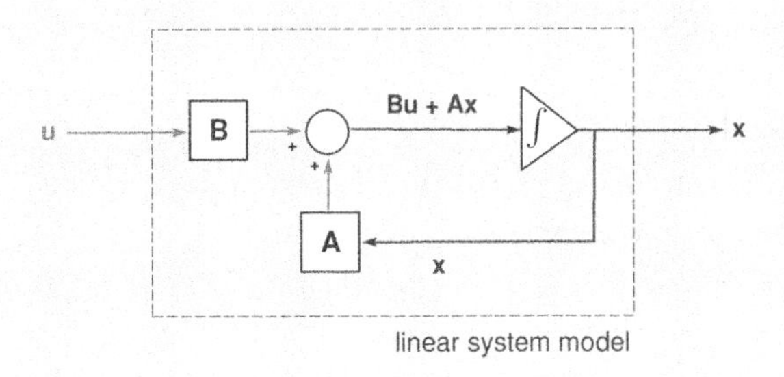

for $t \geq t_0$, then $\boldsymbol{x}^* = \mathbf{0}$ is an equilibrium state of Eq. (3.47).

In the 1960s, Kalman published a series of fundamental papers that revolutionized the entire field of linear feedback control system design. In [44], he states a fundamental control problem as follows:

> *Given any state $\boldsymbol{x}$ of the plant [(3.47)] at time t_0, "generate" a control function $\boldsymbol{u}(t)$, defined for $t \geq t_0$ and depending on $\boldsymbol{x}, t_0$, which causes $\boldsymbol{x}$ to be "transferred" to the equilibrium state $\mathbf{0}$.*

To produce a smooth ***control law***

$$\boldsymbol{u}(t) = \boldsymbol{k}(\boldsymbol{x}(t), t), \qquad \boldsymbol{k} \in C^1 \tag{3.48}$$

"rationally," Kalman states that[44]

> *the integral of a nonnegative function of the state along any motion should be minimized by the choice of* $\mathbf{u}(t)$.

Thus, Kalman equates rational control system design to optimal control. In the engineering systems of the 1960s, it was highly desirable to produce a linear function $\boldsymbol{k}$ so that the control law could be implemented by simple amplifiers. To achieve this linear system design, Kalman suggested a quadratic cost functional of the form

$$J[\boldsymbol{x}(\cdot), \boldsymbol{u}(\cdot), t_f] := \frac{1}{2}\boldsymbol{x}_f^T \boldsymbol{P} \boldsymbol{x}_f + \frac{1}{2}\int_{t_0}^{t_f} \left(\boldsymbol{x}^T(t)\boldsymbol{Q}(t)\boldsymbol{x}(t) + \boldsymbol{u}^T(t)\boldsymbol{R}(t)\boldsymbol{u}(t)\right) dt \tag{3.49}$$

where $\boldsymbol{P}, \boldsymbol{Q}$ and $\boldsymbol{R}$ are design matrices *chosen by the decision maker.* That the choice of the quadratic cost functional generates a linear feedback controller is shown later in Eq. (3.62). Because J is a scalar, $\boldsymbol{P}$ and $\boldsymbol{Q}$ must both be of dimension $N_x \times N_x$ and $\boldsymbol{R}$ must be of dimension $N_u \times N_u$.

Study Problem 3.15

1. *Show that* $\boldsymbol{P}, \boldsymbol{Q}$ *and* $\boldsymbol{R}$ *may be chosen to be symmetric matrices without loss in generality.*
2. *Show that a symmetric formulation reduces the number of design variables from* n^2 *to* $n(n+1)/2$ *where* n *is the size of the square matrix.*

Tech Talk: The "true" concept here is ***multi-objective optimal control*** where the cost function is given by

$$\mathbf{J} := \begin{pmatrix} J_1 \\ J_2 \\ J_3 \end{pmatrix} = \begin{pmatrix} \boldsymbol{x}_f^T \boldsymbol{x}_f \\ \int_{t_0}^{t_f} \boldsymbol{x}^T(t)\boldsymbol{x}(t)\, dt \\ \int_{t_0}^{t_f} \boldsymbol{u}^T(t)\boldsymbol{u}(t)\, dt \end{pmatrix}$$

The weights $\boldsymbol{P}, \boldsymbol{Q}$ and $\boldsymbol{R}$ scalarize the cost function through a linear combination of the individual costs. They can be chosen to explore points in the trade space. Note, however, that their effects may be nonlinear; see Section 4.4.1 on page 266 for an illustrative example.

From the requirement that $J[\boldsymbol{x}(\cdot), \boldsymbol{u}(\cdot), t_f] \geq 0$, it follows that $\boldsymbol{P}, \boldsymbol{Q}$ and $\boldsymbol{R}$ must be chosen to be positive semidefinite. It will be apparent shortly (see Equations (3.52) and (3.66)) that this choice must be tightened a little more to restrict $\boldsymbol{Q}$ and $\boldsymbol{R}$ to be positive definite. Note also that these choices of $\boldsymbol{P}, \boldsymbol{Q}$ and $\boldsymbol{R}$ imply that the value of the global optimal of J is zero.

Equations (3.49) and (3.47) form the centerpiece of ***Kalman's linear-quadratic optimal control problem***. Although Kalman derived his equations using an optimal-control modification of Carathéodory's fundamental equations from the calculus of variations, we can arrive at the same result using Pontryagin's Principle (which was also noted by Kalman). To facilitate a ready application of Pontryagin's Principle, we first organize Kalman's problem into the standard structured format:

$$
(LQ)\left\{\begin{array}{lrl}
\text{Minimize} & J[\boldsymbol{x}(\cdot), \boldsymbol{u}(\cdot), t_f] := & \frac{1}{2}\boldsymbol{x}_f^T \boldsymbol{P} \boldsymbol{x}_f + \frac{1}{2}\int_{t_0}^{t_f}\Big(\boldsymbol{x}^T(t)\boldsymbol{Q}(t)\boldsymbol{x}(t) \\
& & \quad + \boldsymbol{u}^T(t)\boldsymbol{R}(t)\boldsymbol{u}(t)\Big)\,dt \\
\text{Subject to} & \dot{\boldsymbol{x}} = & \boldsymbol{A}(t)\boldsymbol{x} + \boldsymbol{B}(t)\boldsymbol{u} \\
& (\boldsymbol{x}(t_0), t_0) = & (\boldsymbol{x}^0, t^0) \\
& t_f = & t^f
\end{array}\right.
$$

There are no final-time conditions on $\boldsymbol{x}$ despite that it is "desired" to go to zero. For the moment, we shall let t_f be fixed at t^f. In following the *HAMVET* procedure (see page 129), we get

1. *Construct the Hamiltonian:*

$$
H(\boldsymbol{\lambda}, \boldsymbol{x}, \boldsymbol{u}, t) := \frac{1}{2}\left(\boldsymbol{x}^T Q(t)\boldsymbol{x} + \boldsymbol{u}^T R(t)\boldsymbol{u}\right) + \boldsymbol{\lambda}^T\left(\boldsymbol{A}(t)\boldsymbol{x} + \boldsymbol{B}(t)\boldsymbol{u}\right) \tag{3.50}
$$

where $\boldsymbol{\lambda} \in \mathbb{R}^{N_x}$.

2. *Develop the Adjoint Equations:*

$$
-\dot{\boldsymbol{\lambda}} = \frac{\partial H}{\partial \boldsymbol{x}} = \boldsymbol{Q}(t)\boldsymbol{x} + \boldsymbol{A}^T(t)\boldsymbol{\lambda} \tag{3.51}
$$

Note that the adjoint equations are linear in $\boldsymbol{x}$ as a result of the quadratic cost.

3. Minimize the Hamiltonian:

$$(HMC) \quad \begin{cases} \underset{\boldsymbol{u}}{\text{Minimize}} & H(\boldsymbol{\lambda}, \boldsymbol{x}, \boldsymbol{u}, t) \\ \text{Subject to} & \boldsymbol{u} \in \mathbb{R}^{N_u} \end{cases}$$

Because $\boldsymbol{u}$ is unconstrained, we can write $\partial_{\boldsymbol{u}} H = \boldsymbol{0}$; hence, we have

$$\boldsymbol{R}(t)\boldsymbol{u} + \boldsymbol{B}^T(t)\boldsymbol{\lambda} = \boldsymbol{0} \qquad \Rightarrow \qquad \boldsymbol{u} = -\boldsymbol{R}^{-1}(t)\boldsymbol{B}^T(t)\boldsymbol{\lambda} \tag{3.52}$$

Note that if $\boldsymbol{R}$ were to be only positive semidefinite, $\boldsymbol{R}^{-1}$ may not exist. To ensure the existence of its inverse, we needed to tighten the requirement on $\boldsymbol{R}$ to be positive definite.

Substituting Eq. (3.52) in (3.50) we get the minimized or the lower Hamiltonian

$$\begin{aligned} \mathcal{H}(\boldsymbol{\lambda}, \boldsymbol{x}, t) &:= \frac{1}{2}\left(\boldsymbol{x}^T Q(t)\boldsymbol{x} + \boldsymbol{\lambda}^T \boldsymbol{B}(t) R^{-1}(t)\boldsymbol{B}^T(t)\boldsymbol{\lambda}\right) + \boldsymbol{\lambda}^T \boldsymbol{A}(t)\boldsymbol{x} \\ &\qquad - \boldsymbol{\lambda}^T \boldsymbol{B}(t) R^{-1}(t)\boldsymbol{B}^T(t)\boldsymbol{\lambda} \\ &= \frac{1}{2}\left(\boldsymbol{x}^T Q(t)\boldsymbol{x} + 2\boldsymbol{\lambda}^T \boldsymbol{A}(t)\boldsymbol{x} - \boldsymbol{\lambda}^T \boldsymbol{B}(t) R^{-1}(t)\boldsymbol{B}^T(t)\boldsymbol{\lambda}\right) \end{aligned} \tag{3.53}$$

4. Evaluate the Hamiltonian Value Condition:

The first step in establishing the Hamiltonian value condition is to generate the endpoint Lagrangian

$$\overline{E}(\nu, \boldsymbol{x}_f, t_f) := \frac{1}{2}\boldsymbol{x}_f \boldsymbol{P}\boldsymbol{x}_f + \nu(t_f - t^f) \tag{3.54}$$

where ν is the endpoint covector associated with the prescription of a fixed final time. The Hamiltonian Value Condition is now given by

$$\mathcal{H}[@t_f] := -\frac{\partial \overline{E}}{\partial t_f} = -\nu \tag{3.55}$$

This "non-result" is not surprising because the Hamiltonian value condition generates the optimal stopping time condition: If the final time is prescribed, then the prescribed time *is* indeed the optimal stopping time.

Now suppose the final time were free, then we would not have the second term on the right-hand side of Eq. (3.54). In this case, the Hamiltonian value condition generates the optimal stopping time condition:

$$\mathcal{H}[@t_f] := -\frac{\partial \overline{E}}{\partial t_f} = 0 \tag{3.56}$$

Combining equations (3.55) and (3.56), we can write

$$\mathcal{H}[@t_f] = \begin{cases} -\nu & \text{fixed } t_f \\ 0 & \text{free } t_f \end{cases} \tag{3.57}$$

5. Analyze the Hamiltonian Evolution Equation:

From Eq. (3.50), we can write

$$\partial_t H(\boldsymbol{\lambda}, \boldsymbol{x}, \boldsymbol{u}, t) = \frac{1}{2}\left(\boldsymbol{x}^T \frac{d\boldsymbol{Q}(t)}{dt}\boldsymbol{x} + \boldsymbol{u}^T \frac{d\boldsymbol{R}(t)}{dt}\boldsymbol{u}\right) + \boldsymbol{\lambda}^T\left(\frac{d\boldsymbol{A}(t)}{dt}\boldsymbol{x} + \frac{d\boldsymbol{B}(t)}{dt}\boldsymbol{u}\right)$$

If $\boldsymbol{Q}$, $\boldsymbol{R}$, $\boldsymbol{A}$ and $\boldsymbol{B}$ are constants, then

$$\partial_t H(\boldsymbol{\lambda}, \boldsymbol{x}, \boldsymbol{u}, t) = 0$$

Otherwise, $\partial_t H$ will not be zero, in general. With $\boldsymbol{Q}$, $\boldsymbol{R}$, $\boldsymbol{A}$ and $\boldsymbol{B}$ as constants, the Hamiltonian evolution equation becomes

$$\frac{d\mathcal{H}}{dt} = 0$$

6. Determine the Transversality Conditions:

$$\boldsymbol{\lambda}(t_f) = \frac{\partial \overline{E}}{\partial \boldsymbol{x}_f} = \boldsymbol{P}\boldsymbol{x}_f \tag{3.58}$$

Generating a Feedback Solution via RTOC:

A feedback solution can be generated by constructing and solving the BVP, LQ^λ. Collecting all the equations, this BVP can be summarized as

$$\boldsymbol{x} \in \mathbb{R}^{N_x} \quad \boldsymbol{\lambda} \in \mathbb{R}^{N_x}$$

$$(LQ^\lambda) \begin{cases} \text{Find} & (\boldsymbol{x}(\cdot), \boldsymbol{\lambda}(\cdot)) \\ \text{Such that} & \begin{bmatrix} \dot{\boldsymbol{x}} \\ \dot{\boldsymbol{\lambda}} \end{bmatrix} = \begin{bmatrix} \boldsymbol{A}(t) & -\boldsymbol{B}(t)\boldsymbol{R}^{-1}(t)\boldsymbol{B}^T(t) \\ -\boldsymbol{Q}(t) & -\boldsymbol{A}^T(t) \end{bmatrix} \begin{bmatrix} \boldsymbol{x} \\ \boldsymbol{\lambda} \end{bmatrix} \\ & (\boldsymbol{x}_0,\ t_0) = (\boldsymbol{x}^0,\ t^0) \\ & (\boldsymbol{\lambda}_f,\ t_f) = (\boldsymbol{P}\boldsymbol{x}_f,\ t^f) \end{cases}$$

Problem LQ^λ is a *linear* BVP. Solving this problem (using any present-day computational tool) generates the state-costate trajectory $[t^0, t^f] \ni t \mapsto (\boldsymbol{x}, \boldsymbol{\lambda})$. Substituting the costate trajectory for $\boldsymbol{\lambda}$ in Eq. (3.52) produces the open-loop control trajectory

$$\boldsymbol{u}(t) = -\boldsymbol{R}^{-1}(t)\boldsymbol{B}^T(t)\boldsymbol{\lambda}(t)$$

The open-loop control is parameterized by the initial event, $(\boldsymbol{x}^0, t^0)$; hence, it can be rewritten as

$$\boldsymbol{u}(t; \boldsymbol{x}^0, t^0) = -\boldsymbol{R}^{-1}(t)\boldsymbol{B}^T(t)\boldsymbol{\lambda}(t; \boldsymbol{x}^0, t^0) \tag{3.59}$$

From the RTOC principle (see Section 1.5.1 on page 64), Eq. (3.59) generates a time-varying continuous feedback controller $\boldsymbol{u}(t; \boldsymbol{x}, t)$, if $(\boldsymbol{x}^0, t^0)$ is continuously updated. This approach to feedback control using pseudospectral methods is described in [103].

In the 1960s, it was highly desirable to produce feedback controllers using amplifiers, summers and similar electronic circuitry. Solving BVPs online was not a viable option. Furthermore, the feedback principle was limited to producing closed-form solutions. Consequently, Kalman proposed a different procedure for generating a feedback solution.

Kalman's Solution:

Inspired by Eq. (3.58) and motivated by the linearity of the dynamics and co-dynamics, we "guess" $\boldsymbol{\lambda}$ varies linearly with $\boldsymbol{x}$ and write

$$\boldsymbol{\lambda}(t) = \boldsymbol{S}(t)\boldsymbol{x}(t) \tag{3.60}$$

where $\boldsymbol{S}(t)$ is an unknown time-varying square matrix ($N_x \times N_x$) whose value at t_f is given by

$$\boldsymbol{S}(t_f) = \boldsymbol{P} \tag{3.61}$$

This condition is necessary to ensure that Eq. (3.60) satisfies the transversality condition Eq. (3.58).

If we can show that Eq. (3.60) is legitimate, then it can be interpreted as taking the final-time (transversality) condition and pulling it back in time to the initial condition via the "sweep matrix" $\boldsymbol{S}(t)$. Furthermore, the controller is given in terms of a linear feedback law

$$\begin{aligned} \boldsymbol{u} &= -\boldsymbol{R}^{-1}(t)\boldsymbol{B}^T(t)\boldsymbol{\lambda} \\ &= -\boldsymbol{R}^{-1}(t)\boldsymbol{B}^T(t)\boldsymbol{S}(t)\,\boldsymbol{x} \quad = \boldsymbol{K}(t)\,\boldsymbol{x} \end{aligned} \tag{3.62}$$

where $\boldsymbol{K}(t) := -\boldsymbol{R}^{-1}(t)\boldsymbol{B}^T(t)\boldsymbol{S}(t)$.

If our guess of the linearity between the costate and the state is correct, then by differentiating both sides of Eq. (3.60) we get

$$\frac{d\boldsymbol{\lambda}(t)}{dt} = \frac{d\big(\boldsymbol{S}(t)\boldsymbol{x}(t)\big)}{dt} = \dot{\boldsymbol{S}}(t)\boldsymbol{x}(t) + \boldsymbol{S}(t)\dot{\boldsymbol{x}}(t)$$

Substituting the equations for the dynamics and co-dynamics (adjoint equations) we have

$$-\boldsymbol{Q}(t)\boldsymbol{x}(t) - \boldsymbol{A}^T(t)\boldsymbol{\lambda}(t) = \dot{\boldsymbol{S}}(t)\boldsymbol{x}(t) + \boldsymbol{S}(t)\big(\boldsymbol{A}(t)\boldsymbol{x}(t) + \boldsymbol{B}(t)\boldsymbol{u}(t)\big)$$

Substituting Eq. (3.52) in the above equation and rearranging the terms, we get

$$\begin{aligned} &\Big(\dot{\boldsymbol{S}}(t) + \boldsymbol{S}(t)\boldsymbol{A}(t) + \boldsymbol{Q}(t)\Big)\boldsymbol{x}(t) \\ &\qquad + \Big(\boldsymbol{A}^T(t) - \boldsymbol{S}(t)\boldsymbol{B}(t)\boldsymbol{R}^{-1}(t)\boldsymbol{B}^T(t)\Big)\boldsymbol{\lambda}(t) = \boldsymbol{0} \end{aligned} \tag{3.63}$$

Combining equations (3.60) and (3.63), we have

$$\Big(\dot{\boldsymbol{S}}(t) + \boldsymbol{S}(t)\boldsymbol{A}(t) + \boldsymbol{A}^T(t)\boldsymbol{S}(t) + \boldsymbol{Q}(t) \\ - \boldsymbol{S}(t)\boldsymbol{B}(t)\boldsymbol{R}^{-1}(t)\boldsymbol{B}^T(t)\boldsymbol{S}(t)\Big)\boldsymbol{x}(t) = \boldsymbol{0} \tag{3.64}$$

Because Eq. (3.64) must be true for all $\boldsymbol{x}(t)$, the term in parenthesis must be zero; hence, for Eq. (3.60) to hold, we require $\boldsymbol{S}(t)$ to satisfy the *nonlinear* ordinary differential equation:

$$-\frac{d\boldsymbol{S}}{dt} = \boldsymbol{S}\boldsymbol{A}(t) + \boldsymbol{A}^T(t)\boldsymbol{S} - \boldsymbol{S}\boldsymbol{B}(t)\boldsymbol{R}^{-1}(t)\boldsymbol{B}^T(t)\boldsymbol{S} + \boldsymbol{Q}(t) \tag{3.65a}$$

$$\boldsymbol{S}(t_f) = \boldsymbol{P} \tag{3.65b}$$

Equation (3.65a) is known as the ***Riccati equation*** or the *Riccati differential equation.* Kalman's construction of a feedback control system would proceed as follows:

Step 1 Solve off-line the Riccati equation by backward-integrating Eq. (3.65a) using the final-time condition $\boldsymbol{S}(t_f) = \boldsymbol{P}$.

Step 2 Compute (off-line) the Kalman gain matrix $\boldsymbol{K}(t) := -\boldsymbol{R}^{-1}(t)\boldsymbol{B}^T(t)\boldsymbol{S}(t)$ using the solution $\boldsymbol{S}(t)$ generated from Step 1.

Step 3 Store on-board $\boldsymbol{K}(t)$ at some preselected discrete points, $t^0, t^1, \ldots, t^f$.

Step 4 The feedback controller $\boldsymbol{u}(t, \boldsymbol{x}) = \boldsymbol{K}(t)\boldsymbol{x}$ is given by a *gain-scheduled* matrix-vector operation at $t = t^0, t^1, \ldots, t^f$.

The RTOC solution given by Eq. (3.59) is largely equivalent to Kalman's method if the Riccati equation is solved online. A numerical comparison between the two approaches for a specific application problem is discussed in [104].

Special Cases

1. Suppose we choose $\boldsymbol{P} = \boldsymbol{0}$; then, we have $\boldsymbol{\lambda}(t_f) = \boldsymbol{0}$ from the transversality condition. This also implies from Eq. (3.52) that $\boldsymbol{u}[@t_f] = \boldsymbol{0}$. Thus, a choice of $\boldsymbol{P} = \boldsymbol{0}$ automatically turns off the control at the end.

2. Suppose we set t_f to be free in addition to $\boldsymbol{P} = \boldsymbol{0}$. Then, from the Hamiltonian value condition (and the transversality condition) we get

$$\boldsymbol{x}^T(t_f)\boldsymbol{Q}(t_f)\boldsymbol{x}(t_f) = 0 \tag{3.66}$$

See Eq. (3.53). Because we required $\boldsymbol{Q}$ to be positive definite, Eq. (3.66) implies $\boldsymbol{x}(t_f) = \boldsymbol{0}$. That is, the optimal control naturally drives the system to its equilibrium point. Recall that $\boldsymbol{x}_f = \boldsymbol{0}$ was not imposed as an explicit requirement in the formulation of Problem LQ.

3. If $\boldsymbol{A}, \boldsymbol{B}, \boldsymbol{Q}$ and $\boldsymbol{R}$ are constants, then the Riccati equation is time-invariant in its coefficients

$$-\frac{d\boldsymbol{S}}{dt} = \boldsymbol{S}\boldsymbol{A} + \boldsymbol{A}^T\boldsymbol{S} - \boldsymbol{S}\boldsymbol{B}\boldsymbol{R}^{-1}\boldsymbol{B}^T\boldsymbol{S} + \boldsymbol{Q}$$

The equilibrium solution is obtained by setting $\dot{\boldsymbol{S}} = \boldsymbol{0}$:

$$\boldsymbol{S}\boldsymbol{A} + \boldsymbol{A}^T\boldsymbol{S} - \boldsymbol{S}\boldsymbol{B}\boldsymbol{R}^{-1}\boldsymbol{B}^T\boldsymbol{S} + \boldsymbol{Q} = \boldsymbol{0} \tag{3.67}$$

This equation is called the ***algebraic Riccati equation***. Let $\boldsymbol{S}_0$ be the solution to Eq. (3.67); then, Kalman[44] showed that under appropriate conditions $\boldsymbol{S}(t_f) \to \boldsymbol{S}_0$ for t_f sufficiently large.

The importance of Special Case # 3 is that the optimal feedback controller for the linear-quadratic time-invariant problem is given by a constant gain:

$$\boldsymbol{K} = -\boldsymbol{R}^{-1}\boldsymbol{B}^T\boldsymbol{S}_0$$

3.5 The Time-Optimal Double-Integrator

As apparent from Sections 3.3 and 3.4, there are many problems with quadratic cost functionals; see also Section 4.4.1 on page 266. Defining and formulating the "true" cost functional for many practical problems is not necessarily an easy task. Nonetheless, an important performance metric for many problems is the transfer time. Even when the transfer time is not a major consideration in evaluating the performance of an optimal control, it is still important to solve this problem to establish its lower bound. This lower bound then provides a quick way to eliminate infeasible requirements that may be inadvertently placed on the system. For instance, if a control system is required to respond quicker than the minimum time, then it is impossible to meet this requirement.

To illustrate some of the methodology in formulating the right problem, let us formulate a naïve minimum-time problem by simply replacing the cost functional of Problem DQC (discussed in Section 3.3 on page 192) to that of transfer time. This generates the following problem:

$$\boldsymbol{x} := (x, v) \quad \boldsymbol{u} := u \qquad \mathbb{U} = \mathbb{R}$$

$$\left\{ \begin{array}{lrl} \text{Minimize} & J[x(\cdot), v(\cdot), u(\cdot), t_f] = & t_f \\ \text{Subject to} & \dot{x} = & v \\ & \dot{v} = & u \\ & (x_0, v_0) = & (0, 0) \\ & (x_f - 1, v_f) = & (0, 0) \end{array} \right. \tag{3.68}$$

Because this is a minimum-time problem, the event function e does not contain the equality constraint $t_f - 1 = 0$ present in Problem DQC. Imposing this constraint would automatically imply a trivial or infeasible problem.

Computational Tip: In a computational environment like DIDO, it is very easy to add and remove constraints. As a result, a common mistake made by beginners and experts alike is inadvertently imposing constraints like $t_f - 1 = 0$ in a minimum-time problem. In such situations, DIDO will output that the problem as coded is infeasible (e.g., if the time constraint is smaller than the minimum time), or it will generate the solution $t_f = 1$ no matter what. See also Section 4.1.4 on page 253 for other common coding errors.

In any case, proceeding formally to apply Pontryagin's Principle (using the *HAMVET* mnemonic; see page 129) we generate the following:

1. Construct the Hamiltonian:

From the problem formulation, we have $\boldsymbol{x} \in \mathbb{R}^2$; hence, $\boldsymbol{\lambda} \in \mathbb{R}^2$. Choosing deliberate subscripts to tag each component of $\boldsymbol{\lambda}$ to the relevant state variable, we define

$$\boldsymbol{\lambda} := \begin{bmatrix} \lambda_x \\ \lambda_v \end{bmatrix}$$

The Hamiltonian is then given by

$$\begin{aligned} H(\boldsymbol{\lambda}, \boldsymbol{x}, u) &:= \overbrace{0}^{F} + \overbrace{\begin{bmatrix} \lambda_x & \lambda_v \end{bmatrix}}^{\boldsymbol{\lambda}^T} \overbrace{\begin{bmatrix} v \\ u \end{bmatrix}}^{\boldsymbol{f}} \\ &= \lambda_x v + \lambda_v u \end{aligned}$$

2. Develop the Adjoint Equations:

$$-\dot{\lambda}_x := \frac{\partial H}{\partial x} = 0 \qquad \Rightarrow \lambda_x = a, \tag{3.69a}$$

$$-\dot{\lambda}_v := \frac{\partial H}{\partial v} = \lambda_x \qquad \Rightarrow \lambda_v = -at - b \tag{3.69b}$$

where a and b are constants of integration.

3. Minimize the Hamiltonian (with respect to u only):

The static problem, $\min_{u \in \mathbb{R}} H(\boldsymbol{\lambda}, \boldsymbol{x}, u)$, is unconstrained ($\mathbb{U} = \mathbb{R}$); hence, the extremal control is given by

$$\frac{\partial H}{\partial u} = 0 \quad \Rightarrow \lambda_v = 0 \tag{3.70}$$

That is, the Hamiltonian Minimization Condition produces no solution for the control. This essentially means that we have a bad problem formulation. Since the Hamiltonian is linear with respect to the control u, one could have easily guessed (see Fig. 3.21) that it takes on the minimum value of $-\infty$ at either $u =$

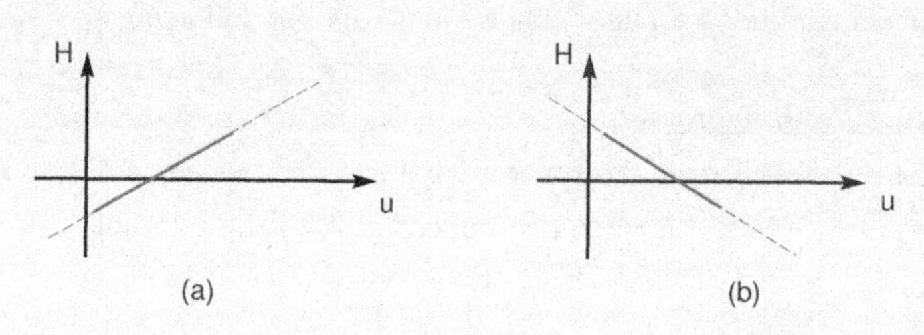

Figure 3.21: The Hamiltonian function $(u \mapsto H(\boldsymbol{\lambda}, \boldsymbol{x}, u))$ for the problem given by Eq. (3.68) is unbounded because u is unbounded.

$-\infty$ (as in Fig. 3.21(a)) or at $u = +\infty$ (as in Fig. 3.21(b)). This implies that we cannot solve the problem with $\mathbb{U} = \mathbb{R}$ (i.e., unbounded u) and must constrain the control space, $\mathbb{U}$. *This is yet another illustration of how Pontryagin's Principle provides clues for sniffing out a bad problem formulation.*

Study Problem 3.16

Why did $\mathbb{U} = \mathbb{R}$ not create a feasibility problem in Problem DQC?

Tech Talk: One could have guessed that u must be impulses in the preceding problem. Although this is technically correct, recall that $\boldsymbol{u}(\cdot)$ is Lebesgue measurable in Pontryagin's Principle. To allow u to be impulses, we need to expand the space of admissible functions to incorporate atomic measures. A simpler approach might be to choose a "homotopy" path where $\mathbb{U}$ is compact and examine the sequence of solutions resulting from a relaxation of $\mathbb{U}$ all the way up to a limiting case of $\mathbb{U} = \mathbb{R}$. This approach was adopted by Lawden[57] in developing his *primer vector theory* for space flight.

Recognizing that the problem given by Eq. (3.68) was badly posed, we reformulate it to include a control "saturation" constraint, $|u| \leq 1$. In addition, since we also wish to seek feedback solutions by way of the RTOC principle, we pose this problem as that of transferring the state of the system from an arbitrary given point (rather than $(0,0)$) to a rest position in minimum time; thus, we have:

$$\boldsymbol{x} := (x, v) \quad \boldsymbol{u} := u \quad \mathbb{U} = \{u : |u| \leq 1\}$$

$$(\text{DMT}) \begin{cases} \text{Minimize} & J[x(\cdot), v(\cdot), u(\cdot), t_f] = t_f \\ \text{Subject to} & \dot{x} = v \\ & \dot{v} = u \\ & (x_0, v_0) = (x^0, v^0) \\ & (x_f, v_f) = (0, 0) \end{cases} \tag{3.71}$$

Note that we have chosen $x_f = 0$ and not $x_f - 1 = 0$, which is a valid problem formulation since we left x_0 arbitrary. There is no loss of generality here since if $x_0 = 0 = v_0$, the problem is automatically solved. This problem formulation is essentially the problem of driving the system to the origin from some arbitrary initial conditions.

It is easy to verify that Steps *1.* and *2.* are the same as before; hence, continuing on to a modified Step *3.*, we get:

3. Minimize the Hamiltonian:

$$(HMC) \begin{cases} \underset{u}{\text{Minimize}} & H(\boldsymbol{\lambda}, \boldsymbol{x}, u) := \lambda_x v + \lambda_v u \\ \text{Subject to} & u \in \mathbb{U} := \{u \in \mathbb{R} : -1 \leq u \leq 1\} \end{cases}$$

This is a box-constrained minimization problem; see Sec. 2.5, and in particular, Eqs. (2.37) and (2.38) on page 118.

The KKT conditions for this problem are obtained by constructing the Lagrangian for Problem HMC. Because u is a scalar, we have the path covector $\boldsymbol{\mu} := \mu$; hence, the Lagrangian of the Hamiltonian is given by

$$\overline{H}(\mu, \boldsymbol{\lambda}, \boldsymbol{x}, u) = \lambda_x v + \lambda_v u + \mu u$$

From the stationarity and complementarity conditions, we have

$$\frac{\partial \overline{H}}{\partial u} = \lambda_v + \mu = 0 \tag{3.72a}$$

$$\mu \begin{cases} \leq 0 & \textit{if} \quad u = -1 \\ = 0 & \textit{if} \quad -1 < u < 1 \\ \geq 0 & \textit{if} \quad u = 1 \end{cases} \tag{3.72b}$$

Although the stationarity condition provides no information on choosing u, the complementarity condition does; it can be rewritten as

$$u = \begin{cases} -1 & \textit{if} \quad \mu \leq 0 \\ \text{any value } \in (-1, +1) & \textit{if} \quad \mu = 0 \\ +1 & \textit{if} \quad \mu \geq 0 \end{cases} \tag{3.73}$$

Equation (3.73) can be visualized as shown in Fig. 3.22: At each instant of time t, we check the sign of $\mu(t)$ and use formula (3.73) to determine $u(t)$. When $\mu(t) = 0$, the control switches ... and in the process takes all values between $(-1, +1)$. This "interpretation" of a switch is consistent with ***nonsmooth calculus***[21, 23], and no handwringing is necessary to define the value of a function at a switch point. In this spirit, we can write the application of Eq. (3.73) for all time in the compact form

$$u(t) = \text{sgn}(\mu(t)) \tag{3.74}$$

where sgn is the sign function (consistent with Eq. (3.73)).

The control given by Eq. (3.74) is called a ***bang-bang control*** — that is, a controller that operates at its extremes: full-on forward, $u(t) = 1$ or full-on reverse, $u(t) = -1$. In this case, the function $t \mapsto \mu$ is called a ***switching function***, a term quite apparent from Fig. 3.22.

Notice that if the control space were defined by the binary set $\mathbb{U} = \{-1, 1\}$,

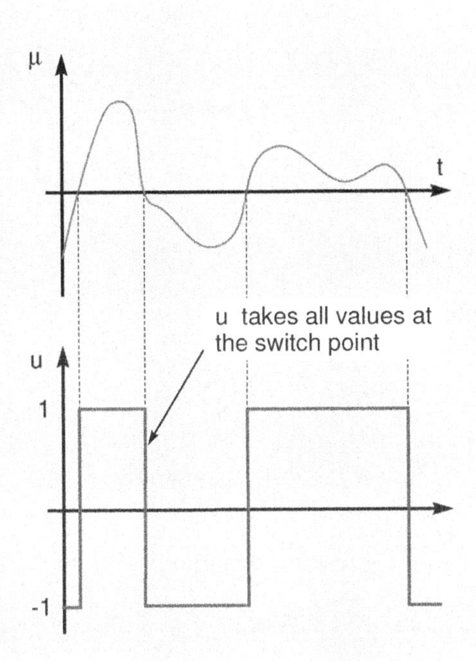

Figure 3.22: Illustrating one possible outcome of applying Eq. (3.73) at each instant of time.

the result would be unchanged! Pontryagin's Principle applies even in this case: Problem HMC would be addressed by integer programming (as opposed to nonlinear programming). The bang-bang solution follows simply by inspection and enumeration of the discrete HMC.

4. Evaluate the Hamiltonian Value Condition:

By inspection, the endpoint function $\boldsymbol{e}$ is given by

$$\boldsymbol{e}(\boldsymbol{x}_f, t_f) := \begin{bmatrix} x_f \\ v_f \end{bmatrix}$$

Hence, we must choose $\boldsymbol{\nu} \in \mathbb{R}^2$ to construct the endpoint Lagrangian:

$$\overline{E}(\boldsymbol{x}_f, t_f) := \overbrace{t_f}^{E} + \overbrace{\begin{bmatrix} \nu_1 & \nu_2 \end{bmatrix}}^{\boldsymbol{\nu}^T} \overbrace{\begin{bmatrix} x_f \\ v_f \end{bmatrix}}^{e}$$

$$= t_f + \nu_1 x_f + \nu_2 v_f$$

The Hamiltonian Value Condition is now given by

$$\mathcal{H}[@t_f] = -\frac{\partial \overline{E}}{\partial t_f} = -1 \tag{3.75}$$

This is a standard condition for minimum-time problems.

5. *Analyze the Hamiltonian Evolution Equation:*

Because

$$\frac{\partial H}{\partial t} = 0$$

we have

$$\dot{\mathcal{H}}(\boldsymbol{\lambda}, \boldsymbol{x}) = 0$$

Hence, the minimized (or lower) Hamiltonian is a constant with respect to time. The function $\mathcal{H}$ can be obtained as follows:

- First produce $\boldsymbol{u}$ as a function of $\boldsymbol{\lambda}$ by eliminating $\boldsymbol{\mu}$ from the KKT conditions associated with Problem HMC. From Eq. (3.72a) we have $\mu = -\lambda_v$. Substituting this in Eq. (3.74) we get

$$u(t) = \operatorname{sgn}(-\lambda_v(t)) = -\operatorname{sgn}(\lambda_v(t)) \tag{3.76}$$

- Substitute Eq. (3.76) in the expression for H:

$$\begin{aligned} \mathcal{H}(\lambda_x, \lambda_v, x, v) &:= H(\lambda_x, \lambda_v, x, v, -\operatorname{sgn}(\lambda_v)) \\ &= \lambda_x v - \lambda_v \operatorname{sgn}(\lambda_v) \end{aligned}$$

Constancy of $\mathcal{H}$ implies that the function $t \mapsto \lambda_x(t)v(t) - \lambda_v(t)\operatorname{sgn}(\lambda_v(t))$ is a constant. The value of this constant is -1 by virtue of Eq. (3.75).

Note also that $\mathcal{H}$ is a nonsmooth function; hence, it is not differentiable (in smooth calculus). In contrast, the Pontryagin Hamiltonian, H, is differentiable.

6. Determine the Transversality Conditions:

The terminal transversality conditions are obtained by simply differentiating the endpoint Lagrangian:

$$\lambda_x(t_f) := \frac{\partial \overline{E}}{\partial x_f} = \nu_1$$

$$\lambda_v(t_f) := \frac{\partial \overline{E}}{\partial v_f} = \nu_2$$

Obviously, we get no new *analytical* information (although it does provide useful computational information for V & V, as discussed in Section 3.2.1).

The Fine Print ... In Case You Missed It:

We can also legitimately interpret Eq. (3.70) as a condition that demands the Hamiltonian be "horizontal", as illustrated in Fig. 3.23. In this case, u can take any value. This condition is known as "singular," and the resulting control

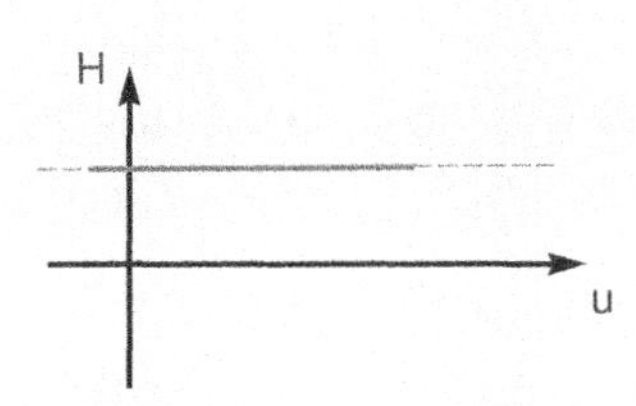

Figure 3.23: If the Hamiltonian function ($u \mapsto H(\boldsymbol{\lambda}, \boldsymbol{x}, u)$) is "horizontally flat," then u can take any value.

that makes the Hamiltonian "horizontal" is called a ***singular control***. This situation can also happen in Fig. 3.22 if μ is zero over a nonzero time interval (as opposed to merely crossing the t-axis). Both cases are characterized by the condition

$$\frac{\partial H}{\partial u} = 0 \qquad \forall\, t \in (t_1, t_2) \subseteq [t_0, t_f]$$

Because this equality is presumed true for all t in some nonzero time interval, we have

$$\frac{d^k}{dt^k}\left(\frac{\partial H}{\partial u}\right) = 0 \qquad \forall\, k = 0, 1, 2, \ldots \tag{3.77}$$

Equation (3.77) can be used to find u if it exists. Applying this condition to Problem DMT, we get

$$\begin{aligned}
& & \lambda_v &= 0 & &(k = 0)\\
& & \dot{\lambda}_v &= 0 & &(k = 1)\\
&\Rightarrow & \lambda_x &= 0 & &\text{(from Eq. (3.69))}\\
&\Rightarrow & H(\boldsymbol{\lambda}, \boldsymbol{x}, u) &\equiv 0 & &\\
&\Rightarrow & \mathcal{H}(\boldsymbol{\lambda}, \boldsymbol{x}) &\equiv 0 & &(\text{and hence}, \neq -1)
\end{aligned}$$

As a result of this contradiction, we can conclude that a singular control is not possible for Problem DMT.

Although the possibility of a singular control is eliminated for this problem by elementary arguments, its exclusion or inclusion requires the application of a "higher order" version of Pontryagin's Principle formulated by Krener[55]. The possibility of a singular control is not as rare as is sometimes implied in the literature. See Section 4.4.4 on page 272 for a simple example.

Incidentally, the absence of a singular control can also be obtained directly (i.e., without invoking Eq. (3.77)) for this particular example problem as follows: Because Eq. (3.70) is assumed to be true for the nonzero time interval (t_1, t_2), we have

$$\begin{aligned}
& & \lambda_v(t) &= 0 & &\forall\, t \in (t_1, t_2)\\
&\Rightarrow & a &= 0 = b & &\text{(from Eq. (3.69))}\\
&\Rightarrow & \lambda_x(t) &= 0 & &\\
&\Rightarrow & H(\boldsymbol{\lambda}, \boldsymbol{x}, u) &\equiv 0 & &
\end{aligned}$$

Generating a Feedback Solution via RTOC:

To generate a feedback controller using the same procedure as the one described in Section 3.3, we need to produce an analytical solution to the differential equation

$$\dot{v}^*(t) = \operatorname{sgn}(at + b) \tag{3.78}$$

Care must be taken to integrate this differential equation since it has a discontinuous right-hand side. Integrating Eq. (3.78) twice, we get

$$
\begin{aligned}
v^*(t) &= \frac{|at+b|}{a} + c \\
x^*(t) &= \frac{(at+b)\,|at+b|}{2a^2} + ct + d
\end{aligned}
$$

Not only are these equations nonlinear (in the coefficients a and b), but they are also ***nonsmooth***. Hence, finding the coefficients, a, b, c, d and final time t_f, for arbitrary values of x_0, v_0 and t_0 requires solving the nonsmooth equations

$$
\begin{aligned}
\frac{(at_0+b)\,|at_0+b|}{2a^2} + ct_0 + d &= x_0 &\quad (3.79a)\\
\frac{|at_0+b|}{a} + c &= v_0 &\quad (3.79b)\\
\frac{(at_f+b)\,|at_f+b|}{2a^2} + ct_f + d &= 0 &\quad (3.79c)\\
\frac{|at_f+b|}{a} + c &= 0 &\quad (3.79d)\\
|at_f+b| &= 1 &\quad (3.79e)
\end{aligned}
$$

where the last equation follows from the Hamiltonian value condition (Eq. (3.75)). In principle, a feedback solution can be obtained by solving Eq. (3.79) in real time. This generates the function

$$(t_0, x_0, v_0) \mapsto (a, b)$$

and an RTOC feedback controller can be implemented in much the same way as shown in Fig. 3.16 on page 199.

While it is conceivable that we can produce good algorithms to solve Eq. (3.79) in real time, it is instructive to discuss the far simpler approach taken by Pontryagin et al[75].

Pontryagin's Solution:

Since the system at hand is only two-dimensional, Pontryagin et al[75] provided a feedback solution to this problem via a graphical approach. They noted that because the optimal controls are either $u = +1$ or $u = -1$, the entire collection of extremals can be characterized as parabolas in the two-dimensional state space $\mathbb{X} = \mathbb{R}^2$; see Table 3.1 and Fig. 3.24.

$u = +1$	$u = -1$
$\ddot{x} = 1$	$\ddot{x} = -1$
$v = t + a_1$	$v = -t - b_1$
$x = t^2/2 + a_1 t + a_2$	$x = -t^2/2 - b_1 t - b_2$
$v^2 = 2x + c_1$	$v^2 = -2x + c_2$

Table 3.1: Extremals for Problem DMT.

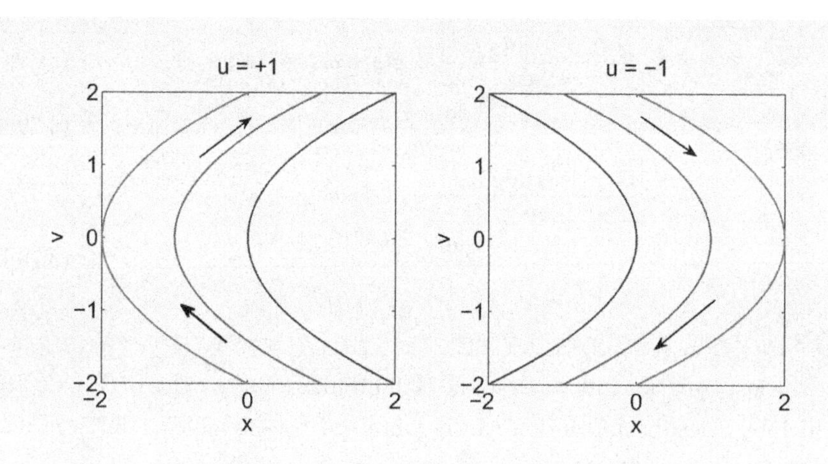

Figure 3.24: Extremals for Problem DMT.

There are only two parabolas that pass through the target point $(x_f, v_f) = (0, 0)$, and only half of each parabola approaches the origin; the other halves depart from it as shown in Fig. 3.25. Thus, the final segment of the optimal trajectory must be on the curve AOB. *Now, because $\lambda_v(t)$ is a linear function of time, the control has at most one switch.* Consequently, if the initial point (x_0, v_0) is not on AOB, it must traverse the appropriate parabola to get to AOB and switch at the junction point to ride AOB to the origin. This notion is illustrated in Fig. 3.26 for two cases:

- $(x_0, v_0) = (-1, -1)$ and
- $(x_0, v_0) = (1, 1)$

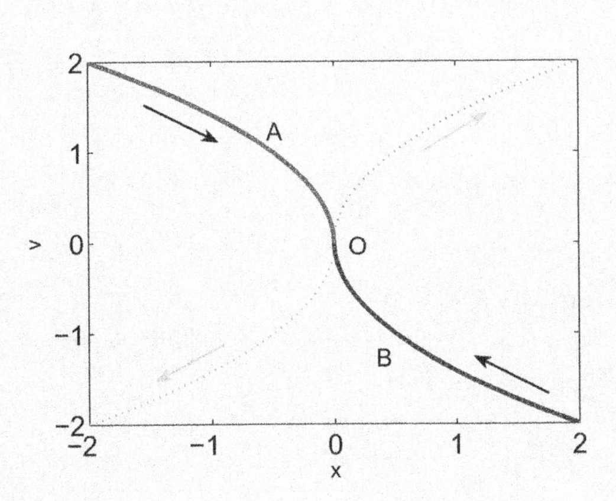

Figure 3.25: The final segment of the optimal trajectory must lie on the curve AOB.

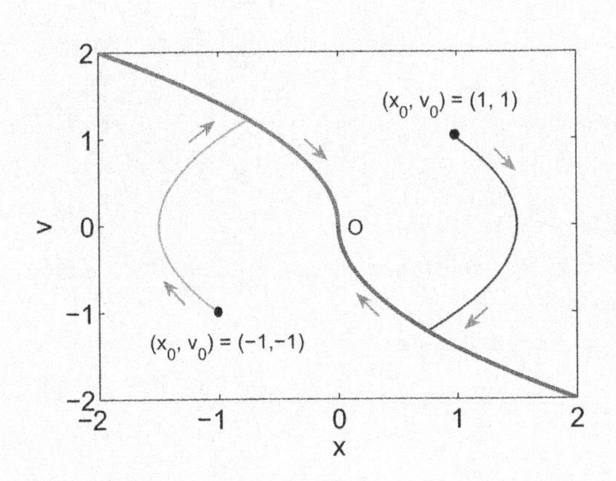

Figure 3.26: Optimal trajectories for two sample initial conditions.

The feedback control for implementing this solution can be generated as follows: It can be easily verified that the equation $v^2\text{sgn}(v) + 2x = 0$ defines AOB; that is,

$$AOB = \left\{(x, v) \in \mathbb{R}^2 : \ v^2\text{sgn}(v) + 2x = 0\right\} \tag{3.80}$$

The feedback solution is then given by the ***nonsmooth control law***

$$u = -\mathrm{sgn}(S(x,v)) \quad \textit{where} \quad S(x,v) := v^2\mathrm{sgn}(v) + 2x \tag{3.81}$$

Since $v\,\mathrm{sgn}(v) = |v|$, we can also write $S(x,v) = v\,|v| + 2x$.

Study Problem 3.17

Except for a dissipative term, the following problem is identical to Problem DMT.

$$\boldsymbol{x} := (x,v) \in \mathbb{R}^2, \quad \boldsymbol{u} := u \in \mathbb{U} := \{u \in \mathbb{R} : |u| \le 1\}$$

$$\left\{\begin{array}{lrl} \text{Minimize} & J[x(\cdot),v(\cdot),u(\cdot),t_f] = & t_f \\ \text{Subject to} & \dot{x} = & v \\ & \dot{v} = & -v + u \\ & (x_0, v_0) = & (x^0, v^0) \\ & (x_f, v_f) = & (0,0) \end{array}\right.$$

Develop both open and closed-loop solutions to this problem.

Study Problem 3.18

Suppose that the cost function in Problem DMT is replaced by the L^1-norm of the control,

$$\int_{t_0}^{t_f} |u(t)|\, dt$$

1. *Discuss why the problem is not well-posed (Hint: can t_f be free?)*
2. *Show that the optimal control for this problem is bang-off-bang.*

3.6 The Powered Explicit Guidance Problem

In space applications, a control system usually implies an attitude control system. An attitude control system is considered to be the inner-loop of a *guidance and control* system; see Section 1.5.1 (page 67). For inner loops, PD or Kalman controllers work extremely well. This is, in part, because, as shown in Sections 3.3 and 3.4, a PD controller can be a near-optimal feedback controller for a double integrator (see Section 3.3.2 page 201), while a Kalman controller is optimal for linear systems (with quadratic cost functionals). The techniques developed in these examples to obtain feedback solutions are not "portable" to guidance problems because of nonlinear dynamics and non-quadratic cost functionals. A standard engineering trick to address this issue is to solve a simplified problem and explore its effectiveness on the "full" problem by extensive simulations (e.g., Monte Carlo simulations). The expectation is that feedback would take care of the model mismatches in addition to plant uncertainties and external disturbances. This approach has been used by aerospace engineers quite effectively since the 1960s. In this context, consider the schematic shown in Fig. 3.27 for a fundamental exo-atmospheric space guidance problem.

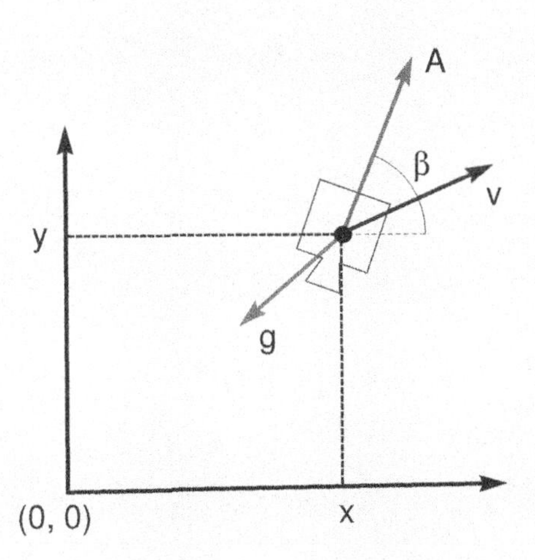

Figure 3.27: Schematic for a fundamental space vehicle guidance problem.

The dynamics of this system are given by

$$\ddot{x} = g_x + A\cos\beta, \qquad \ddot{y} = g_y + A\sin\beta$$

where A is the thrust acceleration, β is the thrust steering angle, and g_x and g_y are *constant* gravitational accelerations in the x- and y-directions. Both A and β are the outer-loop control variables.

Because propellant is a premium for space applications, the driving cost functional for engineering feasibility is the magnitude (L^1-norm) of the acceleration. Thus, a basic space guidance problem can be posed (in the format of Problem B) as

$$\boldsymbol{x} = (x, y, v_x, v_y) \in \mathbb{R}^4 \quad \boldsymbol{u} = (A, \beta)$$

$$\mathbb{U} = \left\{\boldsymbol{u} \in \mathbb{R}^2 : 0 \le A \le A_{max}\right\}$$

$$\left\{\begin{array}{lrl} \text{Minimize} & J[\boldsymbol{x}(\cdot), \boldsymbol{u}(\cdot), t_f] & = \displaystyle\int_{t_0}^{t_f} A(t)\, dt \\ \text{Subject to} & \dot{x} & = v_x \\ & \dot{y} & = v_y \\ & \dot{v}_x & = g_x + A\cos\beta \\ & \dot{v}_y & = g_y + A\sin\beta \\ & t_0 & = t^0 \\ & \boldsymbol{x}_0 & = (x^0, y^0, v_x^0, v_y^0) \\ & \boldsymbol{x}_f - \left(x^f, y^f, v_x^f, v_y^f\right) & = \boldsymbol{0} \\ & t_f - t^f & = 0 \end{array}\right. \tag{3.82}$$

This problem is nonlinear because of the trigonometric terms; see Section 4.4.2 on page 268 for a linear version of this problem.

Following the *HAMVET* mnemonic (see page 129) to analyze this problem, we begin by constructing the Hamiltonian:

1. Construct the Hamiltonian:

Because $\boldsymbol{x} \in \mathbb{R}^4$, we have $\boldsymbol{\lambda} \in \mathbb{R}^4$. Choosing deliberate subscripts on the components of $\boldsymbol{\lambda}$

$$\boldsymbol{\lambda} := \begin{bmatrix} \lambda_x \\ \lambda_y \\ \lambda_{v_x} \\ \lambda_{v_y} \end{bmatrix}$$

we tag each costate to the relevant state variable. The Hamiltonian is then given by

$$\begin{aligned} H(\boldsymbol{\lambda}, \boldsymbol{x}, \boldsymbol{u}) &:= \overbrace{A}^{F} + \overbrace{\begin{bmatrix} \lambda_x & \lambda_y & \lambda_{v_x} & \lambda_{v_y} \end{bmatrix}}^{\boldsymbol{\lambda}^T} \overbrace{\begin{bmatrix} v_x \\ v_y \\ g_x + A\cos\beta \\ g_y + A\sin\beta \end{bmatrix}}^{\boldsymbol{f}} \\ &= A + \lambda_x v_x + \lambda_y v_y + \lambda_{v_x} A\cos\beta + \lambda_{v_y} A\sin\beta + \lambda_{v_x} g_x + \lambda_{v_y} g_y \end{aligned}$$

2. Develop the Adjoint Equations:

$$-\dot{\lambda}_x = \frac{\partial H}{\partial x} = 0 \quad \Rightarrow \quad \lambda_x = a \tag{3.83a}$$

$$-\dot{\lambda}_y = \frac{\partial H}{\partial y} = 0 \quad \Rightarrow \quad \lambda_y = b \tag{3.83b}$$

$$-\dot{\lambda}_{v_x} = \frac{\partial H}{\partial v_x} = \lambda_x \quad \Rightarrow \quad \lambda_{v_x} = -at - c \tag{3.83c}$$

$$-\dot{\lambda}_{v_y} = \frac{\partial H}{\partial v_y} = \lambda_y \quad \Rightarrow \quad \lambda_{v_y} = -bt - d \tag{3.83d}$$

3. Minimize the Hamiltonian (with respect to $\boldsymbol{u}$ only):

$$(HMC) \begin{cases} \underset{A,\beta}{\text{Minimize}} & H(\boldsymbol{\lambda}, \boldsymbol{x}, [A, \beta]) \\ \text{Subject to} & 0 \le A \le A_{max} \end{cases}$$

Study Problem 3.19

1. *Minimize H with respect to A and show that the candidate optimal thrust acceleration is either 0 or A_{max}. (Hint: See Section 3.5, particularly the discussions on page 222.)*
2. *Determine the number of switches in the candidate optimal control trajectory $t \mapsto A(t)$. (Hint: Differentiate the switching function and examine its sign.)*

The minimization of H with respect to β is an unconstrained minimization problem; hence, from $\partial_\beta H = 0$, we have

$$A\left(-\lambda_{v_x} \sin\beta + \lambda_{v_y} \cos\beta\right) = 0$$

If $A \neq 0$, we have

$$\lambda_{v_x} \sin\beta - \lambda_{v_y} \cos\beta = 0$$

In addition, if $\lambda_{v_x} \neq 0$, we have

$$\tan\beta = \frac{\lambda_{v_y}}{\lambda_{v_x}} \tag{3.84}$$

Study Problem 3.20

Show that in addition to Eq. (3.84), a necessary condition for optimality is

$$\frac{\lambda_{v_y}}{\sin\beta} \leq 0 \qquad (\textit{for } \beta \neq 0)$$

(Hint: Consider $\partial^2_\beta H$.)

4. Evaluate the Hamiltonian Value Condition:

By inspection, the endpoint function $\boldsymbol{e}$ is given by

$$\boldsymbol{e}(\boldsymbol{x}_f, t_f) := \begin{bmatrix} x_f - x^f \\ y_f - y^f \\ v_{x_f} - v_x^f \\ v_{y_f} - v_y^f \\ t_f - t^f \end{bmatrix}$$

Hence, we must choose $\boldsymbol{\nu} \in \mathbb{R}^5$ to construct the endpoint Lagrangian:

$$\overline{E}(\boldsymbol{x}_f, t_f) := \overbrace{0}^{E} + \overbrace{\begin{bmatrix} \nu_1 & \nu_2 & \nu_3 & \nu_4 & \nu_5 \end{bmatrix}}^{\boldsymbol{\nu}^T} \overbrace{\begin{bmatrix} x_f - x^f \\ y_f - y^f \\ v_{x_f} - v_x^f \\ v_{y_f} - v_y^f \\ t_f - t^f \end{bmatrix}}^{\boldsymbol{e}}$$

$$= \nu_1(x_f - x^f) + \nu_2(y_f - y^f) + \nu_3(v_{x_f} - v_x^f) + \nu_4(v_{y_f} - v_y^f) + \nu_5(t_f - t^f)$$

The Hamiltonian Value Condition is now given by

$$\mathcal{H}[@t_f] = -\frac{\partial \overline{E}}{\partial t_f} = -\nu_5$$

Clearly, it provides no useful analytical information; however, recall that it provides valuable computational information for V & V (see Section 3.2.1).

5. Analyze the Hamiltonian Evolution Equation:

Because

$$\frac{\partial H}{\partial t} = 0$$

we have

$$\dot{\mathcal{H}} = 0$$

Hence, the minimized (or lower) Hamiltonian is a constant with respect to time.

Study Problem 3.21

Produce the function $\mathcal{H}(\boldsymbol{\lambda}, \boldsymbol{x})$ using the results from Study Problem 3.19 and Eq. (3.84).

6. Determine the Transversality Conditions:

The terminal transversality conditions are obtained by simply differentiating the endpoint Lagrangian:

$$\lambda_x(t_f) := \frac{\partial \overline{E}}{\partial x_f} = \nu_1 \qquad \lambda_y(t_f) := \frac{\partial \overline{E}}{\partial y_f} = \nu_2$$
$$\lambda_{v_x}(t_f) := \frac{\partial \overline{E}}{\partial v_{x_f}} = \nu_3 \qquad \lambda_{v_y}(t_f) := \frac{\partial \overline{E}}{\partial v_{y_f}} = \nu_4$$

Obviously, we get no new *analytical* information (although it does provide useful computational information for V & V, as discussed in Section 3.2.1).

Generating a Feedback Solution via RTOC:

The procedure for generating a feedback solution to this problem — and along more general lines than discussed here — was developed in the classic 1979 paper by McHenry et al[65]. It turns out that their procedure, called ***powered explicit guidance, or PEG***, is a specific application of RTOC. What is more impressive about the ensuing results is that PEG was implemented onboard the Space Shuttle, and with just a few KBs of RAM!

The details of PEG viewed through the prism of RTOC are as follows: Substituting Eqs. (3.83c) and (3.83d) in Eq. (3.84), we get

$$\tan\beta(t) = \frac{bt + d}{at + c} \tag{3.85}$$

This equation is the well-known[16] *bilinear tangent law.*

Study Problem 3.22

1. *Show that if $x(t_f)$ was unspecified (i.e., x_f is free), the bilinear*

tangent law simplifies to a ***linear tangent law:***

$$\tan\beta(t) = c_1 t + c_2 \tag{3.86}$$

2. *Using Eq. (3.86), evaluate the following integrals:*

$$\begin{aligned} I_1(c_1, c_2, t_0, t) &:= \int_{t_0}^{t} \cos\beta(\tau)\, d\tau \\ I_2(c_1, c_2, t_0, t) &:= \int_{t_0}^{t} \sin\beta(\tau)\, d\tau \\ I_3(c_1, c_2, t_0, t) &:= \int_{t_0}^{t} \left(\int_{t_0}^{\tau} \sin\beta(s)\, ds \right) d\tau \end{aligned} \tag{3.87}$$

The integrals indicated in Eq. (3.87) are carried out in [65] through a combination of careful assumptions and quadrature techniques. These integrals are used as follows: Over periods where $t \mapsto A$ is a constant, $v_x(t), v_y(t)$ and $y(t)$ can be written as

$$v_x(t) = v_x(t_0) + g_x(t - t_0) + A \underbrace{\int_{t_0}^{t} \cos\beta(\tau)\, d\tau}_{I_1(c_1, c_2, t_0, t)} \tag{3.88a}$$

$$v_y(t) = v_y(t_0) + g_y(t - t_0) + A \underbrace{\int_{t_0}^{t} \sin\beta(\tau)\, d\tau}_{I_2(c_1, c_2, t_0, t)} \tag{3.88b}$$

$$\begin{aligned} y(t) &= y(t_0) + \int_{t_0}^{t} v_y(\tau)\, d\tau \\ &= y(t_0) + v_y(t_0)(t - t_0) + \frac{g_y}{2}(t - t_0)^2 + A I_3(c_1, c_2, t_0, t) \end{aligned} \tag{3.88c}$$

It is possible to separately produce a "shutdown algorithm," to determine the switch points for A by using the final-time conditions[65]; see also Study Problem 3.19 on page 234. Substituting these "burn-out" conditions in Eq. (3.88) generates the following system of nonlinear equations for the key unknowns c_1

and c_2

$$v_x^b - v_x(t_0) - g_x(t^b - t_0) - AI_1(c_1, c_2, t_0, t^b) = 0 \qquad (3.89a)$$
$$v_y^b - v_y(t_0) - g_y(t^b - t_0) - AI_2(c_1, c_2, t_0, t^b) = 0 \qquad (3.89b)$$
$$y^b - y(t_0) - v_y(t_0)(t^b - t_0) - \frac{g_y}{2}(t^b - t_0)^2 - AI_3(c_1, c_2, t_0, t^b) = 0 \qquad (3.89c)$$

where the superscript b refers to the burn-out conditions. Thus, we can generate the map

$$\Big(v_x(t_0), v_y(t_0), y(t_0), t_0\Big) \mapsto (c_1, c_2)$$

by solving Eq. (3.89). This is done using a predictor-corrector algorithm as part of the PEG formulation[65]. The details are a little different than the ones described here; however, the essence of the concepts is the same. Replacing t_0 by the current time t, we arrive at a feedback solution for the linear tangent law:

$$\tan\beta = c_1\Big(v_x(t), v_y(t), y(t), t\Big)t + c_2\Big(v_x(t), v_y(t), y(t), t\Big) \qquad (3.90)$$

A basic schematic for PEG as an RTOC box is shown in Fig. 3.28.

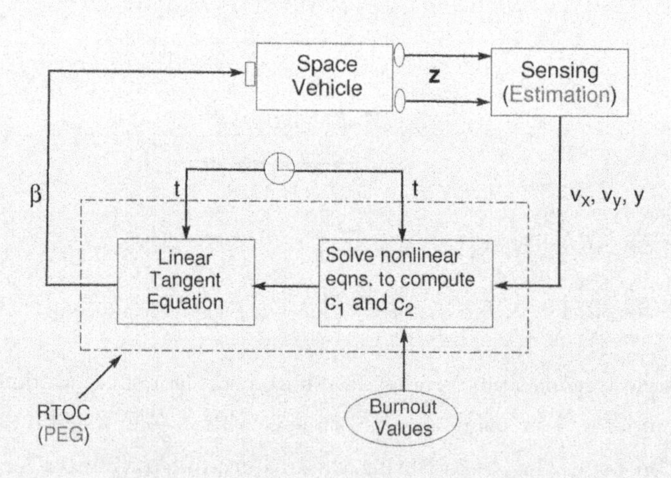

Figure 3.28: Basic block diagram for Powered Explicit Guidance (PEG). The burn-out conditions flow in from an engine shutdown algorithm; see Study Problem 3.19 on page 234.

Recall that β is not really a control variable and that the guidance system is an outer loop of an inner-loop system. Taking these issues into account, a schematic for PEG as an outer loop of a guidance and control system is shown in Fig. 3.29.

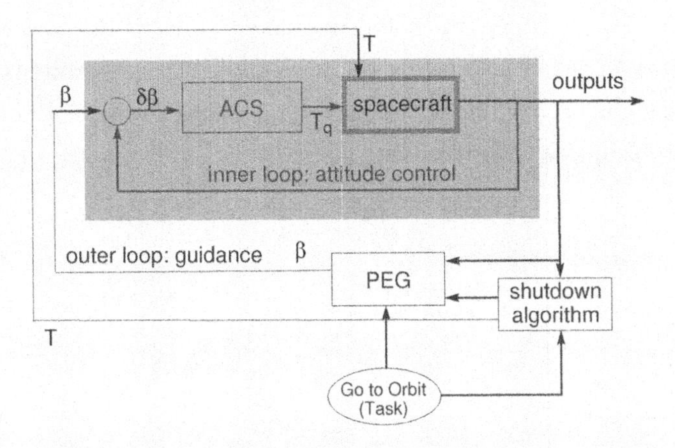

Figure 3.29: PEG is the outer or guidance loop; the inner loop is the attitude control system (ACS) that may be a PD controller.

Study Problem 3.23

In the actual PEG algorithm[65], gravity is not assumed to be a constant, but the linear tangent law is still used. Investigate the implications of using an inverse-square gravity model. Can you produce an algorithm that can run in real time (and in under 64KB as was done onboard the Space Shuttle)? No, this is not a typo; it is Kilo Bytes!

3.7 A Myth-Busting Example Problem

In the 1990s, there was widespread belief that "indirect" methods were more accurate than "direct" methods. Refer back to Section 2.9.2 on page 157 for a recalibration of perspectives on so-called direct and indirect methods. As discussed in Section 2.9.2, this belief was largely due to the absence of the *covector mapping principle* (which was discovered in 2001). Despite facts to the contrary, this folklore was hard to shake. As a means to dispel the myth, the following problem was introduced and discussed in [77]:

$$\boldsymbol{x} := x \in \mathbb{R} \quad \boldsymbol{u} := u \in \mathbb{U} := \{0, 1\}$$

$$(MB) \begin{cases} \text{Minimize} & J[x(\cdot), u(\cdot)] := \dfrac{(x_f - r)^2}{2} \\ \text{Subject to} & \dot{x}(t) = u(t) \\ & (x_0, t_0) = (0, 0) \\ & t_f = 1 \end{cases}$$

where r is a positive irrational number less than 1. Problem MB is an inspired "smooth" modification of a counterexample first designed in [71].

After applying Pontryagin's Principle, we arrive at the BVP:

$$\boldsymbol{x} := x \in \mathbb{R} \quad \boldsymbol{\lambda} := \lambda \in \mathbb{R}$$

$$(MB^{\lambda}) \begin{cases} \text{Find} & [\boldsymbol{x}(\cdot), \boldsymbol{\lambda}(\cdot)] \\ \text{Such that} & \dot{x}(t) = \begin{cases} 0 & if \quad \lambda(t) > 0 \\ 1 & if \quad \lambda(t) < 0 \\ 0 \ or \ 1 & if \quad \lambda(t) = 0 \end{cases} \\ & \dot{\lambda}(t) = 0 \\ & (x_0, t_0) = (0, 0) \\ & t_f = 1 \\ & \lambda(t_f) = x_f - r \end{cases}$$

Study Problem 3.24

Apply Pontryagin's Principle to Problem MB and show that it generates Problem MB^{λ}.

From the adjoint and transversality equations, we have

$$\lambda(t) = x(t_f) - r, \quad \text{a constant}$$

By examining the three possibilities $x_f \lesseqgtr r$, it is apparent that we cannot get an extremal state unless $x_f = r$. Thus, any feasible solution that satisfies $x(t_f) = r$ is an extremal.

Study Problem 3.25

Explain why the nontriviality condition is not violated in the application of the Pontryagin's Principle to generate the solutions shown in Fig. 3.30.

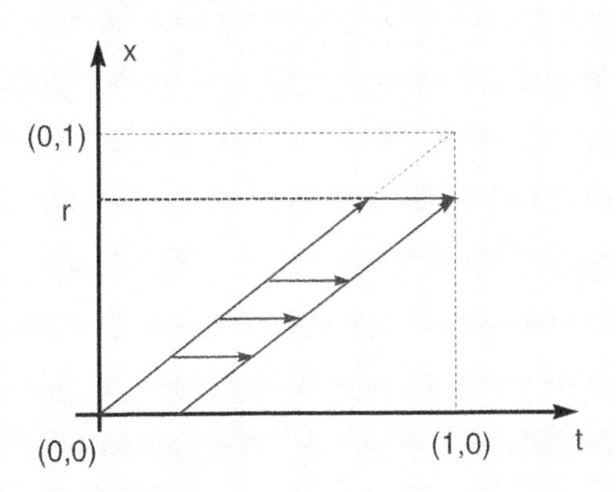

Figure 3.30: Illustrating several of the infinite set of globally optimal solutions to Problem MB.

By inspecting the cost functional it is obvious that all extremals are globally optimal. Some of the infinite set of globally optimal solutions is shown in Fig. 3.30.

Let us now apply an "indirect" method to solve Problem MB. To this end, we discretize the time interval $[0, 1]$ to a uniform grid of "diameter" $h = 1/N$ where $N \in \mathbb{N}$. As $N \to \infty$, we hope to get the "exact" solution. This generates the following problem:

$$X^N := (x_0, x_1, \ldots, x_N), \quad \Lambda^N := (\lambda_0, \lambda_1, \ldots, \lambda_N)$$
$$t_k = kh, \quad h = \frac{1}{N}$$

$$(MB^{\lambda N}) \begin{cases} \text{Find} & (X^N, \Lambda^N) \\ \text{Subject to} & x_{k+1} = x_k + h \begin{cases} 0 & if\ \lambda_k > 0 \\ 1 & if\ \lambda_k < 0 \\ 0\ or\ 1 & if\ \lambda_k = 0 \end{cases} \\ & \qquad k = 0, 1, \ldots, (N-1) \\ & x_0 = 0 \\ & \lambda_N = x_N - r \\ & \lambda_{k+1} = \lambda_k \end{cases}$$

Note that the constraints $t_0 = 1$ and $t_N = 1$ are automatically included in Problem $MB^{\lambda N}$ through the grid construction ($t_k = kh,\ h = 1/N$).

From the discretized transversality condition and adjoint equations, we have

$$\lambda_k = x_N - r, \ \forall\ k$$

Now, since $h = 1/N$ is rational and $x_0 = 0$, it is clear that all feasible discrete states $x_k,\ k = 0, 1, \ldots, N$ are rational (as a result of the Eulerian integration). Thus, for all step sizes $h > 0$ no matter how small, all feasible states x_k are rational. Consequently, $x_N \neq r$; hence, we have

$$\lambda_k \neq 0, \forall\ k$$

This means that we have one of only two possibilities

$$\forall\ k \quad \lambda_k \begin{cases} > 0 & if\ x_N - r > 0 \\ < 0 & if\ x_N - r < 0 \end{cases}$$

with no prospect of a switch. Suppose $\lambda_k > 0$. In this case, we get $x_N = 0$ from the discrete state dynamics. Hence, we have $x_N - r < 0$, which contradicts the assumption $x_N - r > 0$. Thus, we have

$$\lambda_k \not> 0$$

Similarly, with $\lambda_k < 0$ we get $x_N = 1 \Rightarrow x_N - r > 0$, which contradicts the assumption that $x_N - r < 0$; hence, we have

$$\lambda_k \not< 0$$

Thus, there is no solution to Problem $MB^{\lambda N}$ for any h no matter how small. *That is, an "indirect" method generates the wrong answer, despite that the problem has an infinite set of globally optimal solutions!*

It is important to note that Problem $MB^{\lambda N}$ was analyzed completely "by hand" without resorting to a computer. In addition, Problem $MB^{\lambda N}$ was solved exactly! Interestingly, had we resorted to solving this problem with a digital computer, we would have found the answer!! This is because a digital computer can only encode r approximately via a rational approximation.

Study Problem 3.26

1. *Investigate the cases:*
   i) $r > 1$ *and remains irrational.*
   ii) $0 \leq r \leq 1$ *and rational.*
   iii) $0 < r < 1$, r *is irrational, and* $\mathbb{U} = [0, 1]$.
   iv) $0 < r < 1$, r *is irrational,* h *is nonuniform and passes over irrational points in the time domain,* $[0, 1]$.
2. *Explain precisely the meaning of an exact solution. Based on your explanation, compute the exact value of* π.

Let us now consider a "direct" method to solve Problem MB. Applying the same discretization to Problem MB, we get

$$X^N := (x_0, x_1, \ldots, x_N) \quad U^N := (u_0, u_1, \ldots, u_N)$$

$$u_k \in \mathbb{U} := \{0, 1\} \qquad \forall\, k = 0, 1, \ldots, N$$

$$t_k = kh, \quad h = \frac{1}{N}$$

$$(MB^N) \begin{cases} \text{Minimize} & J^N[X^N, U^N] := \dfrac{(x_N - r)^2}{2} \\ \text{Subject to} & x_{k+1} = x_k + h\, u_k \\ & \qquad k = 0, 1, \ldots, (N-1) \\ & x_0 = 0 \end{cases}$$

It is apparent that x_N can be written as

$$x_N = \frac{\sum_{k=0}^{N-1} u_k}{N}$$

Thus, a globally optimal solution is simply that (rational and reachable) value of x_N that is nearest to r. For N sufficiently large, there is more than one globally optimal solution to Problem MB^N; see Fig. 3.31. Thus, Problem MB^N

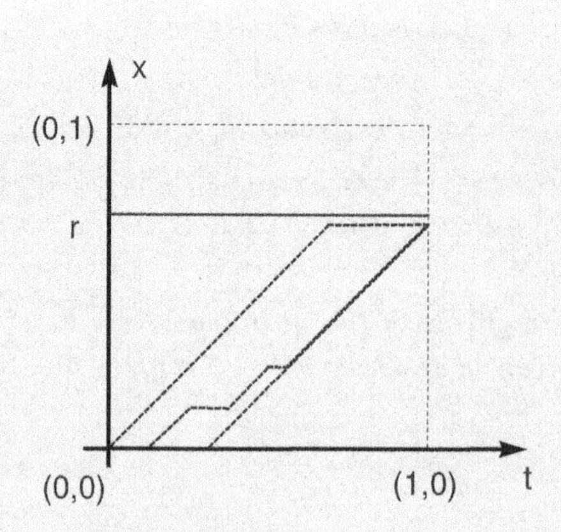

Figure 3.31: Illustrating several of the infinite set of globally optimal solutions to Problem MB^N.

solves Problem MB up to precision h, where we note again that this result is independent of a computer.

The preceding discussions also highlight a well-known precept in advanced optimal control[69, 70, 84]: *Solve approximate problems approximately!*

Study Problem 3.27

Pick any recent journal or conference paper that contains discussions on "numerical" methods for optimal control. Critique the discussion based on the revelations of this section.

Study Problem 3.28

Develop Problem $MB^{N\lambda}$. *(Hint: Apply the KKT conditions over fixed feasible values of* u_k*.)*

Chapter 4

Exercise Problems

Life is an exercise in optimal control.
What is your trajectory?

The scope of optimal control is truly phenomenal: Many problems in economics, engineering, environment, epidemiology, management, mathematics, physics, social sciences and other disciplines can be formulated as optimal control problems. Knowledge of the discipline is a necessary condition to formulate such problems. See also Fig. 1.50 on page 84 to appreciate how problem-analysis and problem-solving are connected to formulating the correct problem.

The following problems are from disparate fields. The first few sections are agnostic to the applications; hence, no knowledge of any particular discipline is necessary to analyze the problems via an application of Pontryagin's Principle.

4.1 One-Dimensional Problems

One-dimensional problems have the advantage that they can be frequently analyzed by simple graphical methods. For maximizing the learning outcome in minimum time, we recommend a student begin by first applying Pontryagin's Principle in the systematic manner discussed in Chapter 2. Thereafter, the resulting equations can be analyzed through graphical methods for a better understanding of the particular problem and Pontryagin's Principle itself.

:ar-Quadratic Exercise Problems

and 4.2 defined below can be solved by following the same process developed in Section 3.3, page 191. Begin with

$$\boldsymbol{x} := x \in \mathbb{R}, \quad \boldsymbol{u} := u \in \mathbb{R}$$

$$\left\{\begin{array}{lrl} \text{Minimize} & J[x(\cdot), u(\cdot)] = & \dfrac{x_f^2}{2} + \dfrac{1}{2}\displaystyle\int_{t_0}^{t_f} u^2(t)\, dt \\ \text{Subject to} & \dot{x} = & x + u \\ & (x_0, t_0) = & (1, 0) \\ & t_f = & 1 \end{array}\right. \tag{4.1}$$

by showing that the adjoint equations and transversality conditions reduce to

$$\dot{\lambda} = -\lambda \qquad \lambda(t_f) = x(t_f)$$

Then, develop both open- and closed-loop solutions by showing that

$$u = -\lambda$$

Validate your analytical results via a numerical simulation.

For further understanding of the concepts developed in this book, run a Monte Carlo simulation of the open- and closed-loop system for $x_0 \sim \mathcal{N}(1, 0.1^2)$. Perform a "big data" analysis. (This is really "small data" but it is bigger than without the Monte Carlo simulation.)

Next, repeat all of the preceding steps for the following "modified" problem:

$$\boldsymbol{x} := x \in \mathbb{R}, \quad \boldsymbol{u} := u \in \mathbb{R}$$

$$\left\{\begin{array}{lrl} \text{Minimize} & J[x(\cdot), u(\cdot)] = & \dfrac{1}{2}\displaystyle\int_{t_0}^{t_f} u^2(t)\, dt \\ \text{Subject to} & \dot{x} = & x + u \\ & (x_0, t_0) = & (1, 0) \\ & (x_f, t_f) = & (0, 1) \end{array}\right. \tag{4.2}$$

Compare and contrast the differences. How are these problems different and similar to Kalman's linear-quadratic problem discussed in Section 3.4?

4.1.2 A Simple Exercise in BVPs

1. Show that Problem (4.2) generates a classic BVP:

$$\begin{aligned} \dot{x} &= x - \lambda \qquad && x(0) = 1 \quad x(1) = 0 \\ \dot{\lambda} &= -\lambda \end{aligned} \tag{4.3}$$

2. To solve this BVP, design a ***"home-brewed" shooting algorithm*** (hbshoot), using the following recipe:

   (a) First, construct an initial value problem solver (IVPsolver) for a generic differential equation

   $$\dot{\boldsymbol{y}} = \boldsymbol{g}(\boldsymbol{y}, t), \qquad \boldsymbol{y}(t^0) = \boldsymbol{y}^0$$

   using Euler's method

   $$\boldsymbol{y}_{k+1} = \boldsymbol{y}_k + h\,\boldsymbol{g}(\boldsymbol{y}_k, t_k)$$

   where, h is the step size.

   (b) Next, let $\boldsymbol{y} = (x, \lambda) \in \mathbb{R}^2$. Using the IVPsolver, produce the function generator

   $$g : \lambda_0 \mapsto x(1), \qquad x(1) = g(\lambda_0)$$

   The value of λ_0 for which $x(1) = 0$ is the solution to the BVP given by Eq. (4.3). Find this (approximate) value of λ_0 by a graphical process.

   (c) Construct a root-finder for the generic equation

   $$\boldsymbol{g}(\boldsymbol{z}) = \boldsymbol{0}$$

   by using the simple "trick," $\boldsymbol{z} = \boldsymbol{g}(\boldsymbol{z}) + \boldsymbol{z}$. This method is called a *fixed-point iteration*. Construct a root-finder (zeroFinder) using the simpler iteration:

   $$\boldsymbol{z}_{k+1} = \boldsymbol{g}(\boldsymbol{z}_k) + \boldsymbol{z}_k \tag{4.4}$$

(d) Finally, put everything together to make hbshoot by using the IVPsolver to produce $g(z)$, and feeding the resulting function $g(z)$ as the input for zeroFinder to find the missing initial conditions.

3. Solve the BVP (4.3) using hbshoot. In case you experience difficulties, you may "cheat" by using the solution obtained by the graphical process as a guess.

4. In addition to hbshoot, use a professional shooting code (*Warning*: Most freeware are not professional codes; they are only worth their price.)

5. Evaluate the performance of all methods used to solve the BVP using at least two figures of merit: (i) Sensitivity and (ii) Relative run-time. That is, systematically change the values of the guess (e.g., λ_0 for the shooting method) from large negative numbers to large positive numbers and record the outcomes under the figures of merit.

6. Redo the preceding numerical tests by experimenting with different values of step sizes. Does a smaller step-size (and hence a "better" integrator) produce a "better" algorithm?

7. Produce "new improved" variants of hbshoot by:

   (a) Replacing IVPsolver by a (professional) Runge-Kutta method and/or

   (b) Replacing zeroFinder by a (damped) Newton's method (your own or out of the box).

   How do these replacements affect the performance of hbshoot? Do these enhancements alter any of your conclusions so far?

8. Solve the BVP given by Eq. (4.3) using any collocation method: Your own or a professional code (e.g., from MATLAB's optimization toolbox). If you experience difficulties, you may "cheat" by using the solution from hbshoot as a guess.

9. Can the results from all these numerical experiments be used to make general conclusions about shooting and/or collocation methods? Use the curse of sensitivity for a perspective. (Recall Section 2.9 on page 150.)

10. Do the conclusions made in 9. pertain to the methods (i.e., shooting vs collocation) or to the software? (*Hint*: If you say "both," your answer may be incorrect. Read Section 2.9 again! Warning: You may have to read and understand the portions marked with the dangerous bends.)

11. Repeat the entirety of the preceding analysis for the BVP generated by Problem (4.1):

$$\begin{aligned} \dot{x} &= x - \lambda & \qquad x(0) &= 1 \\ \dot{\lambda} &= -\lambda & \qquad \lambda(1) &= x(1) \end{aligned} \tag{4.5}$$

12. Based on the results of these experiments, answer the following:

    (a) Critique the statement that a simple algorithm is required for flight implementation.

    (b) What requirements would you impose on flight algorithms?

    (c) Would these requirements be different if you were limited to a microprocessor technology of the 1980s? If so, explain (very briefly).

    (d) Depending upon your your answer to the above question, explain if these requirements would change based on projections of future technology.

    Where applicable, use the following industry-standard figures of merit:

    i) Number of lines of code.

    ii) Number of iterations.

    iii) *Relative* run times (anything coded in MATLAB can be made faster by compiling).

13. Flight implementations require certain mathematical guarantees. Based on this overarching principle, what "new" figures of merit would you require for a flight implementation? (*Hint*: Research notions of convergence, consistency, etc.) Which of the "old" figures of merit would you discard?

4.1.3 An Example Problem of Problems

Small changes to a problem formulation can have a big impact on the solution. To see this, consider the very simple one-dimensional linear dynamical system

$$\dot{x} = u$$

with event conditions given by:

$$(x_0, t_0) = (0, 0)$$
$$(x_f, t_f) = (0, 1)$$

The form of the cost functional

$$J[x(\cdot), u(\cdot)] := \frac{1}{2}\int_{t_0}^{t_f} x^2(t)\, dt$$

is the same for all of the following problems:

1. Let $\mathbb{U} := [-1, 1]$ and minimize J.

   (a) Show that the Hamiltonian Minimization Condition generates:

   $$u = \begin{cases} -1 & \text{if } \lambda \geq 0 \\ [-1, 1] & \text{if } \lambda = 0 \\ +1 & \text{if } \lambda \leq 0 \end{cases} \tag{4.6}$$

   (b) Show that $x(\cdot) = \Theta = u(\cdot)$ satisfies Pontryagin's Principle.

   (c) Is $x(\cdot) = \Theta = u(\cdot)$ a globally optimal solution?

   (d) Refine the condition $u \in [-1, 1]$ if $\lambda = 0$ in Eq. (4.6), by considering the the costate trajectory $\lambda(\cdot) = \Theta$ and all its time derivatives. That is, show that $\ddot{\lambda} \equiv 0$ and hence, $u(\cdot) = \Theta$.*

   (e) Are there any other solutions that satisfy Pontryagin's Principle?

*This solution is called a singular control because $\partial^2 H/\partial u^2$ is zero (singular when u is a vector of dimension two or more). The Hamiltonian is "locally horizontal" with respect to u; see Fig. 3.23 on page 225. Hence, $d^n/dt^n(\partial H/\partial u) = 0$ for $n = 0, 1, 2, \ldots$.

2. Let $\mathbb{U} := [-1, 1]$ and maximize J.

   (a) Show that $x(\cdot) = \Theta = u(\cdot)$ satisfies Pontryagin's Principle.

   (b) Is $x(\cdot) = \Theta = u(\cdot)$ at least a locally optimal solution?

   (c) Are there any other solutions that satisfy Pontryagin's Principle? Is any one of them a globally optimal solution?

3. From the results of these analyses, criticize the statement that a solution to the BVP (Problem B^λ) generates an optimal solution to Problem B.

4. Let $\mathbb{U} := \{-1, 1\}$ and minimize J.

   (a) Determine the optimal solution. (*Hint*: Consider Lebesgue measures)

   (b) Attempt to develop necessary conditions for this problem using the calculus of variations.

   (c) Discuss what happens if $\mathbb{U} := \{-1, 0, 1\}$.

4.1.4 Illustrating Think Twice, Code Once

An unbelievably common mistake made by many users of DIDO is to code a problem before performing even a rudimentary analysis. The following very simple problem[†] illustrates why a code should be the last step of the first iteration (see Fig. 1.50 on page 84 as well as the discussions on page 177) in solving a problem:

$$\boldsymbol{x} := x \in \mathbb{R}, \quad \boldsymbol{u} := u \in \mathbb{R}$$

$$\left\{\begin{array}{lrl} \textsf{Minimize} & J[x(\cdot), u(\cdot)] = & \displaystyle\int_{t_0}^{t_f} t\, u^2(t)\, dt \\ \textsf{Subject to} & \dot{x} = & u \\ & (x_0, t_0) = & (0, 0) \\ & (x_f, t_f) = & (1, 1) \end{array}\right.$$

Obviously this problem looks very benign; however, coding it without analysis will generate countless hours of unnecessary trouble. Analyze this problem and see why.

[†]This problem was inspired by a similar one designed by W. Kang.

4.1.5 A Medley of Thought-Provoking Problems

Many of the following problems are optimal control versions of problems from the calculus of variations. They were designed by the cited authors to stimulate or answer some interesting questions. *Beware of the dangerous bends!*

A 1-D Linear Control Problem

Show that the globally optimal solution to the following problem[35]

$$\boldsymbol{x} := x \in \mathbb{R}, \quad \boldsymbol{u} := u \in \mathbb{U} = \{u \in \mathbb{R} : \ u \geq -1\}$$

$$\begin{cases} \text{Minimize} & J[x(\cdot), u(\cdot)] := & x(t_f) \\ \text{Subject to} & \dot{x}(t) = & u(t) \\ & (x_0, t_0) = & (0, 0) \\ & t_f = & 1 \end{cases}$$

is given by $u(t) = 1, \ \forall \ t \in [0, 1]$.

A Cubic Cost Problem

The following problem from Dreyfus[28] is also discussed in [16] and [64]:

$$\boldsymbol{x} := x \in \mathbb{R}, \quad \boldsymbol{u} := u \in \mathbb{R}$$

$$\begin{cases} \text{Minimize} & J[x(\cdot), u(\cdot)] = & \displaystyle\int_{t_0}^{t_f} u^3(t)\, dt \\ \text{Subject to} & \dot{x} = & u \\ & (x_0, t_0) = & (0, 0) \\ & (x_f, t_f) = & (1, 1) \end{cases}$$

Show that:

1. $(x(t), u(t)) = (t, 1)$ is a feasible solution that satisfies $\partial_u H = 0$ and $\partial_u^2 H > 0$.

2. This problem has no optimal solution. (*Hint*: Sketch the graph of $u \mapsto H(\lambda, x, u)$.)

Kang's Problem[‡]

Analyze the optimality of the candidate solution $x^*(t) = (-t+1)^3$ for the following problem:

$$\boldsymbol{x} := x \in \mathbb{R}, \quad \boldsymbol{u} := u \in \mathbb{R}$$

$$\begin{cases} \text{Minimize} & J[x(\cdot), u(\cdot)] = \displaystyle\int_{t_0}^{t_f} \left(9\,x^2(t) - \frac{u^3(t)}{6}\right) dt \\ \text{Subject to} & \dot{x} = u \\ & (x_0, t_0) = (1, 0) \\ & (x_f, t_f) = (0, 1) \end{cases}$$

A Chattering Problem

In his very entertaining book[105], L. C. Young defines the following problem:

$$\boldsymbol{x} := x \in \mathbb{R}, \quad \boldsymbol{u} := u \in \mathbb{R}$$

$$\begin{cases} \text{Minimize} & J[x(\cdot), u(\cdot)] = \displaystyle\int_{t_0}^{t_f} \left(1 + x^2(t)\right)\left(1 + [u^2 - 1]^2\right) dt \\ \text{Subject to} & \dot{x} = u \\ & (x_0, t_0) = (0, 0) \\ & (x_f, t_f) = (0, 1) \end{cases}$$

1. Show that $u(t) = 0$ is not an optimal solution to this problem. (*Hint*: Consider $\partial_u^2 H$.)
2. Show that $u(t) = \pm 1$ satisfies all the necessary conditions for optimality, and that the cost of this solution is half the cost of $J[0, 0]$.
3. Sketch the optimal solution.

‡Communicated by W. Kang.

A Pathological Problem

The following very interesting problem is from [2]:

$$\boldsymbol{x} := x \in \mathbb{R}, \quad \boldsymbol{u} := u \in \mathbb{R}$$

$$\begin{cases} \text{Minimize} & J[x(\cdot), u(\cdot)] = \displaystyle\int_{t_0}^{t_f} \left(u^2(t) - x^2(t)\right) dt \\ \text{Subject to} & \dot{x} = u \\ & (x_0, t_0) = (0, 0) \\ & t_f > \pi \\ & x_f = 0 \end{cases}$$

Use Pontryagin's Principle to analyze the optimality of the system trajectory given by $x(\cdot) = \Theta = u(\cdot)$.

A State-Constrained Problem

The following problem is from [24]:

$$\boldsymbol{x} := x \in \mathbb{R}, \quad \boldsymbol{u} := (u_1, u_2) \in \mathbb{R}^2$$

$$\begin{cases} \text{Minimize} & J[x(\cdot), \boldsymbol{u}(\cdot)] = -x_f \\ \text{Subject to} & \dot{x} = u_1 \\ & (x_0, t_0) = (-1, 0) \\ & t_f = 1 \\ & x(t) + u_2(t) \leq 0 \\ & u_1 u_2 \geq 0 \\ & |u_i| \leq 2, \quad i = 1, 2 \end{cases}$$

Show that the solution to this problem is given by:

$$x(t) = t - 1$$

$$\big(u_1(t), u_2(t)\big) = \begin{cases} (1,0) & \textit{if} \quad t \in [0,1) \\ (0,-1) & \textit{if} \quad t = 1 \end{cases}$$

Discuss what happens if the path constraint, $x(t) + u_2(t) \leq 0$ is dropped as a condition. (*Hint*: Incorporate the nontriviality condition in Pontryagin's Principle.)

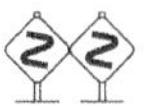 **The Ball-Mizel Problem**

The following problem[4] is connected to many deep mathematical results in existence and regularity of solutions:

$$\boldsymbol{x} := x \in \mathbb{R}, \quad \boldsymbol{u} := u \in \mathbb{R}$$

$$\begin{cases} \text{Minimize} & J[x(\cdot), u(\cdot)] = \displaystyle\int_{t_0}^{t_f} \left[\left(t^2 - x^3(t)\right)^2 u^{14}(t) + \varepsilon\, u^2(t) \right] dt \\ \text{Subject to} & \dot{x} = u \\ & (x_0, t_0) = (0,0) \\ & (x_f, t_f) = (k,1) \end{cases}$$

where, $\varepsilon > 0$ and $k > 0$ are given.

Show that:

1. $x(t) = k\, t^{2/3}$ solves the problem, provided

$$\varepsilon = \left(\frac{2k}{3}\right)^{12} (1 - k^3)(13k^3 - 7)$$

2. The costate has the form $\lambda(t) = \alpha_k\, t^{-1/3}$ where, $\alpha_k \neq 0$ is a quantity that depends upon k.

4.2 A Suite of Test Problems and Solutions

Test Problem # 1 (from [86])

$$\boldsymbol{x} := x \in \mathbb{R}, \quad \boldsymbol{u} := u \in \mathbb{R}$$

$$\left\{\begin{array}{lll} \text{Minimize} & J[\boldsymbol{x}(\cdot), \boldsymbol{u}(\cdot)] & := \displaystyle\int_{-1}^{1} (x(t) - \sin(\pi t))^2 \, dt \\ \text{Subject to} & \dot{x} & = \sin(4\pi x) + u \\ & x(-1) & = 0 \end{array}\right. \tag{4.7}$$

Solution:

$$\begin{aligned} x(t) &= \sin(\pi t) \\ u(t) &= \pi \cos(\pi t) - \sin(4\pi x(t)) \\ \lambda(t) &\equiv 0 \end{aligned}$$

Why does $\lambda(t) \equiv 0$ not violate the nontriviality condition?

Test Problem # 2 (from [34])

$$\boldsymbol{x} := (x_1, x_2) \in \mathbb{R}^2, \quad \boldsymbol{u} := u \in \mathbb{R}$$

$$\left\{\begin{array}{lll} \text{Minimize} & J[\boldsymbol{x}(\cdot), \boldsymbol{u}(\cdot)] & := 4x_1(2) + x_2(2) + 4 \displaystyle\int_{0}^{2} u^2(t) \, dt \\ \text{Subject to} & \dot{x}_1 & = x_2^3 \\ & \dot{x}_2 & = u \\ & \boldsymbol{x}_0 & = (0, 1) \end{array}\right. \tag{4.8}$$

Solution:

$$
\begin{aligned}
x_1(t) &= \frac{2}{5} - \frac{64}{5(2+t)^5} \\
x_2(t) &= \frac{4}{(2+t)^2} \\
u(t) &= -\frac{8}{(2+t)^3} \\
\lambda_1(t) &= 4 \\
\lambda_2(t) &= \frac{64}{(2+t)^3}
\end{aligned}
$$

Test Problem # 3 (from [48])

$$
\boldsymbol{x} := (x_1, x_2) \in \mathbb{R}^2, \quad \boldsymbol{u} := u \in \mathbb{R}
$$

$$
\left\{
\begin{array}{lrcl}
\text{Minimize} & J[\boldsymbol{x}(\cdot), \boldsymbol{u}(\cdot)] & := & \displaystyle\int_0^{\pi} \left(1 - x_1(t) + x_1(t)x_2(t) + x_1(t)u(t)\right)^2 dt \\
\text{Subject to} & \dot{x}_1 & = & -x_1^2 x_2 \\
& \dot{x}_2 & = & -1 + \dfrac{1}{x_1} + x_2 + \sin t + u \\
& \boldsymbol{x}_0 & = & (1,\ 0) \\
& \boldsymbol{x}_f & = & \left(\dfrac{1}{\pi+1},\ 2\right)
\end{array}
\right.
\tag{4.9}
$$

Solution:

$$
\begin{aligned}
x_1(t) &= \frac{1}{1 - \sin t + t} \\
x_2(t) &= 1 - \cos t \\
u(t) &= -(t+1) + \sin t + \cos t \\
J[\boldsymbol{x}^*(\cdot), \boldsymbol{u}^*(\cdot)] &= 0
\end{aligned}
$$

4.3 RTOC for Nonbelievers

The following problems are quite simple to code and study. Notwithstanding their simplicity, they generate a powerful set of ideas for practical implementation. Embedded in the following problems are certain concepts and principles in "how to" and "how not to." In fact, the how-to's that are apparent in the following problems are used in the operation of certain advanced industrial systems including flight implementations of pseudospectral optimal control techniques[87].

4.3.1 RTOC for Problem DQC

Refer to Section 3.3 on page 191 for specifics and Section 1.5.1 on page 64 for the general principles.

1. Construct an open-loop simulation of the optimal control ($u = at + b$) as shown in Fig. 4.1. This specific construction prepares the next step of closing the loop via RTOC as shown in Fig. 3.16 on page 199.

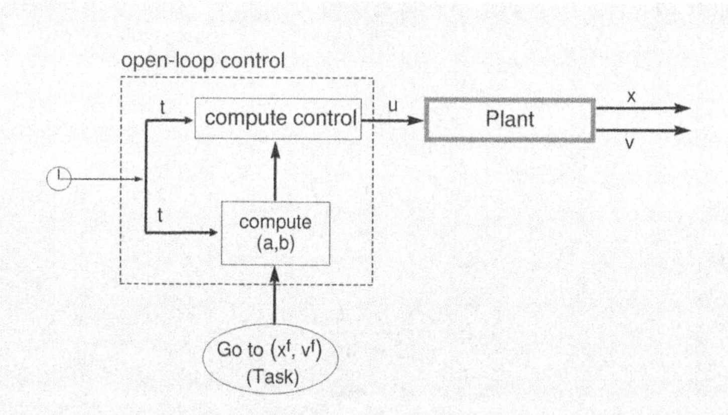

Figure 4.1: Conceptual block diagram for an optimal open-loop simulation. This construction is as an initial step for the RTOC implementation shown in Fig. 3.16 on page 199.

2. Test the construction of Fig. 4.1 using the results of Section 3.3. This is part of a verification and validation (V & V) procedure.

3. Add the additional "wiring" to Fig. 4.1 to complete the RTOC implementation shown in Fig. 3.16 on page 199. *Do not invert the $\boldsymbol{T}$-matrix.* Test this modified implementation. What are the metrics for a successful test?

4. Suppose that the plant is uncertain with an uncertainty parameter m that affects the dynamics linearly:

$$m\ddot{x} = u$$

Assume that m has a uniform (non-Gaussian) distribution of $\pm 5\%$ around a mean value of $\mu_m = 1$. Perform a Monte Carlo simulation and evaluate the performance of the RTOC controller.

5. Inject Gaussian sensor noise in the simulation with $\mathcal{N}(0, 0.1^2)$ and redo the the Monte Carlo analysis. Compare your final result with the one shown in Fig. 4.2 produced by M. Karpenko.

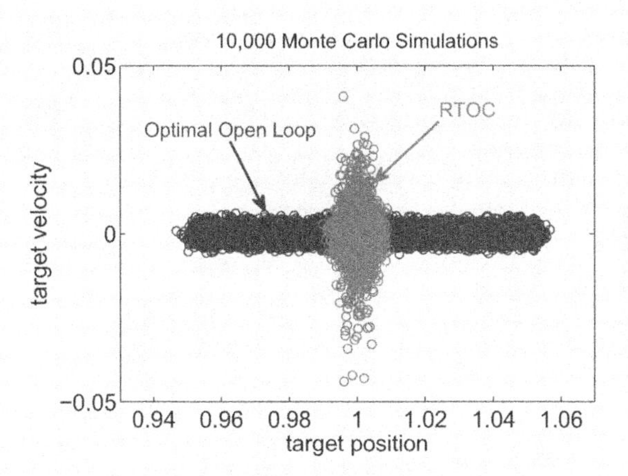

Figure 4.2: Performances of RTOC and the optimal open loop controller for the figure of merit of acquiring the targeted state (1,0).

Explain why the optimal open-loop controller produces smaller errors than RTOC in the velocity component.

6. Next, refer to Fig. 3.17 on page 202. Design a (constant gain) PD controller. Describe your design criteria.

7. Does the PD controller achieve the target end condition (i.e. $x(1) = 1$, $v(1) = 0$)? Evaluate the quadratic-control cost for the PD controller,

$$\begin{aligned}\mathcal{J}_{PD} &:= \frac{1}{2}\int_{t_0}^{t_f} u_{PD}^2(x(t), v(t))\, dt \\ &= \frac{1}{2}\int_{t_0}^{t_f} \left(k_x\, x(t) + k_v\, v(t)\right)^2 dt\end{aligned}$$

Compare your results to that of the optimal.

8. Perform a Monte Carlo simulation for mass uncertainty only (see previous question 4) and compare the performance of your PD controller with that of RTOC.

9. Redo the Monte Carlo simulations and analysis with the inclusion of Gaussian sensor noise, as in question 5, and compare your results with the one shown in Fig. 4.3 produced by M. Karpenko.

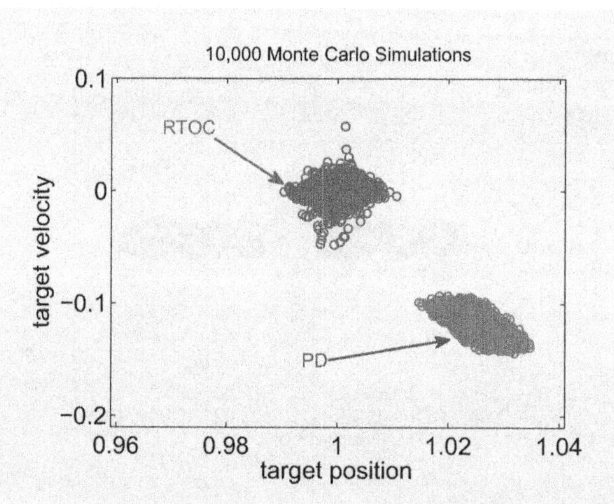

Figure 4.3: Performances of PD and RTOC controllers for the figure of merit of acquiring the targeted state (1,0).

Karpenko's PD controller was based on the "standard" design of a damping ratio of $\zeta = 0.7$ with settle time $t_s = t_f = 1$.

(a) Compare your results with Fig. 4.3. Does your PD controller have an "offset" or error from the target condition of $(1, 0)$? Explain.

(b) The cost $\mathcal{J}_{PD}$ for Karpenko's controller was about 35. Compare the cost of your PD-controller with this value vis-à-vis the target-error value in question (a).

(c) Is it possible to reduce the target error by ***gain tuning***? What happens to the value of the $\mathcal{J}_{PD}$ as you change the gains?

10. Redo the entirety of the preceding steps by performing the following alternative implementations of closed-loop optimal controllers:

    (a) Update a and b by inverting the $\boldsymbol{T}$-matrix online.

    (b) Implement "time-varying PD" controller given by Eq. (3.39). Recall that this is obtained by an off-line inversion of the $\boldsymbol{T}$-matrix.

    Compare and contrast the performances of all three implementations of the optimal closed-loop controllers.

4.3.2 Strap-On Optimal Guidance

Refer to Section 1.5.1 on page 67 for the general concept of "strapping on" a guidance block over a legacy inner-loop control system. In particular refer to Figs. 1.39 and 1.40. Specifics of these concepts for the double integrator plant are shown in Fig. 4.4. Develop this architecture in the following phases (called

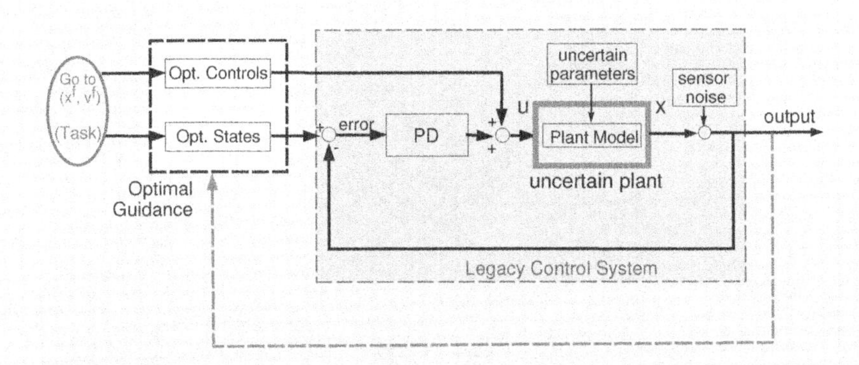

Figure 4.4: Strap-on optimal guidance over a legacy (PD) control system: This construction straps on Fig. 3.16 over Fig. 3.17; see pages 199–202.

a "spiral path" in industry-speak). Include testing and analysis at the end of each phase.

1. Start with the inner-loop. Turn off all the uncertainties. Inject only the optimal states to form an open-loop guidance system. Analyze, test and validate this system.

2. Inject the optimal controls (in addition to the optimal states) and repeat the entire battery of tests. This architecture completes the open-loop guidance block shown in Fig. 4.4.

3. Close the guidance loop. Turn on all uncertainties and repeat the tests.

Note that, by construction, the process includes a test and analysis of several failure modes of the guidance and control system concept. During each phase of the construction, the uncertainties must be turned on and off for testing. Include the following figures of merit in your analysis:

1. Satisfaction/deviation from the targeted conditions (in the presence of uncertainties).

    (a) Compare your Monte Carlo results with the one shown in Fig. 4.5 produced by M. Karpenko.

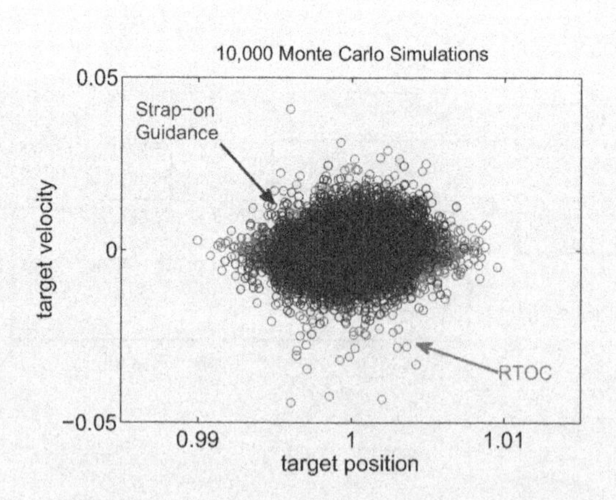

Figure 4.5: Performances of a strap-on open-loop guidance system (with PD inner-loop) and RTOC controller for the figure of merit of acquiring the targeted state (1,0).

(b) Explain why an open-loop strap-on optimal guidance architecture does just as well, or better, than the RTOC approach. This aspect of the performance of the two architectures is more apparent in the distribution plots generated from the Monte Carlo simulations as shown in Fig. 4.6.

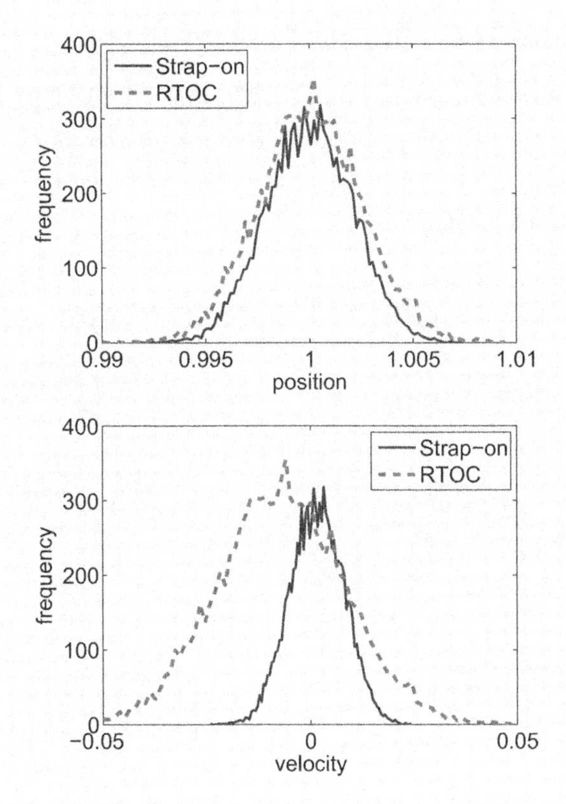

Figure 4.6: Distributions at the target state generated from 10,000 Monte Carlo simulations for two control system architectures.

2. Value of cost (this is the integral of the square of the control).

3. Transfer time.

4.4 Double-Integrator-Type Problems

As noted in Sections 3.3 and 3.5, double integrator systems have been extensively studied since the 1950s because they provide an excellent initial model for the study of electromechanical systems. The following are a small selection of such problems motivated by some physical consideration.

4.4.1 A Modified Quadratic Control Problem

This problem is nearly identical to Problem DQC that was formulated and solved in Section 3.3, pages 191–208. As noted throughout this book, it is always helpful to study the *problem of problems*. In this spirit, consider the following modification to Problem DQC,

$$\boldsymbol{x} := (x, v) \in \mathbb{R}^2 \qquad \boldsymbol{u} := u \in \mathbb{R} \qquad \mathbb{U} = \mathbb{R}$$

$$(\mathsf{mDQC}) \begin{cases} \textsf{Minimize} & J[x(\cdot), v(\cdot), u(\cdot)] = P_x \dfrac{(x_f - 1)^2}{2} + P_v \dfrac{v_f^2}{2} \\ & \qquad + \dfrac{1}{2} \displaystyle\int_{t_0}^{t_f} u^2(t)\, dt \\ \textsf{Subject to} & \dot{x} = v \\ & \dot{v} = u \\ & (x_0, v_0, t_0) = (x^0, v^0, t^0) \\ & t_f - 1 = 0 \end{cases}$$

This problem and its many variants are extensively discussed in Bryson and Ho[16]. From Section 3.4, page 211, it is apparent that this problem is a special case of Kalman's linear-quadratic problem.

The main difference between Problem DQC and mDQC is that there are no terminal state constraints on the latter problem; however, the cost function contains endpoint "penalty" terms. The weights, $P_x > 0$ and $P_v > 0$, are somewhat arbitrary but are based on the intuitive notion that a larger number means a more important term.

1. Apply Pontryagin's Principle to Problem mDQC and show that it generates the same equations as Problem DQC except for the terminal transver-

sality conditions which are given by

$$\begin{aligned}\lambda_x(t_f) &= P_x(x_f - 1)\\ \lambda_v(t_f) &= P_v v_f\end{aligned}$$

2. Show that the open-loop optimal control is given by

$$u = P_x(x_f - 1)(t - 1) - P_v v_f \tag{4.10}$$

3. Following the same procedure as the one discussed in pages 195–199, show that the closed-loop optimal control can be obtained by solving the 2×2 linear matrix equation

$$\underbrace{\begin{pmatrix} P_x \dfrac{T_0^3}{6} + 1 & T_0 - P_v \dfrac{T_0^2}{2} \\ P_x \dfrac{T_0^2}{2} & 1 - P_v T_0 \end{pmatrix}}_{\boldsymbol{T}_2} \underbrace{\begin{pmatrix} x_f - 1 \\ v_f \end{pmatrix}}_{\boldsymbol{p}_2} = \underbrace{\begin{pmatrix} x_0 - 1 \\ v_0 \end{pmatrix}}_{\boldsymbol{q}_2} \tag{4.11}$$

where $T_0 := t_0 - 1$. Note that $\boldsymbol{T}_2$ is not singular when $T_0 = 0$.

4. Solve Eq. (4.11).
   (a) Is it possible for $(x_f - 1)$ to be equal to zero?
   (b) Is it possible for v_f to be zero?
   (c) Substitute the solution in Eq. (4.10) and rearrange the terms as in Eq. (3.39) on page 202. Show that the gains, and hence the controls do not go to infinity as $t_{go} \to 0$.
5. (a) Is the relationship between P_x and $(x_f - 1)$ linear?
   (b) Is the relationship between P_v and v_f linear?
   (c) Is there a relationship between P_x and v_f?
   (d) Is there a relationship between P_v and $(x_f - 1)$?
6. Based on your answer to the preceding question, critique the use of P_x and P_v as tuning knobs to *independently* control $(x_f - 1)$ and v_f respectively.
7. Choose $(x_0, v_0, t_0) = \mathbf{0}$, and quantitatively compare the performance of the solution of Problem mDQC to that of DQC for the following cases:

(a) $P_x = P_v = 1$,

(b) $P_x = 0.1P_v = 1$, and

(c) $P_x = 10P_v = 1$.

8. Discuss what happens for the following limiting cases:

(a) $P_x \to \infty, P_v \to 0$, and

(b) $P_x \to \infty, P_v \to \infty$.

Compare your limiting solutions to those obtained by directly solving Problem DQC; see Section 3.3.

4.4.2 An L^1-Optimal Control Problem

Consider the following L^1-optimal control problem for the double integrator:

$$\boldsymbol{x} := (x, v) \in \mathbb{R}^2 \quad \boldsymbol{u} := u \in \mathbb{U} := \{u \in \mathbb{R} : \ |u| \leq 6\}$$

$$(\text{L1P}) \left\{ \begin{array}{lrcl} \text{Minimize} & J_1[\boldsymbol{x}(\cdot), \boldsymbol{u}(\cdot)] = & \displaystyle\int_0^1 & |u(t)|\, dt \\ \text{Subject to} & \dot{x} = & v & \\ & \dot{v} = & u & \\ & (x_0, v_0) = & (0, 0) & \\ & (x_f, v_f) = & (1, 0) & \end{array} \right. \tag{4.12}$$

Show that the solution to Problem $L1P$ is given by:

$$u_1(t) = \left\{ \begin{array}{rl} 6 & t \in \Delta_1 \\ 0 & t \in \Delta_2 \\ -6 & t \in \Delta_3 \end{array} \right.$$

$$
\begin{aligned}
x_1(t) &= \begin{cases} 3t^2 & t \in \Delta_1 \\ 3\Delta(2t - \Delta) & t \in \Delta_2 \\ 6(t + \Delta - \Delta^2) - 3(1 + t^2) & t \in \Delta_3 \end{cases} \\
v_1(t) &= \begin{cases} 6t & t \in \Delta_1 \\ 6\Delta & t \in \Delta_2 \\ 6(1 - t) & t \in \Delta_3 \end{cases} \\
\lambda_{x_1}(t) &= \frac{2}{2\Delta - 1} \\
\lambda_{v_1}(t) &= \frac{1 - 2t}{2\Delta - 1}
\end{aligned}
$$

where, Δ_i, $i = 1, 2, 3$ are three subintervals of $[0, 1]$ defined by

$$
\Delta_1 = [0, \Delta], \quad \Delta_2 = [\Delta, 1 - \Delta], \quad \Delta_3 = [1 - \Delta, 1]
$$

and

$$
\Delta = \frac{1}{2} - \sqrt{\frac{1}{12}} \simeq 0.211
$$

A plot of the L^1-optimal control along with the linear-quadratic optimal control discussed in Section 3.3 is shown in Fig. 4.7. Further details on this problem

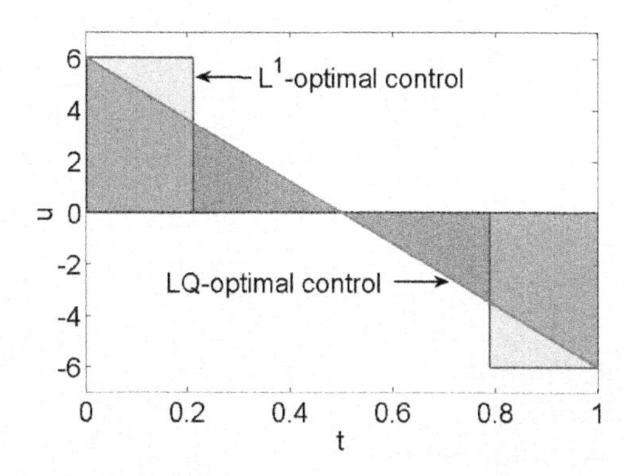

Figure 4.7: L^1 vs linear-quadratic optimal controls; figure is from [79].

and how it compares with Problem DQC (see Section 3.3 and page 192) are discussed in [79].

Show that

$$J_1[\boldsymbol{x}_1(\cdot), \boldsymbol{u}_1(\cdot)] := \int_0^1 |u_1(t)|\, dt = 12\Delta \simeq 2.536 \tag{4.13a}$$

$$J_1[\boldsymbol{x}_Q(\cdot), \boldsymbol{u}_Q(\cdot)] := \int_0^1 |u_Q(t)|\, dt = 3.0 \tag{4.13b}$$

where, $\boldsymbol{x}_Q$ and $\boldsymbol{u}_Q$ are the state and control variables of Problem DQC respectively.

Equation (4.13) implies that if a quadratic cost is used as a proxy for fuel consumption, the performance penalty incurred is about 18%. In orbital applications, the limit of this number is 50%; see [79].

4.4.3 Breakwell's Problem

Like many great problems, Breakwell's problem[16] is deceptively simple; it is given by:

$$\boldsymbol{x} := (x, v) \in \mathbb{R}^2 \quad \boldsymbol{u} := u \in \mathbb{R}$$

$$\left\{\begin{array}{lrcl} \text{Minimize} & J[\boldsymbol{x}(\cdot), \boldsymbol{u}(\cdot)] & = & \dfrac{1}{2}\displaystyle\int_{t_0}^{t_f} u^2(t)\, dt \\ \text{Subject to} & \dot{x} & = & v \\ & \dot{v} & = & u \\ & (x_0, v_0, t_0) & = & (0, 1, 0) \\ & (x_f, v_f, t_f) & = & (0, -1, 1) \\ & x(t) & \leq & \ell \end{array}\right. \tag{4.14}$$

where $\ell > 0$ is a given number.

1. Show that the adjoint equations are given by

$$-\dot{\lambda}_x = \mu, \qquad -\dot{\lambda}_v = \lambda_x$$

where, $\mu \dagger x$; that is, μ is complementary to x.

2. Show that

$$u = -\lambda_v$$

and hence, prove that if $x = \ell$ over a nonzero time interval, then

$$0 = u = \lambda_v = \lambda_x = \mu$$

3. Because $\mu = 0$ if $x < \ell$, and $\mu = 0$ over $x = \ell$, does this mean $t \mapsto \lambda_x(t) =$ a constant? (Refer to page 139 and Fig. 2.18.)

4. Assume the switching structure consists of three segments:

   (a) $x < \ell$ for $0 \leq t \leq t_1$,

   (b) $x = \ell$ for $t_1 \leq t \leq t_2$, and

   (c) $x > \ell$ for $t_2 \leq t \leq 1$.

   Prove that $\lambda_x > 0$ in segment (a) and $\lambda_x < 0$ in (c); and hence, the control trajectory consists of a linear segment with positive slope, followed by $u = 0$ and ending with a linear segment of negative slope.

5. Show that $t_1 = 3\ell$ and $\lambda_x = 2/9\ell^2$ in segment (a). (*Hint*: Use the following conditions: $v(0) = 1$, $x(0) = 0$, $u(t_1) = 0$, $v(t_1) = 0$, $x(t_1) = \ell$.)

6. Show that $t_2 = 1 - 3\ell$. Discuss what happens if $\ell - 1/6$. From this justify the assumption of the switching structure for $\ell < 1/6$.

7. Redo the analysis for the case when $\ell > 1/6$. Explain what happens when $\ell \geq 1/4$.

8. Prove that $t \mapsto \lambda_v$ is continuous. (*Note: In [16] a different multiplier theory is used which generates a discontinuous* λ_v.)

4.4.4 Gong's Motion Planning Test Problem

Consider the deceptively simple problem of moving a block of unit mass from points A to B with the least amount of "work;" see Fig. 4.8. A particular version

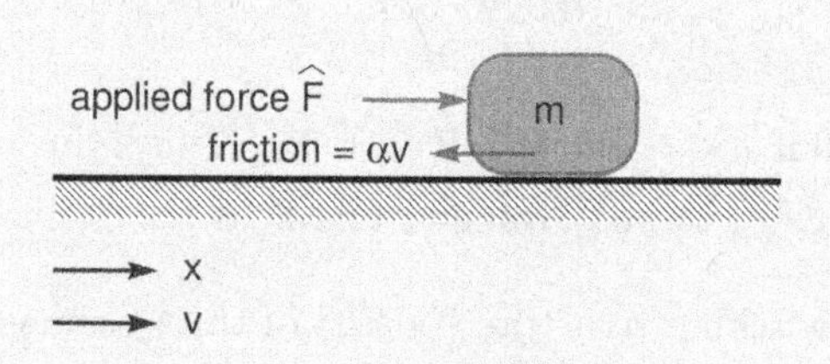

Figure 4.8: Schematic for Gong's motion planning problem.

of this problem as formulated by Gong[34] is given by:

$$\boldsymbol{x} \in \mathbb{X} := \left\{(x, v) \in \mathbb{R}^2 : v \geq 0\right\}, \quad \boldsymbol{u} \in \mathbb{U} := \left\{\widehat{F} \in \mathbb{R} : 0 \leq \widehat{F} \leq 2\right\}$$

$$(G) \left\{ \begin{array}{lrl} \text{Minimize} & J[\boldsymbol{x}(\cdot), \boldsymbol{u}(\cdot)] = & \displaystyle\int_{t_0}^{t_f} v(t)\widehat{F}(t)\, dt \\ \text{Subject to} & \dot{x}(t) = & v(t) \\ & \dot{v}(t) = & -v(t) + \widehat{F}(t) \\ & (x_0, v_0, t_0) = & (0, 1, 0) \\ & (x_f, v_f, t_f) = & (1, 1, 1) \end{array} \right.$$

Show that the solution to Problem G is given by

$$x(t) = t, \qquad v(t) = 1, \qquad \widehat{F}(t) = 1$$

and that the the costates satisfy

$$\lambda_x(t) = -2, \qquad \lambda_v(t) = -1$$

Despite that the solution to this problem is very simple, many computational techniques and associated software fail to produce the correct solution. For instance, an Euler discretization produces a chattering control as shown in

Fig. 4.9. A higher-order method (see Section 2.9.2 on page 157) does not

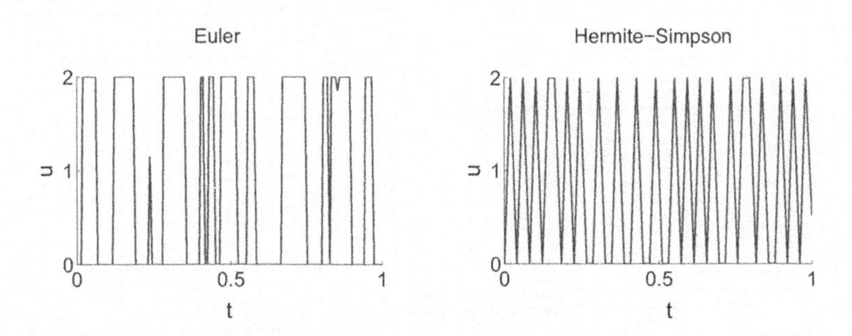

Figure 4.9: Popular Runge-Kutta collocation methods fail to converge to the simple solution of Problem G; figure adapted from [34].

solve the problem. The solution generated by a popular 4th-order Runge-Kutta method (known as a Hermite-Simpson discretization[6, 36], and encoded in many commercial software packages) is also shown in Fig. 4.9. Furthermore, none of these solutions converge with mesh refinements; in fact, they make it worse[34].

Such problems with numerical methods have been known since the 1990s; hence, it contributed to the folklore that optimal control problems were just hard. Similar problems have been generally recognized as "known hard" problems because they are "singular." That is, problems where the ***Hessian of the Hamiltonian***

$$\frac{\partial^2 H}{\partial \boldsymbol{u}^2} \qquad (an\ N_u \times N_u\ matrix)$$

is singular. In Problem G, because u is a scalar, we have $\partial^2 H/\partial u^2 = 0$.

Many (not all!) of these problems started to disappear with the advent of pseudospectral (PS) optimal control theory[87]. A PS solution for Problem G is shown in Fig. 4.10 along with the costates generated by DIDO. That a PS solution should indeed produce this result is part of the mathematical theory initiated by Gong et al in [34]. In very broad terms, pseudospectral theory abandons the "equation centric" approach and embraces a "solution centric" philosophy[78]. For example, in a PS technique, the differential equations are not integrated as in Runge-Kutta methods; instead, the equations are simply viewed as supplying the requisite tangent vector information (see Fig. 1.7 on

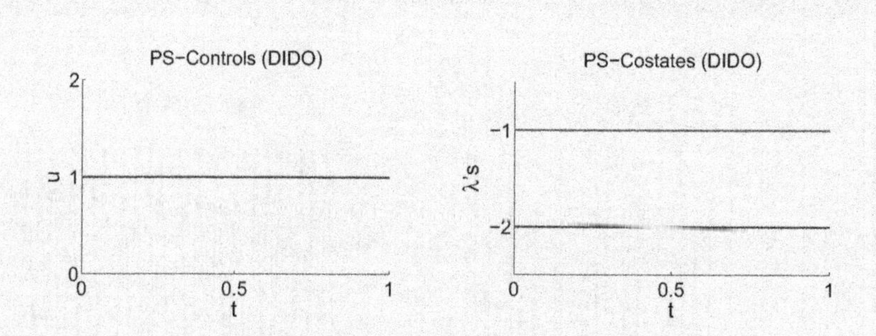

Figure 4.10: DIDO solution to Gong's problem.

page 9) to a candidate state *solution* $t \mapsto \boldsymbol{x}$ represented by a generalized Fourier series. See Section 2.9.2 on page 157 for a quick introduction to these concepts. Additional details on this paradigm shift are described in [78] and [84].

A major challenge with PS theory is that it has the same idiosyncrasy as Pontryagin's Principle: *Easy to use, hard to prove.* The ease of using PS methods has had some unfortunate consequences: It has generated a perceptible amount of misinformed research. A naïve use of PS theory can lead to spectacular failures. PS solutions to Problem G using ***Radau and Gauss points*** are shown in Fig. 4.11. Obviously, these PS controls chatter in much the same

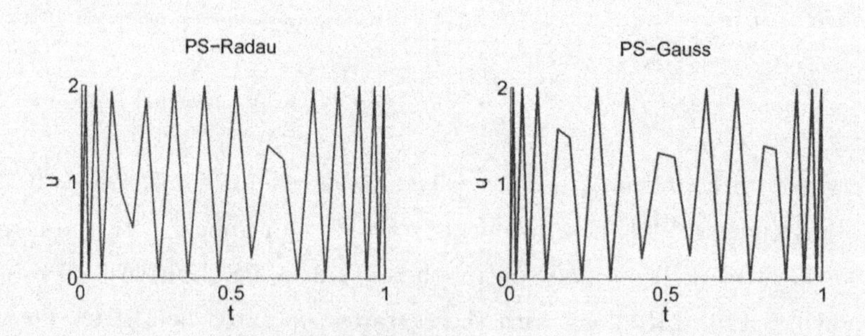

Figure 4.11: A naïve implementation of pseudospectral theory, as implemented in some DIDO clones, produces convergence problems similar to those of Fig. 4.9; figure is adapted from [30].

manner as those generated by Runge-Kutta collocation methods.[§] What is more "shocking" about this result is that, as shown in Fig. 4.12, the costates converge! Clearly, a naïve use of PS theory can also be used to add fuel to the fire that

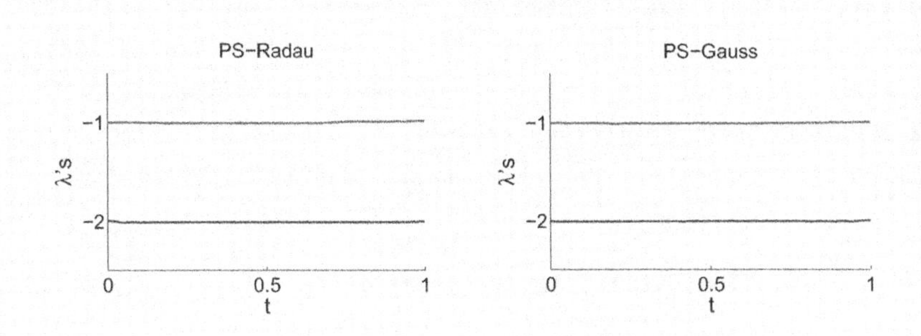

Figure 4.12: Convergence of costates does not imply convergence of controls; figure adapted from [30].

solving an optimal control problem is just hard.

From the fundamentals of PS theory[87], these results are not at all shocking; particularly since the optimal cost for Problem G computed over Radau and Gauss points are lower than the theoretical optimal! This structural error indicates that a lower cost does not imply a better method; see [30].

There is no doubt that optimal control theory and problem solving can be made to look very hard by misapplications. Misusing theory or technique does not equate to hard problems. While many "hard" problems of the past are indeed solvable today, a growing number of new problems and challenges have emerged. See [3, 14, 23, 62, 84, 87] and Sections 1.5 and 1.6 of this book for a quick sample of these emerging research areas.

[§]Although Radau points generate convergence problems for PS discretization of finite-horizon optimal control problems, they are the correct set of points for infinite horizon problems; see [29] where these concepts were first introduced.

Distinctions Without A Difference: Ever since (2006) NASA's historic flight test of pseudospectral (PS) optimal control theory — as implemented in DIDO — there have been many attempts to "improve" or redefine PS methods with different adjectives. Regardless of any "re-branding," what matters in the end is if a solution can be verified and validated via Pontryagin's Principle as outlined in Section 3.2.1 on page 180. In any event, as shown by Gong et al[35], there are very clear mathematical reasons why all roads lead to the ***Legendre*** PS method with ***Chebyshev*** being a close second. This is why these two techniques have come to be known as the "big two" methods.

4.4.5 A Minimum Energy Problem

Gong's problem (see Section 4.4.4) is a special case of a minimum energy problem. A more general problem can be formulated as:

$$\boldsymbol{x} \in \mathbb{X} := \left\{(x, v) \in \mathbb{R}^2\right\}, \quad \boldsymbol{u} \in \mathbb{U} := \left\{\widehat{F} \in \mathbb{R} : \left|\widehat{F}\right| \leq \widehat{F}_{max}\right\}$$

$$(G)\begin{cases} \text{Minimize} & J[\boldsymbol{x}(\cdot), \boldsymbol{u}(\cdot)] = \displaystyle\int_{t_0}^{t_f} \left|v(t)\widehat{F}(t)\right| dt \\ \text{Subject to} & \dot{x}(t) = v(t) \\ & \dot{v}(t) = -\alpha\, v(t) + \widehat{F}(t) \\ & (x_0, v_0, t_0) = (0, v^0, 0) \\ & (x_f, v_f, t_f) = (x^f, v^f, 1) \end{cases}$$

The quantities $\widehat{F}_{max} > 0$ and $\alpha > 0$ are given numbers. See Fig. 4.8 on page 272.

Pontryagin's Principle is not directly applicable to this problem because the running cost is not differentiable with respect to a state variable (v). A direct approach is to apply Clarke's nonsmooth calculus[21, 22]. Another approach is to transform the problem to a smooth one and then apply Pontryagin's Principle. Perform such an analysis by carrying out the following steps:

1. Refer to Section 1.4.4 on page 55. Transform the L^1 cost functional to a smooth problem by introducing additional variables and constraints.

2. Map the transformed problem to functions and variables of Problem P discussed in Section 1.4.2 on page 46.

3. Refer to Section 2.7 on page 131. Apply Pontryagin's Principle to the transformed problem.

4.4.6 Fuller's Problem

The following problem[32] results in the famous Fuller's phenomenon:

$$\boldsymbol{x} := (x, v) \in \mathbb{R}^2, \quad \boldsymbol{u} := u \in \mathbb{U} = \{u \in \mathbb{R} : \ |u| \leq 1\}$$

$$\left\{\begin{array}{lrl} \text{Minimize} & J[\boldsymbol{x}(\cdot), \boldsymbol{u}(\cdot), t_f] & = \displaystyle\int_{t_0}^{t_f} x^2(t)\, dt \\ \text{Subject to} & \dot{x} & = v \\ & \dot{v} & = u \\ & t_0 & = 0 \\ & \boldsymbol{x}(t_0) & = \boldsymbol{x}^0 \neq \boldsymbol{0} \\ & \boldsymbol{x}(t_f) & = \boldsymbol{0} \end{array}\right. \tag{4.15}$$

4.5 Optimal Control For Nonbelievers

Optimal control can address and solve many problems in disparate fields: Artificial intelligence, economics, epidemiology, management, social sciences etc. Despite its engineering origins, many engineers view the field with some cynicism. Their qualms are largely traceable to an improper understanding of the fundamental notion of optimality. The following collection of problems illustrates one of these aspects.

The "bang-bang" control generated in Section 3.5 is famously misunderstood. Nonbelievers claim that the optimal solution is a bad one because it

(a) stresses the actuators too much,

(b) generates vibration modes,

(c) creates passenger discomfort, or

(d) insert your favorite complaint here.¶

¶Ergo, they say, optimality is not desirable; "good" feasible solutions are quite sufficient.

The problem here is with the problem formulation, not the optimal solution. This is why solving ***the*** problem may involve solving an iterative set of problems that converge to the correct problem formulation which properly models the previously-unmodeled issues. This is part of the problem-of-problems work-flow illustrated in Fig. 1.50 on page 84.

Let us begin by assuming that the switch in the control solution to Problem DMT (discussed on page 221) was undesirable (because of other systems-engineering considerations). The new requirement is to produce a smooth solution. If "smooth" and "time-optimal" are not defined precisely, then, one simple "trick" is to add a mollifier to the cost function.

4.5.1 A Mollified Minimum-Time Problem

An "indirect" approach to producing a "smooth" time-optimal control is to modify the cost function by adding a mollifier:

$$\boldsymbol{x} := (x, v) \in \mathbb{R}^2, \quad \boldsymbol{u} := u \in \mathbb{U} = \{u \in \mathbb{R} : \ |u| \leq 1\}$$

$$\left\{\begin{array}{lrl} \text{Minimize} & J[\boldsymbol{x}(\cdot), \boldsymbol{u}(\cdot), t_f] & = t_f + \dfrac{M}{2}\displaystyle\int_{t_0}^{t_f} u^2(t)\, dt \\ \text{Subject to} & \dot{x} & = v \\ & \dot{v} & = u \\ & (x_0, v_0, t_0) & = (x^0, v^0, t^0) \\ & (x_f, v_f) & = (0, 0) \end{array}\right. \tag{4.16}$$

1. Show that for for M sufficiently large ($M > 0$), the optimal control to Problem 4.16 is smooth.

2. Discuss the issues with this problem formulation.

3. What additional criticisms can you level at the solution to this problem? Discuss new problem formulations to address these additional criticisms.

4.5.2 A Jerk-Limited Minimum-Time Problem

A "direct" approach to producing a smoother time-optimal control is to limit the "control rate." Recall that in optimal control theory, any decision variable with a derivative is a state variable; see the discussions on page 6. Consequently, adding a rate-limit to the control changes its description to a state variable. Furthermore, because controls in an optimal control framework have no "inertia", the control variable in the modified system is the *jerk* variable (= derivative of acceleration).

The new problem formulation is given by||

$$\boldsymbol{x} := (x, v, a) \in \mathbb{X} \subset \mathbb{R}^3, \quad \boldsymbol{u} := j \in \{j \in \mathbb{R} : |j| \leq j_{max}\}$$

$$\left\{\begin{array}{llll} \text{Minimize} & J[\boldsymbol{x}(\cdot), \boldsymbol{u}(\cdot), t_f] & = t_f \\ \text{Subject to} & \dot{x} & = v \\ & \dot{v} & = a \\ & \dot{a} & = j \\ & (x_0, v_0, t_0) & = (x^0, v^0, t^0) \\ & (x_f, v_f) & = (0, 0) \\ & |a| & \leq 1 \end{array}\right. \tag{4.17}$$

where, j_{max} is the maximum value of allowable jerk. The original control u from Problems DMT and 4.16 is now the state variable a; hence, it is constrained to the original control space $[-1, 1]$, which is now part of the state space. As a result, we end up with a state-constrained problem; hence, Pontryagin's Principle for Problem P needs to be invoked for answering the following questions:

1. What is the structure of the optimal jerk profile?

2. Is the optimal acceleration smooth?

3. Are the costates continuous?

4. Consider the problem to be parameterized by the requirement on the upper bound for the value of jerk. Discuss the relative "loss" in performance (t_f) as j_{max} is made increasingly smaller. See [31] for a detailed discussion on this problem as it applies to a practical spacecraft.

||The dynamics of this system is known as a triple integrator; hence, the concept of adding a "du/dt-term" is known as adding an integrator.

5. What additional criticisms can you level at the solution to this problem? Discuss new problem formulations to address these additional criticisms.

4.5.3 A Rate-Limited Minimum-Time Problem

In certain situations, the engineering issue or operational constraint may be with the "rate" (v) and not the switch in the control solution to Problem DM I (see page 221). In this case, the problem can be easily modified to limit the velocity to a maximum allowable value v_{max} as follows:

$$\boldsymbol{x} := (x, v) \in \mathbb{R}^2, \quad \boldsymbol{u} := u \in \mathbb{U} = \{u \in \mathbb{R} : \ |u| \leq 1\}$$

$$\left\{\begin{array}{lrl} \text{Minimize} & J[\boldsymbol{x}(\cdot), \boldsymbol{u}(\cdot), t_f] := & t_f \\ \text{Subject to} & \dot{x} = & v \\ & \dot{v} = & u \\ & (x_0, v_0, t_0) = & (x^0, v^0, t^0) \\ & (x_f, v_f) = & (0, 0) \\ & |v| \leq & v_{max} \end{array}\right. \tag{4.18}$$

1. Develop the optimal control solution to this problem. Does this conform with your intuition of driving a car (in a speed-limited zone)?

2. Are the costates continuous?

3. Determine a lower bound for the minimum-time solution using the value of v_{max}.

4.5.4 A Constrained Minimum-Time Problem

Finally, suppose that requirements are imposed that constitute an assortment of constraints: rate-limits, jerk-limits, vague smoothness requirements, etc. Such constraints are quite typical in practical and flight applications; see for instance [52]. Then, as an example, the following problem may be constructed:

$$\boldsymbol{x} := (x, v, a) \in \mathbb{X} \subset \mathbb{R}^3, \quad \boldsymbol{u} := j \in \{j \in \mathbb{R} : \ |j| \leq j_{max}\}$$

$$\left\{\begin{array}{lrl} \textsf{Minimize} & J[\boldsymbol{x}(\cdot), \boldsymbol{u}(\cdot), t_f] := & t_f + \dfrac{M}{2}\displaystyle\int_{t_0}^{t_f} u^2(t)\, dt \\ \textsf{Subject to} & \dot{x} = & v \\ & \dot{v} = & a \\ & \dot{a} = & j \\ & (x_0, v_0, t_0) = & (x^0, v^0, t^0) \\ & (x_f, v_f) = & (0, 0) \\ & |v| \leq & v_{max} \\ & |a| \leq & 1 \end{array}\right. \tag{4.19}$$

1. Develop the optimal control solution to this problem.
2. Compare the optimal solution of this problem with those of the previous problems. Identify which of the constraints have the most dominant impact on optimality.

On The Practice of Optimal Control:

1. Many practical optimal control problems come with an overhead of imprecise or unclear notion of optimality and constraints. The task of a practitioner is to clear the clutter through a multitude of problem formulations and solutions.
2. A practical optimal control problem usually involves a collection of constraints. The unconstrained problem may provide a false impression of optimality, while the overly-constrained problem may produce nothing valuable.
3. An over-simplified dynamical model (like a double-integrator) may produce overly-conservative solutions to the system it represents.
4. In many engineering systems, false constraints are unwittingly imposed because of a misunderstanding of requirements. The practice of optimal control requires one to trace the origins of suspected false constraints in order to formulate the right problem.

4.6 Queen Dido and the Badlands

In his May 12th address of 1893, Lord Kelvin introduced his lecture as follows[97],

> The first isoperimetrical problem known in history was practically solved by Dido, a clever Phoenician princess, who left her Tyrian home and emigrated to North Africa, with all her property and a large retinue, because her brother Pygmalion murdered her rich uncle and husband Acerbas, and plotted to defraud her of the money which he left. On landing in a bay about the middle of the north coast of Africa she obtained a grant from Hiarbas, the native chief of the district, of as much land as she could enclose with an ox-hide. She cut the ox-hide into an exceedingly long strip, and succeeded in enclosing between it and the sea a very valuable territory on which she built Carthage.

A word-formulation of an isoperimetric (*iso* = same, *perimeter* = perimeter) problem can be stated as: Given a perimeter L, determine the curve that encloses the greatest area. Queen Dido (B.C. 800 or 900 [97]) formulated and solved this ***first optimal control problem***. For well over two millennia, this problem has inspired some of the greatest mathematicians and philosophers from Zenodorus to Hurwitz[8]. See [8] and [96] for well-written historical accounts

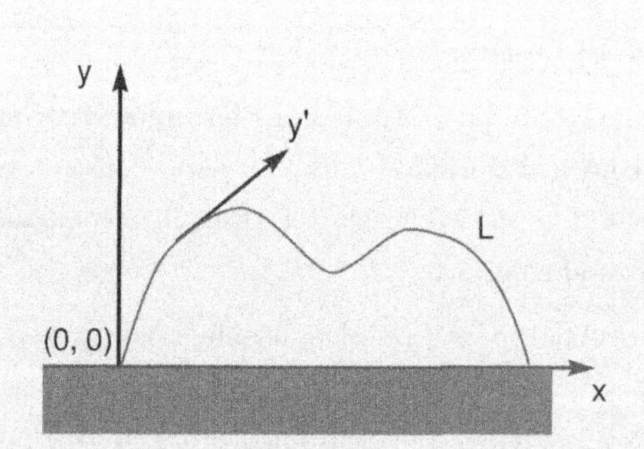

Figure 4.13: Schematic for Queen Dido's problem.

with mathematical details. Tapia[96] goes so far as to argue that this problem is "history's most impactful mathematical problem." The software package DIDO — first created in its object-oriented form in 2001 — is named in honor of Queen Dido.

Using Fig. 4.13, show that Dido's problem can be formulated as

$$\boldsymbol{x} := y \in \mathbb{R}, \quad \boldsymbol{u} := u \in \mathbb{R}$$

$$(D_o) \begin{cases} \text{Maximize} & J[\boldsymbol{x}(\cdot), \boldsymbol{u}(\cdot), x_f] = \int_{x_0}^{x_f} y(x)\, dx \\ \text{Subject to} & y' = u \\ & \int_{x_0}^{x_f} \sqrt{1 + u^2(x)}\, dx \leq L \\ & (y_0, x_0) = (0, 0) \\ & y_f = 0 \\ & x_f < L \end{cases}$$

The independent variable here is x; hence, the dynamics of the state variable y is given by $y' := dy/dx$. In the classical calculus of variations, this problem is formulated by replacing u with y' in the integral constraint. We have also written the integral constraint as an inequality in order to allow the unlikely possibility that we may not need the entirety of the given length, $L > 0$.

Problem D_o is not in the format of Problem B; however, it can be cast in a standard form using the well-known transformational trick (see Section 1.4.4 on page 55) of defining a new state variable z, governed by the dynamics $z' = \sqrt{1 + u^2}$, $\quad z(x_0) = 0$. This trick generates the following problem:

$$\boldsymbol{x} := (y, z) \in \mathbb{R}^2, \quad \boldsymbol{u} := u \in \mathbb{R}$$

$$(D_B) \begin{cases} \text{Minimize} & J[\boldsymbol{x}(\cdot), \boldsymbol{u}(\cdot), x_f] = -\int_{x_0}^{x_f} y(x)\, dx \\ \text{Subject to} & y' = u \\ & z' = \sqrt{1 + u^2} \\ & (y_0, z_0, x_0) = (0, 0, 0) \\ & y_f = 0 \\ & z_f \leq L \\ & x_f < L \end{cases}$$

1. Show that λ_z is a constant and λ_y is linear in x.

2. Derive an equation for a candidate control function $x \mapsto u$.

3. Show that u at $x = 0$ is infinite. (*Hint*: Consider the Hamiltonian evolution equation and the Hamiltonian value condition.) Does this fact imply computational problems? Is the problem of infinite u, a problem with Pontryagin's theory, or a problem with computers, or a problem with the problem formulation itself? See also Section 4.1.4 on page 253.

4. Explain what happens if we set $x_f > 2L/\pi$? (Is this cheating?)

5. Make an attempt to prove that the solution to this problem is an arc of a circle. Discuss the difficulty in arriving at this conclusion. Recall that Queen Dido solved this problem around 800 B.C. without the aid of optimal control theory or even calculus!

6. Bryson and Ho[16] suggest the transformation $u = \tan\theta$. Transform Problem D_B to D_θ where θ is the new control variable. Discuss the pros and cons of solving the transformed problem.

7. Show that the following problem is a transformation of Problem D_B:

$$\boldsymbol{x} := (x, y, z) \in \mathbb{R}^3, \quad \boldsymbol{u} := (u_1, u_2, u_3) \in \mathbb{U} \subset \mathbb{R}^3$$

$$(D_3)\left\{\begin{array}{lrl} \text{Minimize} & J[\boldsymbol{x}(\cdot), \boldsymbol{u}(\cdot), t_f] & = -\displaystyle\int_{t_0}^{t_f} y(t)u_1(t)\,dt \\ \text{Subject to} & \dot{x} & = u_1 \\ & \dot{y} & = u_2 \\ & \dot{z} & = u_3 \\ & (x_0, y_0, z_0, t_0) & = (0, 0, 0, 0) \\ & y_f & = 0 \\ & z_f & \leq L \\ & x_f & < L \\ & u_3^2 - u_1^2 - u_2^2 & = 0 \end{array}\right.$$

8. Discuss why Problem D_3 is a better problem formulation than any of Problems D_o, D_B or D_θ.

9. Show that λ_x and λ_z are constants and that $\dot{\lambda}_y = u_1$.

10. Show that λ_y can also be written as

$$\lambda_y + \frac{y\, u_2}{u_1} = 0 \tag{4.20}$$

(*Hint*: Write down the KKT conditions for Problem HMC.)

11. Using Eq. (4.20) and the adjoint equations, prove that the solution to Queen Dido's problem is an arc of a circle.

12. Determine the radius and center of the arc of the circle using only the given problem data.

13. Explain what happened to the singularities and issues with the previous problem formulations. You may need to refer to Chapter 1 again.

4.7 Zermelo Problems

According to Serres[94], E. Zermelo formulated the following "word problem" in 1931:

> *In an unbounded plane where the wind distribution is given by a vector field as a function of position and time, a ship moves with constant velocity relative to the surrounding air mass. How must the ship be steered in order to come from a starting point to a given goal in the shortest time?*

The differential equations for Zermelo's problem can be written as

$$\begin{aligned} \dot{x} &= W_1(x, y, t) + u_1 \\ \dot{y} &= W_2(x, y, t) + u_2 \end{aligned} \tag{4.21}$$

where, $\mathbf{W}(x, y) := (W_1(x, y), W_2(x, y))$ is the wind vector field and $\boldsymbol{u} = (u_1, u_2)$ is the steering vector normalized by the ship's speed and constrained by

$$u_1^2 + u_2^2 = 1$$

Simplified versions of this problem appear in many textbooks[16, 54, 64, 105], the most common of which is the case when the wind vector field depends linearly on the y-coordinate only. In [89], a nonlinear version of this problem is addressed as described next.

4.7.1 A Classic Nonlinear Zermelo Problem

Let $W_1(x,y,t) = 10y^3$ and $W_2(x,y,t) = 0$; this generates the following problem:

$$\boldsymbol{x} := (x,y) \in \mathbb{R}^2, \quad \boldsymbol{u} \in \mathbb{U} := \left\{(u_1,u_2) \in \mathbb{R}^2 : \ u_1^2 + u_2^2 = 1\right\}$$

$$p = 10$$

$$(Z_D)\left\{\begin{array}{lrcl} \text{Minimize} & J[\boldsymbol{x}(\cdot),\boldsymbol{u}(\cdot),t_f] := & & t_f \\ \text{Subject to} & \dot{x}(t) &=& py^3(t) + u_1(t) \\ & \dot{y}(t) &=& u_2(t) \\ & (\boldsymbol{x}(t_0),t_0) &=& (2.25,1,0) \\ & \boldsymbol{x}(t_f) &=& \boldsymbol{0} \end{array}\right. \tag{4.22}$$

1. Apply Pontryagin's Principle and develop the necessary conditions for optimality.

2. Solve Problem Z_D by any method and compare your answer to the solution shown in Fig. 4.14.

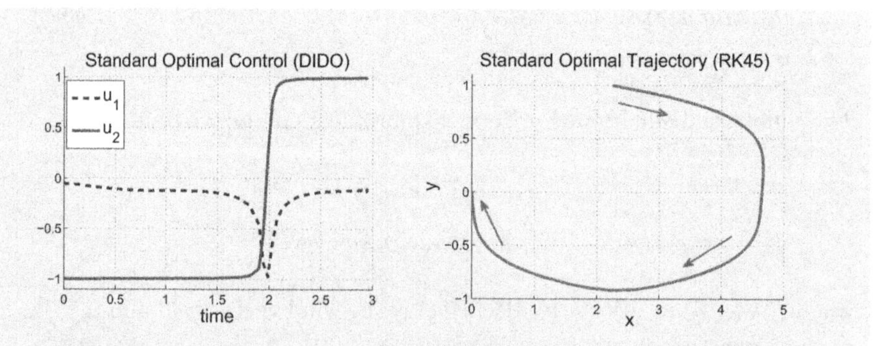

Figure 4.14: A solution to Problem Z_D obtained via the procedure discussed in Section 3.2 on pages 177–187.

The solution shown in Fig. 4.14 was obtained via DIDO by following the steps discussed in Section 3.2 (see pages 177–187). For the purposes of brevity not all the verification and validation steps and processes discussed in Section 3.2 are discussed.

4.7.2 A Tychastic Zermelo Problem

As discussed in Section 1.6 (see page 75), a standard optimal control problem can be used as a generator for tychastic optimal control problems. Using Problem Z_D as a generator, we can formulate a transcendental tychastic optimal control problem as follows:

$$\boldsymbol{x} := (x, y) \in \mathbb{R}^2, \quad \boldsymbol{u} \in \mathbb{U} := \left\{(u_1, u_2) \in \mathbb{R}^2 : \ u_1^2 + u_2^2 = 1\right\}$$

$$p \sim \mathcal{N}(10, 2^2)$$

$$(Z_{tyc}^{\infty}) \begin{cases} \text{Minimize} & J[\boldsymbol{x}(\cdot,\cdot), \boldsymbol{u}(\cdot), t_f] := \ t_f \\ \text{Subject to} & \dot{x}(t,p) = \ p\, y^3(t,p) + u_1(t) \\ & \dot{y}(t,p) = \ u_2(t) \\ & (\boldsymbol{x}(t_0,p), t_0) = \ (2.25, 1, 0) \\ & \boldsymbol{x}(t_f, p) = \ \mathbf{0} \quad \forall\, p \in (-\infty, \infty) \end{cases} \tag{4.23}$$

For theoretical analysis, the Lebesgue-Stieltjes-generalized version of Pontryagin's Principle[91] must be applied to this problem.

As developed in [89, 90], this problem can be "semi-discretized" to a standard optimal control problem. One particular semi-discretization generates an *unscented Zermelo problem*:

$$\boldsymbol{X} := [(x_1, y_1), (x_2, y_2)] \in \mathbb{R}^4, \quad \boldsymbol{u} \in \mathbb{U} := \left\{(u_1, u_2) \in \mathbb{R}^2 : \; u_1^2 + u_2^2 = 1\right\}$$

$$p_1 = 8, \quad p_2 = 12$$

$$(Z_U^\infty) \begin{cases} \text{Minimize} & J[\boldsymbol{X}(\cdot), \boldsymbol{u}(\cdot), t_f] := \; t_f \\ \text{Subject to} & \dot{x}_1(t) = \; p_1 y_1^3(t) + u_1(t) \\ & \dot{y}_1(t) = \; u_2(t) \\ & \dot{x}_2(t) = \; p_2 y_2^3(t) + u_1(t) \\ & \dot{y}_2(t) = \; u_2(t) \\ & t_0 = \; 0 \\ & (x_1(t_0), y_1(t_0)) = \; (2.25, 1) \\ & (x_2(t_0), y_2(t_0)) = \; (2.25, 1) \\ & (x_1(t_f), y_1(t_f)) = \; (0, 0) \\ & (x_2(t_f), y_2(t_f)) = \; (0, 0) \end{cases} \tag{4.24}$$

where $p_1 = 8$ and $p_2 = 12$ are the ***sigma points*** of Julier et al[42, 43].

1. Develop the necessary conditions for optimality for Problem Z_U^∞ by applying Pontryagin's Principle.

2. Based on your results to the preceding exercise, guess a Pontryagin-type Principle for Problem Z_{tyc}^∞. Compare your result with the Riemann-Stieltjes-generalized version of Pontryagin's Principle developed in [91].

3. Solve Problem Z_U^∞ by any method and compare your answer to the solution shown in Fig. 4.15.

4. Choose $(p_1, p_2) \neq (8, 12)$ and solve Problem Z_U^∞ for this modification. Do the new results look the same or different from Fig. 4.15? (This surprising result is due to R. J. Proulx.)

5. The controls shown in Fig. 4.15 do not guarantee a solution to Problem Z_{tyc}^∞; however, they can be numerically tested a posteriori through a Monte Carlo simulation; see also Fig. 1.49 on page 81. The result of this exercise is shown in Fig. 4.16. It is apparent that the ***unscented optimal control***

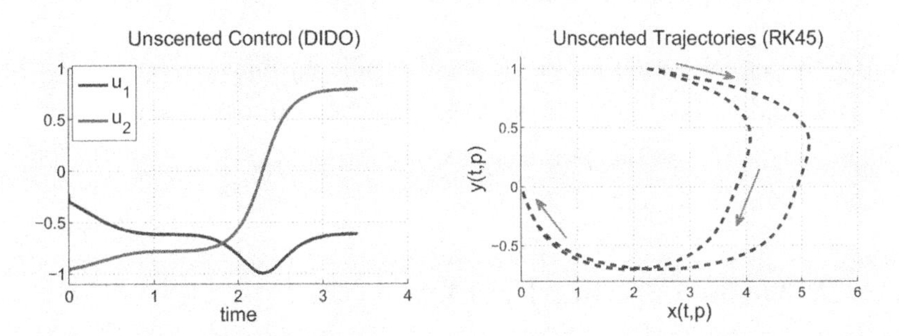

Figure 4.15: A DIDO-solution to Problem Z_U^∞ obtained via the procedure discussed in Section 3.2 on pages 177–187.

does indeed solve Problem Z_{tyc}^∞ within numerical precision. Perform the same exercise for your optimal controls. See [89] and [90] for further details.

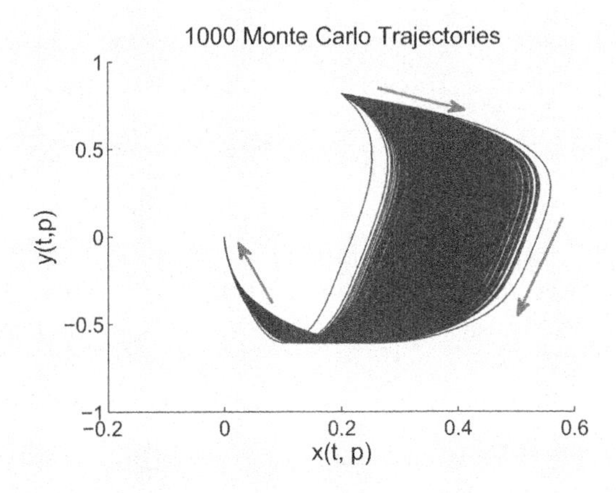

Figure 4.16: Monte Carlo simulations of the candidate unscented optimal control shown in Fig. 4.15.

4.8 Astrodynamic Optimization

Astrodynamic optimization problems are dynamic optimization problems arising in astronautical systems. In addition to the examples discussed in [64], the following sample problems indicate why astronautics is one of the largest "consumers" of optimal control theory.

4.8.1 A Moon-Landing Problem

A simple "one-dimensional" Moon-landing problem can be defined as:

$$\boldsymbol{x} := (h, v, m) \in \mathbb{R}^3, \quad \boldsymbol{u} := T \in \mathbb{U} := \{T \in \mathbb{R} : \; 0 \le T \le T_{max}\}$$

$$(ML_{1D}) \begin{cases} \text{Minimize} & J[\boldsymbol{x}(\cdot), \boldsymbol{u}(\cdot), t_f] := \displaystyle\int_{t_0}^{t_f} \frac{T(t)}{v_e}\, dt \\ \text{Subject to} & \dot{h} = v \\ & \dot{v} = -g + \dfrac{T}{m} \\ & \dot{m} = -\dfrac{T}{v_e} \\ & t_0 = 0 \\ & (h_0, v_0, m_0) = (h^0, v^0, m^0) \\ & (h_f, v_f) = (0, 0) \end{cases}$$

where the quantities $T_{max}, g, v_e, h^0, v^0$ and m^0 are all given numbers. Variants of of this problem are discussed in [67] and the classical text by Kirk[54].

1. Show that the co-altitude (λ_h) is a constant and that the co-velocity (λ_v) is a linear function of time.

2. Is it possible to find the constants associated with λ_h and λ_v (using the data from the problem)?

3. Show that the co-mass (λ_m) and the lower Hamiltonian ($\mathcal{H}$) satisfy the following conditions,

$$\lambda_m(t_f) = 0, \quad \mathcal{H}[@t_f] = 0, \quad \text{and} \quad \Big(t \mapsto \mathcal{H}[@t]\Big) = \text{constant} \tag{4.25}$$

4. Prove that the thrust program ($t \mapsto T$) may switch at most once. (*Hint*: Differentiate the thrust switching function $S := \lambda_v/m - \lambda_m/v_e$ with respect to time.)

5. Scale the problem using canonical units. If necessary, refer back to Section 1.1.4 on page 20. Discuss the implications of choosing canonical units such that

$$m^0 = 1, \quad h^0 = 1, \quad \text{and} \quad g = 1 \tag{4.26}$$

6. Based on the choice of these canonical units, discuss the practical meaning of the following data:

$$T_{max} = 1.227, \quad v_e = 2.349, \quad \text{and} \quad v^0 = -0.783 \tag{4.27}$$

7. Solve the problem by any method and compare your solution to the one shown in Fig. 4.17.

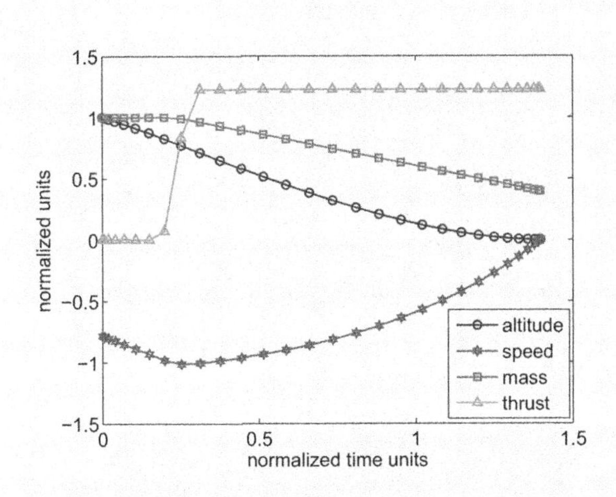

Figure 4.17: A DIDO-solution to Problem ML_{1D} obtained via the procedure discussed in Section 3.2 on pages 177–187.

8. The DIDO-generated costates and Hamiltonian evolution for this problem are shown in Fig. 4.18. Determine what features and values of these plots are checkable against theory. (*Hint*: Start with question # 1).

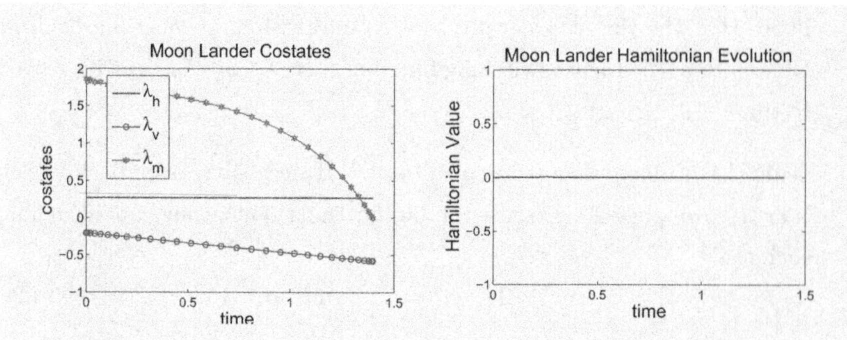

Figure 4.18: Costates and Hamiltonian evolution for Problem ML_{1D} generated by DIDO.

9. Extend the one-dimensional aspect of this problem to two dimensions using the variables indicated in Fig. 4.19. You may assume thrust-steering is provided independently by a "fast inner-loop" attitude control system as discussed on page 68 in Chapter 1.

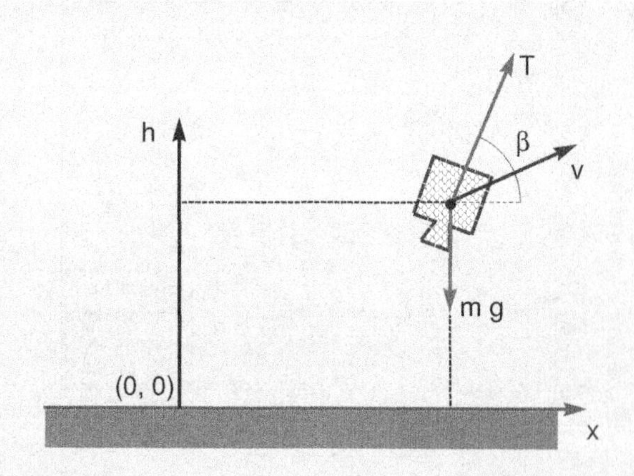

Figure 4.19: Schematic for a two-dimensional moon-landing problem.

4.8.2 Velocity Steering

A simpler version of the powered-explicit-guidance problem (given by Eq. (3.82) on page 232) may be posed as maximizing the final horizontal velocity under continuous acceleration:

$$\boldsymbol{x} := (x, y, v_x, v_y) \in \mathbb{R}^4, \quad \boldsymbol{u} := \beta \in \mathbb{R}$$

$$\left\{\begin{array}{lrl} \text{Minimize} & J[\boldsymbol{x}(\cdot), \boldsymbol{u}(\cdot), t_f] = & -v_x(t_f) \\ \text{Subject to} & \dot{x} = & v_x \\ & \dot{y} = & v_y \\ & \dot{v}_x = & A\cos\beta \\ & \dot{v}_y = & A\sin\beta \\ & (x, y, v_x, v_y)(t_0) = & (0, 0, 0, 0) \\ & (x, v_y)(t_f) = & (1, 0) \end{array}\right. \tag{4.28}$$

Develop the necessary conditions for this problem and compare it to those of Problem 3.82 discussed on page 232.

4.8.3 Max-Energy Orbit Transfer: Cartesian Formulation

See Fig. 4.20. The following problem maximizes the energy of the final orbit under continuous thrusting:

$$\boldsymbol{x} := (x, y, v_x, v_y) \in \mathbb{R}^4, \quad \boldsymbol{u} := \beta \in \mathbb{R}$$

$$\left\{\begin{array}{lrl} \text{Minimize} & J[\boldsymbol{x}(\cdot), \boldsymbol{u}(\cdot), t_f] := & \dfrac{\mu}{\sqrt{x_f^2 + y_f^2}} - \dfrac{v_{x_f}^2 + v_{y_f}^2}{2} \\ \text{Subject to} & \dot{x} = & v_x \\ & \dot{y} = & v_y \\ & \dot{v}_x = & -\dfrac{\mu x}{(x^2+y^2)^{3/2}} + A\cos\beta \\ & \dot{v}_y = & -\dfrac{\mu y}{(x^2+y^2)^{3/2}} + A\sin\beta \\ & t_0 = & 0 \\ & (x_0, y_0, v_{x_0}, v_{y_0}) = & (x^0, 0, 0, \sqrt{\mu/x_0}) \\ & t_f = & t_1\sqrt{x_0^3/\mu} \end{array}\right. \tag{4.29}$$

The quantities A, t_1, μ and x^0 are given positive numbers.

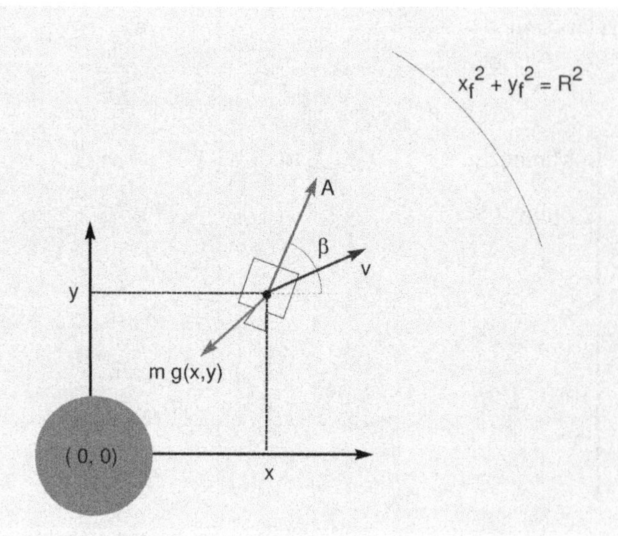

Figure 4.20: Schematic for the orbit transfer problems given by Problems (4.29) and (4.34). The latter problem is discussed on page 304.

1. Show that the adjoint equations are given by

$$\begin{aligned}
-\dot{\lambda}_x &= \lambda_{v_x}\left(-\frac{\mu}{(x^2+y^2)^{3/2}} + \frac{3\mu x^2}{(x^2+y^2)^{5/2}}\right) \\
&\qquad + \lambda_{v_y}\left(\frac{3\mu xy}{(x^2+y^2)^{5/2}}\right) \\
-\dot{\lambda}_y &= \lambda_{v_x}\left(\frac{3\mu xy}{(x^2+y^2)^{5/2}}\right) \\
&\qquad + \lambda_{v_y}\left(-\frac{\mu}{(x^2+y^2)^{3/2}} + \frac{3\mu y^2}{(x^2+y^2)^{5/2}}\right) \\
-\dot{\lambda}_{v_x} &= \lambda_x \\
-\dot{\lambda}_{v_y} &= \lambda_y
\end{aligned} \tag{4.30}$$

and that the adjoint covectors satisfy the transversality conditions

$$\begin{aligned}
\lambda_x(t_f) &= -\frac{\mu x_f}{(x_f^2 + y_f^2)^{3/2}} \\
\lambda_y(t_f) &= -\frac{\mu y_f}{(x_f^2 + y_f^2)^{3/2}} \\
\lambda_{v_x}(t_f) &= -v_{x_f} \\
\lambda_{v_y}(t_f) &= -v_{y_f}
\end{aligned}$$

2. Show that the optimal steering angle, β, must satisfy the equation

$$-\lambda_{v_x} \sin\beta + \lambda_{v_y} \cos\beta = 0 \tag{4.31}$$

3. Discuss the pros and cons of solving for β using either of the following two options:

   (a) $\beta = \tan^{-1}\left(\dfrac{\lambda_{v_y}}{\lambda_{v_x}}\right)$

   (b) Instead of β, use

$$\sin\beta = \frac{\lambda_{v_y}}{\sqrt{\lambda_{v_x}^2 + \lambda_{v_y}^2}}$$

$$\cos\beta = \frac{\lambda_{v_x}}{\sqrt{\lambda_{v_x}^2 + \lambda_{v_y}^2}}$$

   (*Hint*: Explore cases when the denominators go to zero.)

4. Generate the BVP:

   (a) In the minimalist's approach, Eq. (4.31) is used without solving for β. This generates a BVP with an algebraic constraint; hence we get a ***differential-algebraic BVP***.

   (b) In the "standard" approach, β is "eliminated" to arrive at a "simpler" BVP.

5. Using the following canonical units (also known as Schuler units)

$$DU := x^0 \qquad \text{(Distance Unit)}$$
$$VU := \sqrt{\mu/x^0} \qquad \text{(Velocity Unit)}$$
$$TU := VU/DU \qquad \text{(Time Unit)}$$

show that we can set

$$x^0 = 1 \; DU \qquad v_{y0} = 1 \; VU$$
$$\mu = 1 \; DU^3/TU^2 \qquad t_f = t_1 \; TU$$

regardless of the "actual" values of these quantities.

6. Reformulate the BVP in terms of the canonical units.

7. Using $A = 0.01$ and $t_1 = 10$ (all in canonical units) solve (or attempt to solve) the BVP using any numerical method (e.g., shooting, collocation). (*Warning*: You will likely have substantial difficulties, but that is the point of this exercise. *Also, you cannot "cheat" by coordinate transformations, which is also the point of this exercise!*)

   (a) Catalog all the numerical difficulties encountered in solving or attempting to solve this BVP. If you are successful, use the following two figures of merit in evaluating the approach: (i) Sensitivity and (ii) Run-time.

   (b) Take a second look at the denominators of the adjoint equations (and the state equations). Is this the cause of the numerical difficulty (for the collocation method)?

8. A solution to this problem using DIDO is shown in Fig. 4.21. Read-off any value of any variable from these plots and use it as a "guess" for the BVP solver. Does the BVP converge to a solution?

9. Although DIDO generated a solution to this problem without much trouble, when the problem is solved as stated on page 293, two common issues erupt that require a little more nuanced understanding of Pontryagin's Principle:

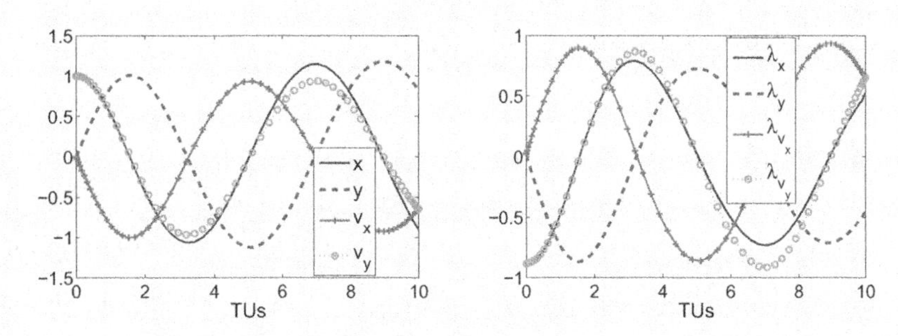

Figure 4.21: DIDO-generated states and costates to Problem 4.29 obtained via the procedure discussed in Section 3.2 on pages 177–187.

(a) First, note that there are no β-dynamics in the problem; that is there is no $\dot{\beta}$ equation. This implies, that β can "jump". ($\beta \in L^\infty$)

(b) Second, β does not directly affect $\dot{\boldsymbol{x}}$; only the sines and cosines of β do. This means that after one revolution, β can reset like the hands of a clock. This is called ***modulo-2π operation***.

Because DIDO and Pontryagin's Principle go hand-in-hand, it generates the control trajectory shown in Fig. 4.22. Note that there is nothing

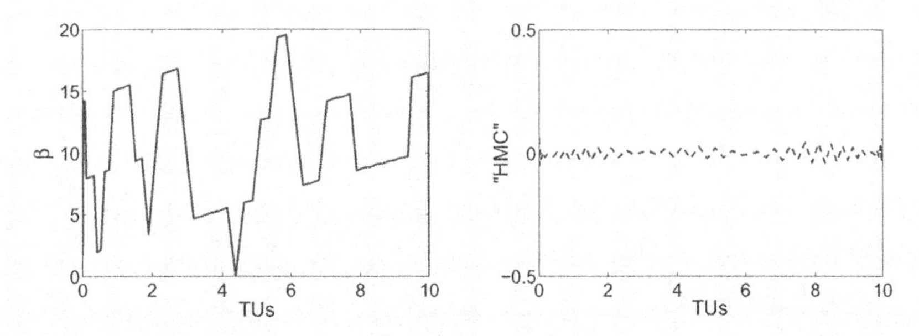

Figure 4.22: The control trajectory $t \mapsto \beta$ exhibits discontinuity and modulo-2π issues while faithfully satisfying the Hamiltonian Minimization Condition (see Eq. (4.31)) as indicated by the "HMC" plot on the right.

wrong with this result! If we insist that $t \mapsto \beta$ must be continuous, then it must be part of the problem formulation!! Refer to Chapter 1 for details on problem formulation and how different problem formulations generate different outcomes.

10. In principle, the modulo-2π issue can be addressed by re-parameterizing the control space $\mathbb{U}$ using two control variables u_x and u_y constrained to lie on a unit circle; see Fig. 1.14 on page 24 and the discussions contained therein (Recall that $\mathbb{U} = S^1$). While this "trick" can be used in many situations, in this problem we are also interested (albeit an unspecified requirement!) in the number of wrappings (or ***windings***) of 2π which cannot be obtained by the u_x-u_y parameters. A new trick is to add an integrator

$$\dot{\beta} = \omega \tag{4.32}$$

with ω taking the role of the control variable while β becomes a state variable. Note that when Eq. (4.32) is appended to Problem 4.29, it is technically a ***new problem***, and Pontryagin's Principle must be reapplied to this new problem to generate a new understanding (β is now in $W^{1,\infty}$). Perform this exercise and and show that $\lambda_\beta \equiv 0$ and ω must be singular! (In other words, singular arcs are not as uncommon as is commonly believed ...and their appearance can actually be a good thing!)

11. Does the new BVP generated through the addition of Eq. (4.32) produce an easier problem than without it? Solving this new problem in DIDO generates the same states and costates shown in Fig. 4.21 with the addition of the extra state (β) and costate (λ_β) shown in Fig. 4.24.

12. It is apparent from Fig. 4.24 that β winds past 2π. This aspect of the trajectory is more apparent in Fig. 4.24. What happens if t_1 is increased to a value greater than 10?

13. A plot of the pair $(\lambda_{v_x}, \lambda_{v_y})$ as a vector emanating from the point (x, y) is shown in Fig. 4.25. Note that this vector appears to be opposite to that of the A-vector shown in Fig. 4.24. This is not an accident; it was discovered by Lawden[57] who named it the ***primer vector***. Also shown in Fig. 4.25 is the adjoint covector projected in the x-y plane; i.e., the pair (λ_x, λ_y) as a vector emanating from the point (x, y). This vector is called the

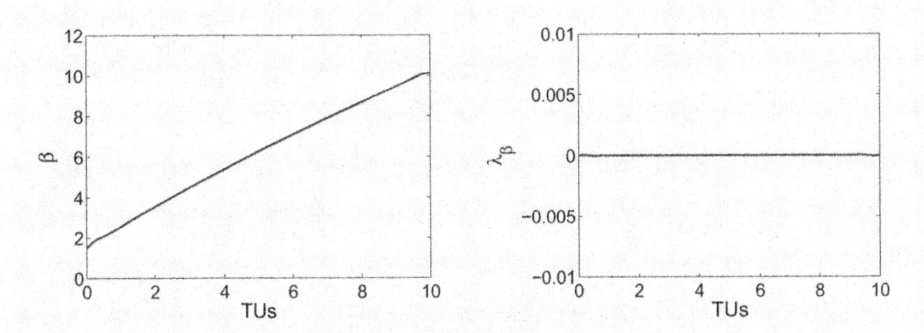

Figure 4.23: The addition of Eq. (4.32) generates a continuous control trajectory $t \mapsto \beta$. The additional costate trajectory $t \mapsto \lambda_\beta$ is identically zero according to DIDO.

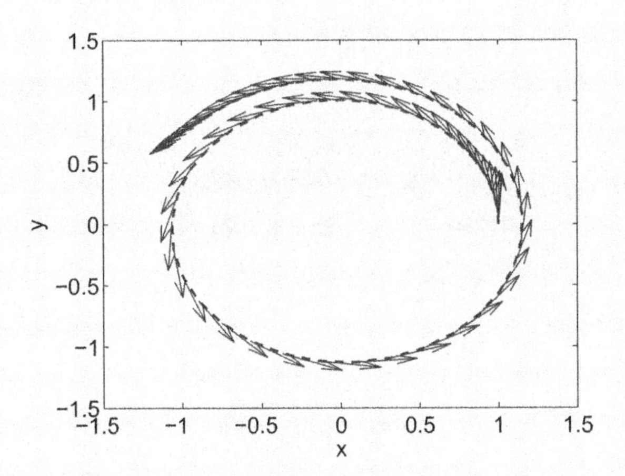

Figure 4.24: The control acceleration vector A, indicated by the arrows along the x-y trajectory. This visual illustrates the the need to track the windings of the system trajectory as well as the control vector.

proxy vector to the (projection of the) adjoint covector. It is connected to the *Riesz representation theorem*; hence, it is also known as a ***Riesz vector***. Refer to Fig. 2.5 on page 93 for a more practical introduction to the meaning of covectors and graphical means to represent them.

14. Discuss what happens when the problem is made more realistic with the

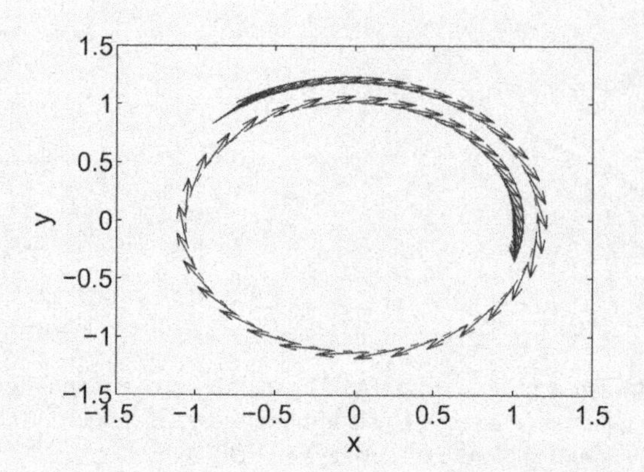

Figure 4.25: The pair $(\lambda_{v_x}, \lambda_{v_y})$, plotted as a vector field along the x-y trajectory is shown on top: It is called the primer vector and has a special significance in orbital maneuvers[64]. The pair (λ_x, λ_y) is plotted similarly below: It is called a proxy vector to the adjoint covector.

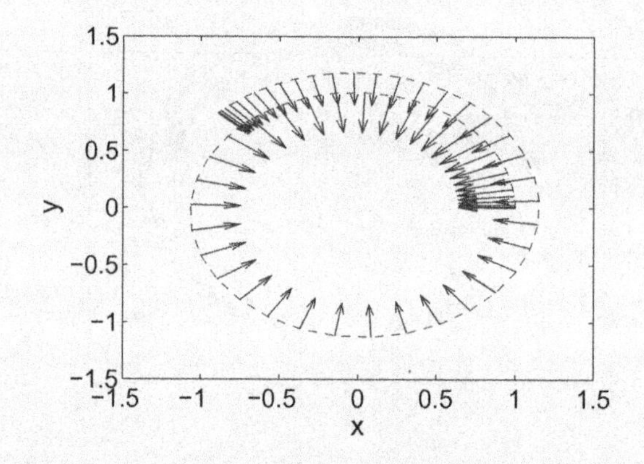

addition of the third "z" coordinate. (Note: Industrial-strength problem have additional differential equations, several path constraints and target sets.)

15. Based on your experience with this simple problem, discuss if a BVP approach is the preferred way to solve optimal control problems. Refer to Section 2.4.4 on page 113 for additional context.

4.8.4 Max-Energy Orbit Transfer: Polar Formulation

See Fig. 4.26. Problem 4.29, discussed in Section 4.8.3, can be reformulated in terms of polar coordinates as:

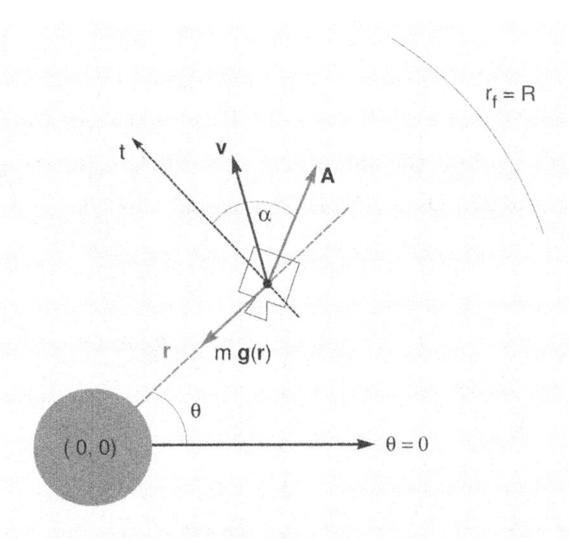

Figure 4.26: Schematic for the orbit transfer problem given by Eq. (4.33).

$$\boldsymbol{x} := (r, \theta, v_r, v_t) \in \mathbb{R}^4, \quad \boldsymbol{u} := \alpha \in \mathbb{R}$$

$$\left\{\begin{array}{lrl}
\text{Minimize} & J[\boldsymbol{x}(\cdot), \boldsymbol{u}(\cdot), t_f] := & \dfrac{\mu}{r_f} - \dfrac{v_{r_f}^2 + v_{t_f}^2}{2} \\
\text{Subject to} & \dot{r} = & v_r \\
 & \dot{\theta} = & \dfrac{v_t}{r} \\
 & \dot{v}_r = & \dfrac{v_t^2}{r} - \dfrac{\mu}{r^2} + A \sin\alpha \\
 & \dot{v}_t = & -\dfrac{v_r v_t}{r} + A\cos\alpha \\
 & t_0 = & 0 \\
 & (r_0, \theta_0, v_{r_0}, v_{t_0}) = & (r^0, 0, 0, \sqrt{\mu/r_0}) \\
 & t_f = & t_1\sqrt{r_0^3/\mu}
\end{array}\right. \tag{4.33}$$

1. Generate a BVP for this problem (in canonical units) by following the procedure discussed in Section 4.8.3.

2. Solve the BVP for the same values of A and t_1 as in Section 4.8.3; i.e., $A = 0.01$ and $t_1 = 10$. Unlike the Cartesian formulation of the previous section where the BVP was substantially difficult, the polar formulation is more amenable to classical computational techniques (like shooting and collocation; see Section 2.9.2 on page 157). Regardless, you may still experience difficulty. If so, you may now "cheat" by using a guess for the initial value of $\lambda(t_0) = (-1, 0, 0, -1)$. If this cheating does not alleviate the problem, read off the values of any variable from the DIDO results shown in Fig. 4.27. Does the BVP converge now? Is λ_θ a constant and

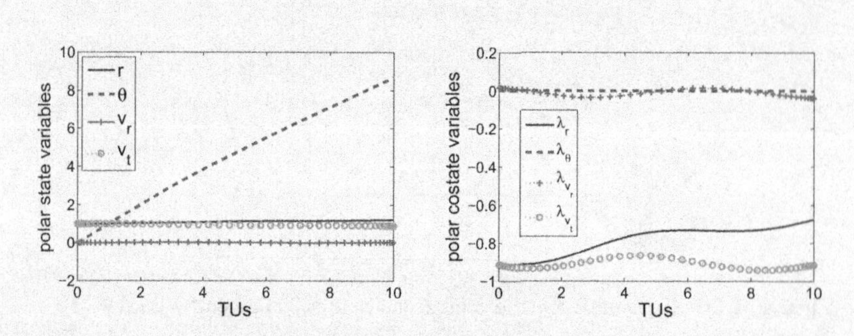

Figure 4.27: DIDO-generated states and costates to Problem 4.33 obtained via the procedure discussed in Section 3.2 on pages 177–187.

equal to zero? Explain.

3. This problem is solved in [37] for $t_1 = 50$ using a 5th-degree Gauss-Lobatto collocation method. Changing the value of t_1 in DIDO and resolving the problem generates the results shown in Fig. 4.28. These results, produced by M. Karpenko, are consistent with those of Herman and Conway[37].

4. Show that the evolution of the Hamiltonian for this problem follows a constant value. A DIDO-generated Hamiltonian for this problem is shown in Fig. 4.29. Figure is courtesy of M. Karpenko. Is the value of the Hamiltonian zero? Explain why or why not.

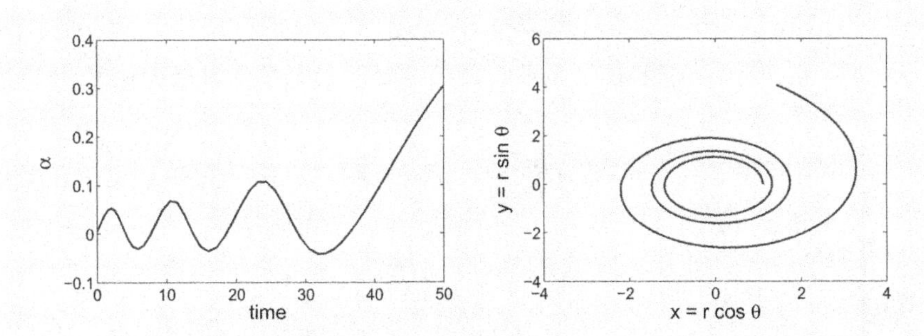

Figure 4.28: DIDO-generated control and a "spiral" orbital trajectory to Problem 4.33 for $t_1 = 50$.

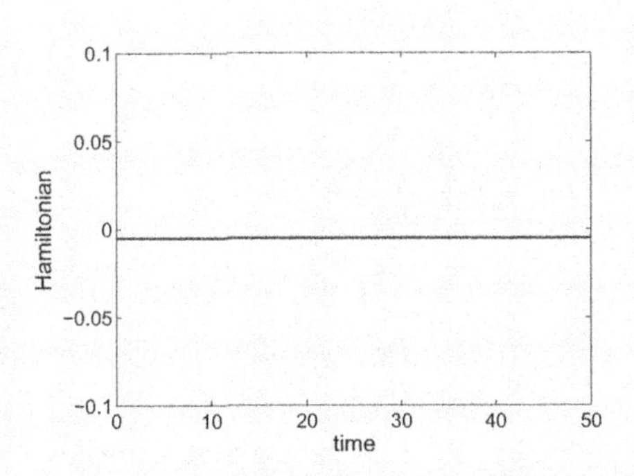

Figure 4.29: DIDO-generated Hamiltonian to Problem 4.33 with $t_1 = 50$ obtained via the procedure discussed in Section 3.2 on pages 177–187.

4.8.5 Minimum-Propellant Orbit Transfer

A dominant issue in orbital maneuvers is propellant. A classic problem in astrodynamics is a circle-to-circle orbit transfer. Refer to Fig. 4.20 on page 294, and replace the symbol A by T. A minimum-propellant circle-to-circle orbit transfer problem can be posed as follows:

$$\boldsymbol{x} := (x, y, v_x, v_y, m) \in \mathbb{R}^5, \quad \boldsymbol{u} := (T, \beta) \in \mathbb{U} = \{(T, \beta) : \ 0 \leq T \leq T_{max}\}$$

$$\left\{\begin{array}{lrl}
\text{Minimize} & J[\boldsymbol{x}(\cdot), \boldsymbol{u}(\cdot), t_f] := & m_0 - m_f \\
\text{Subject to} & \dot{x} = & v_x \\
& \dot{y} = & v_y \\
& \dot{v}_x = & -\dfrac{\mu x}{(x^2 + y^2)^{3/2}} + \dfrac{T}{m} \cos\beta \\
& \dot{v}_y = & -\dfrac{\mu y}{(x^2 + y^2)^{3/2}} + \dfrac{T}{m} \sin\beta \\
& \dot{m} = & -\dfrac{T}{v_e} \\
& t_0 = & 0 \\
& (x_0, y_0, v_{x_0}, v_{y_0}, m_0) = & (x^0, 0, 0, \sqrt{\mu/x_0}, m^0) \\
& x_f^2 + y_f^2 - R^2 = & 0 \\
& v_{x_f}^2 + v_{y_f}^2 - \dfrac{\mu}{R} = & 0 \\
& x_f v_{x_f} + y_f v_{y_f} = & 0
\end{array}\right. \tag{4.34}$$

The quantities $T_{max}, \mu, x^0, m^0, v_e$, and R are given positive numbers.

1. Show that the adjoint equations for Problem (4.34) are the same as (4.29) with the addition of a mass adjoint equation (mass co-flow rate equation) given by

$$\dot{\lambda}_m = \frac{\lambda_{v_x} T \cos\beta}{m^2} + \frac{\lambda_{v_y} T \sin\beta}{m^2} \tag{4.35}$$

2. Next, show that the endpoint Lagrangian can be written as

$$\overline{E}(\boldsymbol{\nu}, \boldsymbol{x}_f, t_f) := m_0 - m_f + \nu_1(x_f^2 + y_f^2 - R^2) + \nu_2\left(v_{x_f}^2 + v_{y_f}^2 - \frac{\mu}{R}\right) + \nu_3\left(x_f v_{x_f} + y_f v_{y_f}\right) \tag{4.36}$$

and hence, the transversality conditions are given by

$$
\begin{aligned}
\lambda_x(t_f) &= 2\nu_1 x_f + \nu_3 v_{x_f} \\
\lambda_y(t_f) &= 2\nu_1 y_f + \nu_3 v_{y_f} \\
\lambda_{v_x}(t_f) &= 2\nu_2 v_{x_f} + \nu_3 x_f \\
\lambda_{v_y}(t_f) &= 2\nu_2 v_{y_f} + \nu_3 y_f \\
\lambda_m(t_f) &= -1
\end{aligned}
$$

3. Show that the optimal steering angle, β, satisfies Eq. (4.31) while the optimal thrust program is given by

$$T = T_{max}\ \mathrm{step}(-S)$$

where, step is the Heaviside step function, defined by

$$\mathrm{step}(x) = \begin{cases} 0 & \text{if } x < 0 \\ [0,1] & \text{if } x = 0 \\ 1 & \text{if } x > 0 \end{cases}$$

and S, called the switching function, is given by

$$S := \frac{\lambda_{v_x}\cos\beta}{m} + \frac{\lambda_{v_y}\sin\beta}{m} - \frac{\lambda_m}{v_e}$$

4. Generate the BVP.

   (a) Document your difficulties in the context of generating the function $\boldsymbol{u} = \arg\min_{\boldsymbol{u}} H(\boldsymbol{\lambda}, \boldsymbol{x}, \boldsymbol{u}, t)$.

   (b) Can the boundary conditions be formulated in a "standard" form; i.e., in forms employed by most professional BVP solvers? (*Hint*: Consider eliminating ν_1, ν_2, ν_3 from the transversality conditions.)

5. Reformulate the BVP in terms of the canonical units and solve it for $T_{max} = 0.05$ and $R = 2$ (all in canonical Schuler units; see page 296). Unlike Problem (4.29), you may "cheat" by coordinate transformations if you experience difficulties. (*Warning*: Your difficulties may not go away. Catalog all the numerical difficulties encountered in solving or attempt-

ing to solve this BVP. Use at least the following two figures of merit: (i) Sensitivity and (ii) Run-time.)

6. Change the cost function to transfer time; i.e., reformulate the problem to a minimum time problem. Do the numerical difficulties go away? See [25, 64, 79, 84, 78, 86] for a variety of solutions to this problem and its problem of problems.

7. Discuss what happens when the problem is made more realistic with the addition of the third "z" coordinate. (Note: Industrial-strength problem have additional differential equations, several path constraints and non-circular target sets.)

4.9 A Rigid-Body Steering Problem

Steering a rigid body is a fundamental control problem in aerospace, robotics, and many other engineering disciplines. An illustrative optimal control problem for "kinematical" steering can be framed as:

$$\boldsymbol{x} := (q_1, q_2, q_3, q_4, \omega_1, \omega_2, \omega_3) \in \mathbb{R}^7 \qquad \boldsymbol{u} := (u_1, u_2, u_3) \in \mathbb{R}^3$$

$$\left\{\begin{array}{lrcl} \textsf{Minimize} & J[\boldsymbol{x}(\cdot), \boldsymbol{u}(\cdot), t_f] &:=& \dfrac{1}{2}\displaystyle\int_{t_0}^{t_f} \left(u_1^2(t) + u_2^2(t) + u_3^2(t)\right) dt \\ \textsf{Subject to} & \dot{q}_1 &=& \frac{1}{2}\left(\omega_1 q_4 - \omega_2 q_3 + \omega_3 q_2\right) \\ & \dot{q}_2 &=& \frac{1}{2}\left(\omega_1 q_3 + \omega_2 q_4 - \omega_3 q_1\right) \\ & \dot{q}_3 &=& \frac{1}{2}\left(-\omega_1 q_2 + \omega_2 q_1 + \omega_3 q_4\right) \\ & \dot{q}_4 &=& \frac{1}{2}\left(-\omega_1 q_1 - \omega_2 q_2 - \omega_3 q_3\right) \\ & \dot{\omega}_1 &=& u_1 \\ & \dot{\omega}_2 &=& u_2 \\ & \dot{\omega}_3 &=& u_3 \\ & (\boldsymbol{q}_0, \boldsymbol{\omega}_0) &=& (\boldsymbol{q}^0, \boldsymbol{0}) \\ & (\boldsymbol{q}_f, \boldsymbol{\omega}_f) &=& (\boldsymbol{q}^f, \boldsymbol{0}) \\ & (t_0, t_f) &=& (0, t^f) \end{array}\right. \tag{4.37}$$

The quantities $\boldsymbol{q}^0, \boldsymbol{q}^f$ and t^f are given numbers such that $\|\boldsymbol{q}^0\|_2^2 = 1 = \|\boldsymbol{q}^f\|_2^2$.

1. Show that the optimal control and the adjoint equations are given by:

$$u_1 = -\lambda_{\omega_1}, \qquad u_2 = -\lambda_{\omega_2}, \qquad u_3 = -\lambda_{\omega_3} \tag{4.38}$$

$$\begin{aligned}
-\dot{\lambda}_{q_1} &= \frac{1}{2}\left(-\lambda_{q_2}\omega_3 + \lambda_{q_3}\omega_2 - \lambda_{q_4}\omega_1\right)\\
-\dot{\lambda}_{q_2} &= \frac{1}{2}\left(\lambda_{q_1}\omega_3 - \lambda_{q_3}\omega_1 - \lambda_{q_4}\omega_2\right)\\
-\dot{\lambda}_{q_3} &= \frac{1}{2}\left(-\lambda_{q_1}\omega_2 + \lambda_{q_2}\omega_1 - \lambda_{q_4}\omega_3\right)\\
-\dot{\lambda}_{q_4} &= \frac{1}{2}\left(\lambda_{q_1}\omega_1 + \lambda_{q_2}\omega_2 + \lambda_{q_3}\omega_3\right)\\
-\dot{\lambda}_{\omega_1} &= \frac{1}{2}\left(\lambda_{q_1}q_4 + \lambda_{q_2}q_3 - \lambda_{q_3}q_2 - \lambda_{q_4}q_1\right)\\
-\dot{\lambda}_{\omega_2} &= \frac{1}{2}\left(-\lambda_{q_1}q_3 + \lambda_{q_2}q_4 + \lambda_{q_3}q_1 - \lambda_{q_4}q_2\right)\\
-\dot{\lambda}_{\omega_3} &= \frac{1}{2}\left(\lambda_{q_1}q_2 - \lambda_{q_2}q_1 + \lambda_{q_3}q_4 - \lambda_{q_4}q_3\right)
\end{aligned}$$

2. Construct the BVP for this problem. Solve the problem via hbshoot (see page 249), its variant or any other BVP solver for the following data:

$$\begin{aligned}
\boldsymbol{q}_0 &= (0, 0, 0, 1)\\
\boldsymbol{q}_f &= (0.06708, -0.1118, 0.2236, 0.9659)\\
t_f &= 20
\end{aligned}$$

3. *If you experience difficulties with question 2, refer to Study Problem 2.16 on page 145.* If you experience continued difficulties, try using the following "clever" guess for the costates:

$$\begin{aligned}
\boldsymbol{\lambda}_{q_0} &= (-0.1628,\ 0.2714,\ -0.5428,\ 0)\\
\boldsymbol{\lambda}_{\omega_0} &= (-0.0406,\ 0.0677,\ -0.1354)
\end{aligned} \tag{4.39}$$

   Results from a version of hbshoot, developed by M. Karpenko, are shown in Fig. 4.30. Your results for $t \mapsto \boldsymbol{\lambda}_q$ may be different than the one shown in Fig. 4.30.

4. From the fourth quadrant of Fig. 4.30, $t \mapsto \boldsymbol{\lambda}_\omega$ appears to be linear. Prove

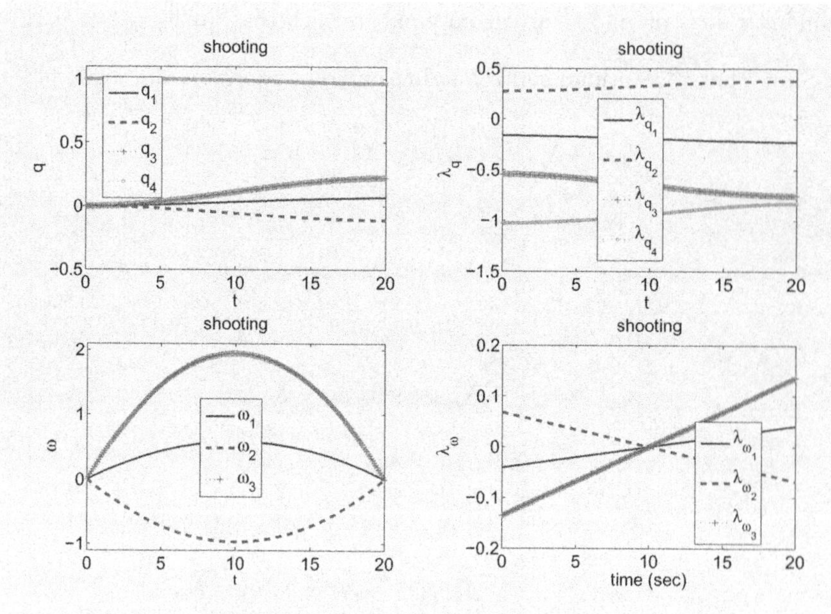

Figure 4.30: Solution to the BVP generated by Problem 4.37. Results generated via a shooting method (hbshoot) that used Eq. (4.39) as a guess and a scale factor of 20.

or disprove this conjecture. (*Hint*: Consult Section 3.3 on page 191, and in particular Fig. 3.14 on page 197.)

5. According to Study Problem 2.16, the function $t \mapsto \boldsymbol{\lambda}_q \cdot \boldsymbol{\lambda}_q$ (where q is a 4-vector) must be a constant. Demonstrate the numerical validity of your solution by computing this integral of motion. Karpenko's results are shown in Fig. 4.31 along with the Hamiltonian evolution.

6. Recall from Section 2.9 discussed on page 150 that the shooting method has an inherent ***curse of sensitivity***. Although the horizon for this problem is relatively short, a shooting method fails to converge if the guess is not very good. The "lucky" guess provided in Eq. (4.39) was not lucky after all; it was generated by solving the problem in DIDO. The DIDO-generated results, produced by M. Karpenko, are shown in Fig. 4.32. Observe that the costate trajectory $t \mapsto \boldsymbol{\lambda}_q$ produced by DIDO is different from the shooting method even though the guess for the shooting method was almost exactly the initial values from DIDO. Explain this

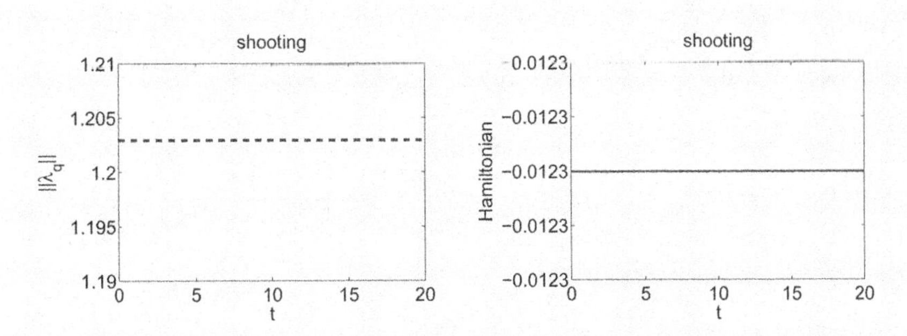

Figure 4.31: Verification and validation of the results from the shooting method via a check on the integrals of motion generated by Study Problem 2.16 discussed on page 145.

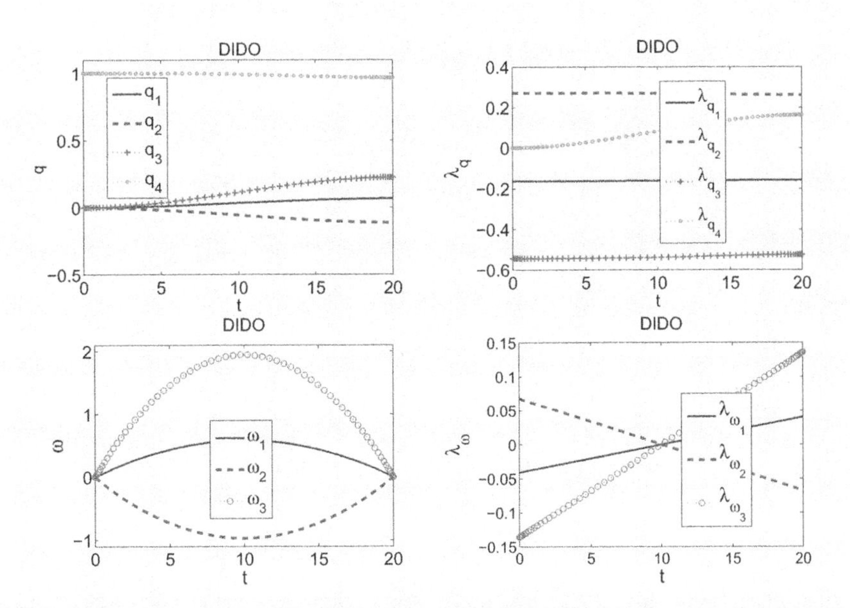

Figure 4.32: DIDO-generated states and costates to Problem 4.37 obtained via the procedure discussed in Section 3.2 on pages 177–187.

phenomenon using the results from Study Problem 2.16. The DIDO companion to Fig. 4.31 is shown in Fig. 4.33.

7. The controls from both DIDO and the shooting method are shown are

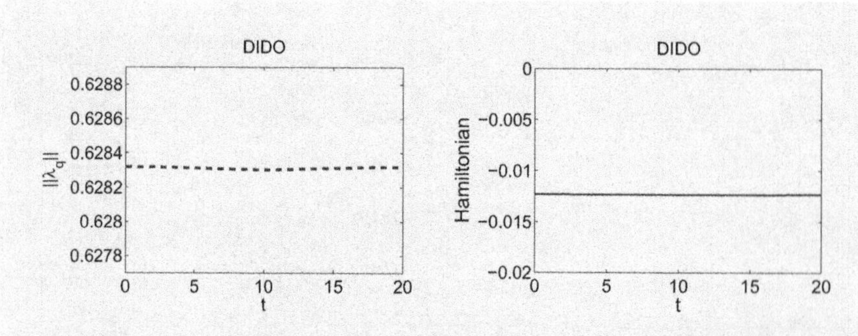

Figure 4.33: Verification and validation of the DIDO results via a check on the integrals of motion generated by Study Problem 2.16 discussed on page 145.

Fig. 4.34. Clearly they match as they should. Note that the controls

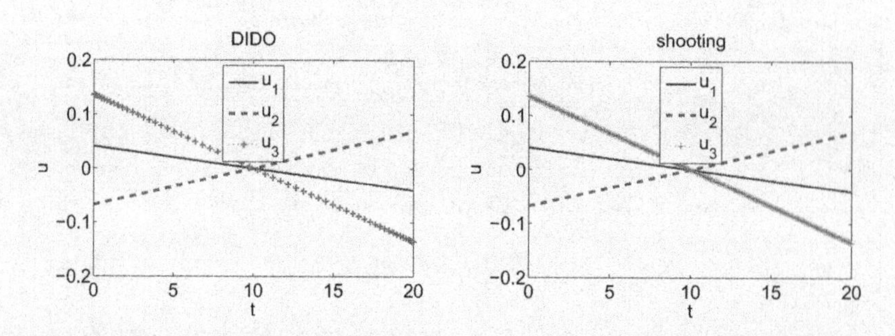

Figure 4.34: DIDO-controls to Problem 4.37 alongside those from the shooting method.

appear to be linear. Prove or disprove this conjecture. (*Hint*: Consult Section 3.3 on page 191, and in particular Fig. 3.14 on page 197.)

8. The solution presented in Fig. 4.34 can be also be verified and validated as discussed in Section 3.2.1 on page 180. Examine the plots for the controls (DIDO or shooting) and determine if they satisfy Eq. (4.38). Also, prove that the Hamiltonian evolution equation generates a constant value. Is it possible to generate the value of this constant by analytical methods?

9. When the BVP is solved using the ***collocation method*** implemented in

MATLAB's bvp4c, the results are exactly the same as before except for $t \mapsto \boldsymbol{\lambda}_q$ as shown in Fig. 4.35. Explain why the co-quaternion trajectories

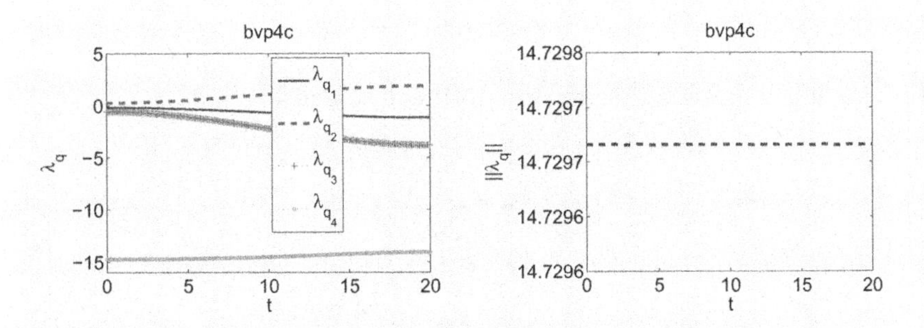

Figure 4.35: Adjoint solution to the BVP generated by Problem 4.37 using the collocation method implemented in MATLAB's bvp4c.

are different using the results from Study Problem 2.16.

10. Change the cost function to transfer time and redo the problem. See [87] and the references contained therein for an extensive discussion of this problem, its origins, and subsequent generalizations that eventually led to its various flight implementations[51] for maneuvering spacecraft for imaging, data dumping, occultation avoidance etc.

4.10 A Coupled Motion-Planning Problem

In motion planning, it is customary to assume the translational motion of a rigid body to be decoupled from its rotational motion. While this is true in many practical systems, it is possible to move objects without such a design. This problem is best addressed in the context of the work flow indicated in Fig. 1.50 on page 84.

A Canonical Problem

Consider the planar motion of a rigid body that is actuated by an off-center force, $\mathbf{F}$, shown in Fig. 4.36. Show that the equations of motion for this system can be written as,

$$\begin{aligned} \ddot{x} &= u \cos\beta \\ \ddot{y} &= u \sin\beta \\ \ddot{\beta} &= \ell u \end{aligned} \tag{4.40}$$

where, $\ell \neq 0$ is a given parameter (proportional to c in Fig. 4.36).

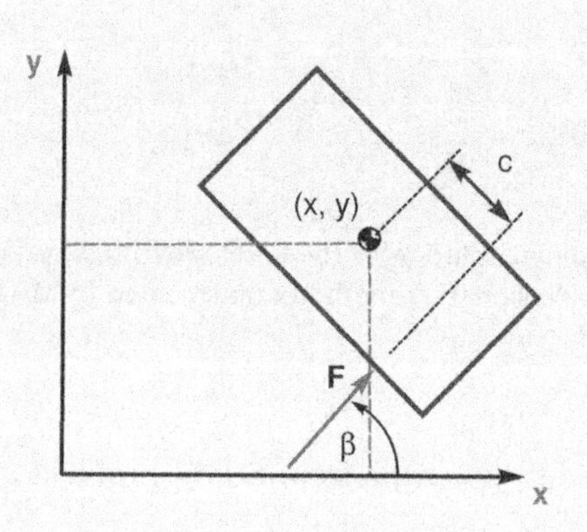

Figure 4.36: Schematic for the dynamics given by Eq. (4.40).

1. Suppose that at $t_0 = 0$, we have $(\dot{x}, \dot{y}, \dot{\beta}) \neq \mathbf{0}$. Is it possible to drive $(\dot{x}, \dot{y}, \dot{\beta}) = \mathbf{0}$ within some time $t_f < \infty$?

2. Construct an optimal control problem to transfer the rigid body from point A to B using the integral of the square of the control as a cost functional

$$J[\boldsymbol{x}(\cdot), \boldsymbol{u}(\cdot), t_f] := \int_{t_0}^{t_f} u^2(t)\, dt$$

   Show that the optimal control is given by

$$u = -(\lambda_{v_x} \cos\beta + \lambda_{v_y} \sin\beta + \lambda_\omega)/2$$

   where $v_x := \dot{x}, v_y := \dot{y}$ and $\omega := \dot{\beta}$. Develop the remainder of the necessary conditions for this optimal control problem.

3. Do the necessary conditions indicate any sign of trouble with the choice of points A and B?

4. Choose $\ell = 1$. Let $t_0 = 0$ and $t_f = 2$. Use any method to solve the problem for the following endpoint conditions:

$$
\begin{aligned}
(x, v_x, y, v_y, \beta, \omega)(t_0) &= \mathbf{0} \\
(x, v_x, y, v_y, \beta, \omega)(t_f) &= (5, 0, 10, 0, \pi, 0)
\end{aligned}
$$

Compare your results with the one shown in Fig. 4.37. It is apparent from this figure, as well as from the physics of the problem, that the solution exhibits a multiscale phenomenon. Figure 4.37 was produced by Q. Gong using a ***multiscale pseudospectral method***.

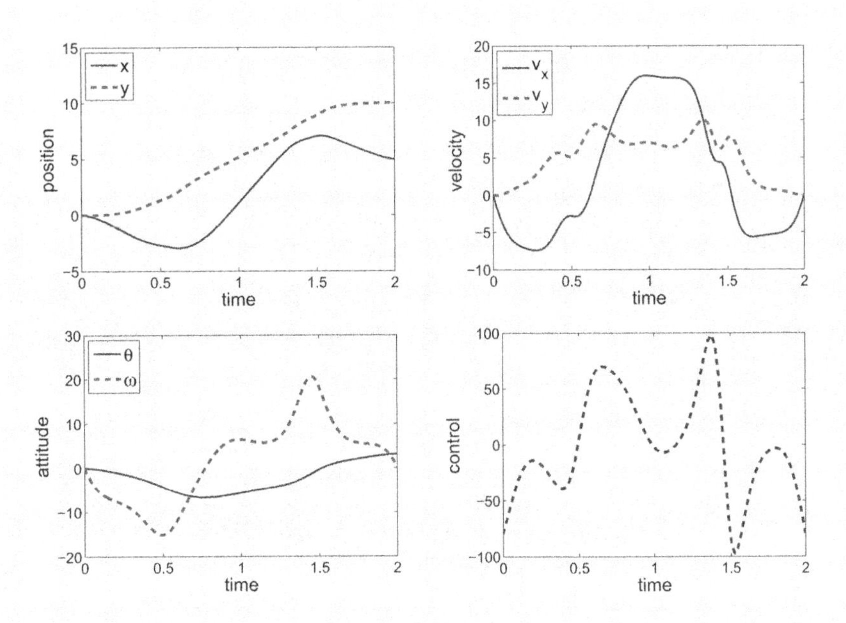

Figure 4.37: A pseudospectral solution to the canonical motion-planning problem.

5. Regardless of your method of choice, show that your solution satisfies Pontryagin's Principle. Compare your solution with Gong's results shown in Fig. 4.38.

   (a) Figure 4.38, quadrants one and two, indicate that λ_x and λ_y are constants. Is this consistent with the necessary conditions developed in question 2?

(b) Figure 4.38, quadrants one and two, indicate that λ_{v_x} and λ_{v_y} are are linear. Is this consistent with the necessary conditions developed in question 2?

(c) Figure 4.38, quadrant four, indicates that the Hamiltonian is a constant. Is this consistent with the Hamiltonian evolution equation?

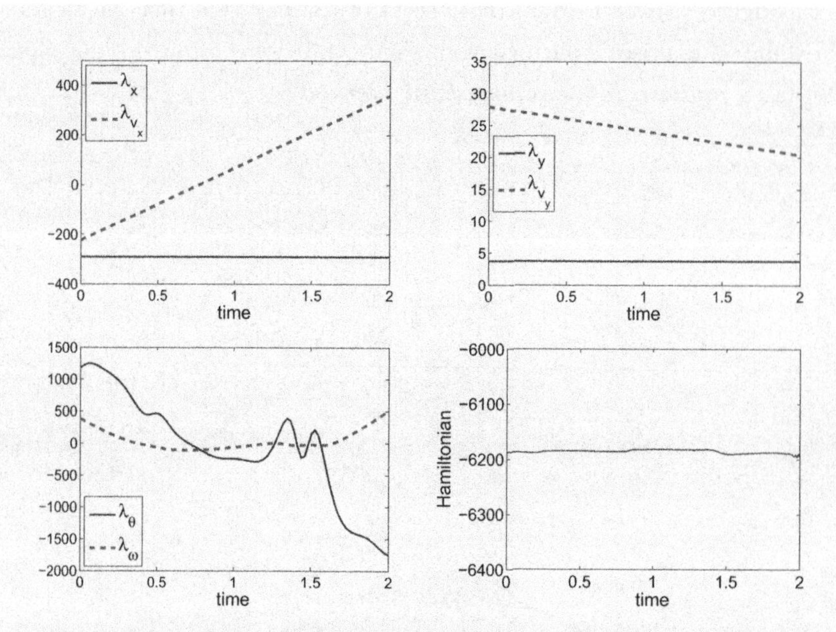

Figure 4.38: Costates and Hamiltonian evolution obtained by an application of the ***Covector Mapping Principle***[87] for the canonical motion-planning problem.

6. Let $u \in [-1, 1]$. Construct a minimum-time problem and analyze the resulting necessary conditions.

7. Rewrite Eq. (4.40) in state-space form and analyze the resulting control-affine system using standard tools from nonlinear control theory. (*Hint*: You may be able to construct the Lie brackets more easily using a symbolic tool box; the analysis, however, is nontrivial.)

8. Design and discuss other interesting questions that may aid understanding this problem.

Self-Retrieval of a Failed Spacecraft

The canonical problem illustrated in Fig. 4.36 was motivated by a problem of retrieving a failed spacecraft in orbit; see Fig. 4.39. Ignoring gravity gradient,

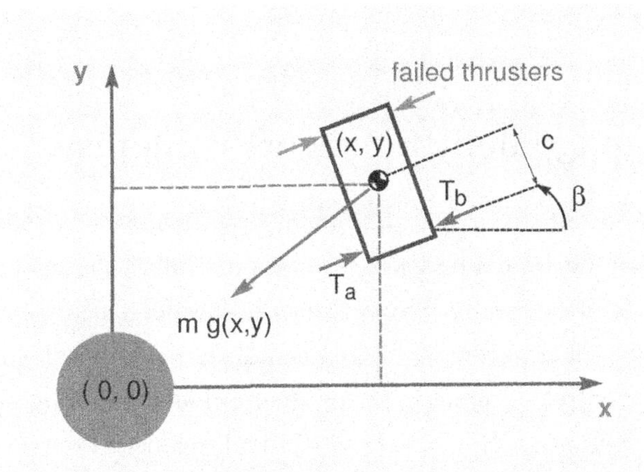

Figure 4.39: Schematic for the dynamics given by Eq. (4.41).

show that the equations of motion for this application are given by

$$\begin{aligned}\ddot{x} &= -\frac{\mu x}{(x^2+y^2)^{3/2}} + u\cos\beta \\ \ddot{y} &= -\frac{\mu y}{(x^2+y^2)^{3/2}} + u\sin\beta \\ \ddot{\beta} &= \ell u\end{aligned} \tag{4.41}$$

where ℓ is proportional to the parameter c indicated in Fig. 4.39. The two working thrusters, $T_a \geq 0$ and $T_b \geq 0$, provide the control $u \in \mathbb{R}$.

A logical cost functional for this problem is the L^1 norm of the control (a proxy for measuring propellant in the absence of including a mass-flow rate equation); see [79] for details.

1. Repeat the entire analysis of the canonical problem for this modified problem.

2. What happens to controllability when the gravity-gradient term is included?

3. Reformulate the problem in polar coordinates, and repeat the analysis.

4. Extend the equations of motion for the full three-dimensions, and repeat the analysis. You may assume a "third" thruster at the "bottom" of spacecraft with the same off-set as T_a and T_b.

4.11 Management Science Problems

In management science, economics and many other fields, one of the main problems is modeling the system under consideration. Unlike physics-based systems where a model can be easily derived from the laws of nature, in many other applications (e.g., pharmacology, environmental science) a mathematical model must be developed based on some set of assumptions, reasoning and empirical proof. *This is not easy!* But once a model is developed, Pontryagin's Principle can be applied systematically (for example, via the *HAMVET* mnemonic described on page 129) to produce answers along the work flow indicated in Fig. 1.50, page 84. Then the answers to the posed problem can be used to reformulate the problem.

A central assumption used in formulating such problems is that the system is describable by means of differential equations. This means that discrete variables are "smoothed out" to allow the use of calculus for modeling the dynamics. The following classical problems illustrate some of these principles.

4.11.1 A Wheat-Trading Problem

Wheat-trading problems are discussed in many books on optimal control. We follow the original version formulated by Ijiri and Thompson[41].

A firm has only two types of assets: cash $c(t)$ measured in dollars and wheat $w(t)$ measured in bushels. Let $m(t)$ denote the rate of buying or selling wheat, expressed in bushels per unit time; then, we can write

$$\dot{w}(t) = m(t)$$

If $m(t)$ is positive, wheat is purchased, negative if sold. Assume $m(t)$ is bound-constrained by

$$-M^{sell} \leq m(t) \leq M^{buy}$$

where $M^{sell} > 0$ and $M^{buy} > 0$ are given numbers which denote the maximum rate at which the firm is permitted to sell or buy wheat. (Why is this necessary?)

Let $h(w)$ be the inventory holding cost that must be paid on an inventory of size w: The more the bushels of wheat to store, the larger the holding cost. This holding cost must be paid out from the firm's cash asset $c(t)$; hence, the firm's cash is depleted at the rate

$$\dot{c}(t) = -h(w(t)) \tag{4.42}$$

The firm's cash is also depleted when it buys wheat at the current market value $p(t) > 0$ measured in dollars per bushel; hence, Eq. (4.42) must be modified to

$$\dot{c}(t) = -h(w(t)) - p(t)m(t) \tag{4.43}$$

Note that Eq. (4.43) automatically accounts for revenue generated through the sale of wheat when $m(t) < 0$. Revenue is also generated through interest received on cash; hence, the cash-balance equation can be completed by

$$\dot{c}(t) = Rc(t) - h(w(t)) - p(t)m(t) \tag{4.44}$$

where $R > 0$ is a constant interest rate.

The firm's objective is to maximize its total asset at $t = t_f$ given by

$$c(t_f) + p(t_f)w(t_f)$$

Collecting all of the relevant equations, the optimal wheat-trading problem can be posed as

$$\boldsymbol{x} := (c, w) \in \mathbb{R}^2, \quad \boldsymbol{u} := m \in \left\{m \in \mathbb{R} : \ -M^{sell} \leq m \leq M^{buy}\right\}$$

$$(WT) \begin{cases} \text{Maximize} & J[\boldsymbol{x}(\cdot), \boldsymbol{u}(\cdot), t_f] := c(t_f) + p(t_f)w(t_f) \\ \text{Subject to} & \dot{c} = Rc - h(w) - p(t)m \\ & \dot{w} = m \\ & (c_0, w_0, t_0) = (c^0, w^0, 0) \\ & t_f = t^f \end{cases}$$

where, c^0, w^0 are given numbers that correspond to the firm's assets at $t = 0$. The final time t^f is a given planning horizon. The model assumes the market value of wheat at any time t is known; hence, the function $t \mapsto p(t)$ is assumed to be known.

1. Develop all the necessary conditions for optimality.

2. Show that the adjoint co-cash variable $\lambda_c(t)$ is given explicitly by

$$\lambda_c(t) = -e^{R(t_f - t)}$$

3. Show that the adjoint co-wheat variable $\lambda_w(t)$ is given by

$$\lambda_w(t) = -p(t_f) + \int_t^{t_f} \partial_w h(w(s))\, e^{R(t_f - s)}\, ds$$

4. Assuming

$$S(t) := e^{R(t_f - t)} p(t) - p(t_f) + \int_t^{t_f} \partial_w h(w(s))\, e^{R(t_f - s)}\, ds \neq 0$$

   show that the optimal trading policy consists of buying and selling wheat at the maximum permissible values of M^{buy} and M^{sell} respectively. This is known as ***bang-bang control***.

5. Consider the special case of $R = 0$ and $h(w) = Qw$ where $Q > 0$. Show that the optimal trading policy is as shown in Fig. 4.40.

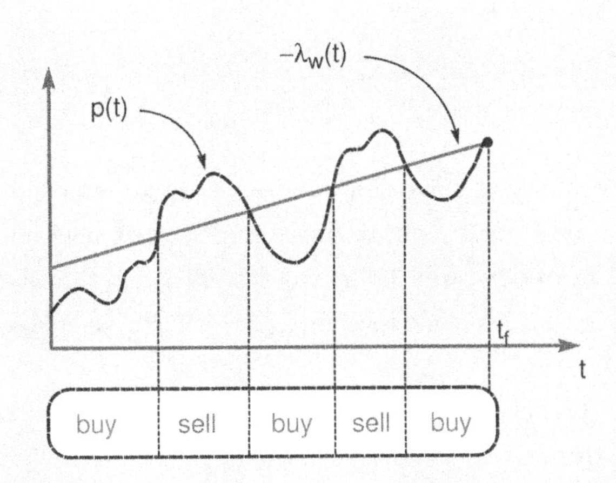

Figure 4.40: The optimal trading policy for buying and selling wheat for a given variation in the market price.

Note that while optimal control theory conforms with the common-sense to buy low and sell high, Fig. 4.40 shows that not all (low-) high-points are (sells) buys.

Problem WT allows the firm to be short in wheat or cash or both. If short-selling of wheat is forbidden, then the state constraint

$$w(t) \geq 0 \tag{4.45}$$

must be added to the problem.

6. Incorporate Eq. (4.45) in Problem WT to create a new problem, say Problem WT_2. Develop the optimal trading policy for this new problem.

4.11.2 Checking-Schedule for Detecting Failures

The following problem is based on Keller[53]; it is also discussed in the classic text by Kamien and Schwartz[47].

Let $n(t)$ be the number of (continuous) checks per unit time of a system, and c the cost per inspection. Then, the accumulated cost of checking over the time interval $[t_0, t]$ is given by

$$c \int_{t_0}^{t} n(\tau) d\tau \tag{4.46}$$

Suppose the system fails at some time, and is detected immediately ($\Delta t = 0$), then the loss incurred is the cost to fix it and bring it back online. In this case, no revenue is lost due to a failure to detect. If a failure is not detected immediately ($\Delta t \neq 0$), then some revenue $L(\Delta t)$ is lost: The longer the Δt, the greater the loss.

The time-interval between checks is $1/n(t)$; hence, this is also the maximum time interval between failure and detection. Setting $\Delta t = 1/n(t)$, the maximum possible lost revenue due to a failure to detect is given by

$$L\left(\frac{1}{n(t)}\right) \tag{4.47}$$

From Eqs. (4.46) and (4.47) it is apparent that as the frequency of checks, $n(t)$, is increased, the cost per check grows but the loss due to failure-detection drops. The economic objective is to address the sum

$$c \int_{t_0}^{t} n(\tau) d\tau + L\left(\frac{1}{n(t)}\right) \tag{4.48}$$

using n as our control variable. Hence, we can define a state variable $x(t)$ as,

$$\dot{x}(t) = n(t)$$

and transform the quantity in Eq. (4.48) to:

$$c\, x(t) + L\left(\frac{1}{n(t)}\right); \qquad x(t_0) = 0$$

This model generates the following optimal control problem:

$$\boldsymbol{x} := x \in \mathbb{R}, \quad \boldsymbol{u} := n \in \mathbb{R}$$

$$(\mathsf{K}') \quad \begin{cases} \text{Minimize} & J[x(\cdot), n(\cdot)] := \displaystyle\int_{t_0}^{t_f} \left(c\, x(t) + L\left(\frac{1}{n(t)}\right) \right) dt \\ \text{Subject to} & \dot{x} = n \\ & (x_0, t_0) = (0, 0) \\ & t_f = t^f \end{cases}$$

1. Discuss any problem-formulation issues with Problem K′.
2. Using Pontryagin's Principle show that the optimal checking schedule ($t \mapsto n$) is obtained by solving the equation

$$c n^2 \left(t_f - t\right) - L'\left(\frac{1}{n}\right) = 0 \tag{4.49}$$

   where L' is the derivative of the function $z \mapsto L(z)$.

3. Explore the ramifications of the optimal schedule given by Eq. (4.49) for the following two cases:
   (a) The loss function $z \mapsto L(z)$ is linear in z.
   (b) The loss function $z \mapsto L(z)$ is quadratic in z; that is, the loss increases at an increasing rate.

Let $F_c(t)$ be the cumulative distribution function (CDF) of failure of the system such that $F_c(t_0) = 0$; that is, the system is surely functioning at t_0. Then, the objective can be framed in terms of the expected cost functional leading to the following modification to Problem K′,

$$\boldsymbol{x} := x \in \mathbb{R}, \quad \boldsymbol{u} := n \in \mathbb{R}$$

$$(\mathsf{K}) \quad \begin{cases} \text{Minimize} & J[x(\cdot), n(\cdot)] := \displaystyle\int_{t_0}^{t_f} \left(c\, x(t) + L\left(\frac{1}{n(t)}\right) \right) \dot{F}_c(t)\, dt \\ \text{Subject to} & \dot{x} = n \\ & (x_0, t_0) = (0, 0) \\ & t_f = t^f \end{cases}$$

where, $\dot{F}_c(t)$ is the probability density function (PDF) obtained by differentiating $F_c(t)$.

1. Show that the optimal schedule is now given by

$$c n^2 \left(F_c(t_f) - F_c(t)\right) - \dot{F}_c(t)\, L'\left(\frac{1}{n}\right) = 0 \tag{4.50}$$

2. Under what conditions do Eqs. (4.50) and (4.49) generate the same result?

4.12 Endnotes

Some of the problems discussed in this chapter have been used by the author as motivating problems along the style of Chapter 1 to construct "new" introductions to optimal control. That is, these problems have been used instead of the Brachistochrone problem as the basis for classroom instruction for different application-specific courses.

Most of the figures in this chapter were produced using standard tools from MATLAB® and DIDO©. Figures 4.2, 4.3, 4.5, 4.6, 4.28–4.35 were generated by M. Karpenko, and Figs. 4.9, 4.37 and 4.38 are due to Q. Gong.

References

[1] A. A. Agrachev and R. V. Gamkrelidze, "The Pontryagin Maximum Principle 50 Years Later," *Proceedings of the Steklov Institute of Mathematics* Vol. 253, No. 1, Supplement pp S4-S12, 2006.

[2] A. V. Arutyunov and R. B. Vinter, "A Simple 'Finite Approximations' Proof of the Pontryagin Maximum Principle, Under Reduced Differentiability Hypotheses," *Journal of Set Valued Analysis*, No. 12, 2004, pp. 5–24.

[3] J.-P. Aubin, A. M. Bayen and P. Saint-Pierre, *Viability Theory: New Directions*, Second Edition, Springer-Verlag Berlin Heidelberg, 2011.

[4] J. M. Ball and V. J. Mizel, "Singular Minimizers for Regular One-Dimensional Problems in the Calculus of Variations," *Bulletin (New Series) of the American Mathematical Society*, Vol. 11, No. 1, July 1984, pp. 143–146.

[5] N. Bedrossian, S. Bhatt, W. Kang and I. M. Ross, "Zero-Propellant Maneuver Guidance," *IEEE Control Systems Magazine*, October 2009, pp. 53–73.

[6] J. T. Betts, *Practical Methods for Optimal Control Using Nonlinear Programming*, SIAM, Philadelphia, PA, 2001.

[7] S. Bhatt, N. Bedrossian, K. Longacre and L. Nguyen, "Optimal Propellant Maneuver Flight Demonstrations on ISS," *AIAA Guidance, Navigation, and Control Conference*, August 19-22, 2013, Boston, MA. AIAA 2013-5027.

[8] V. Blåsjö, "The Evolution of ... The Isoperimetric Problem," *The American Mathematical Monthly*, June-July 2005, pp. 526–566.

[9] G. A. Bliss, "A Note on Functions of Lines," *Proceedings of the National Academy of Sciences*, Vol. 1, No. 2, 1915, pp. 173–177.

[10] K. P. Bollino, *High-Fidelity Real-Time Trajectory Optimization for Reusable Launch Vehicles*, Ph.D. Dissertation, Naval Postgraduate School, December 2006.

[11] K. P. Bollino and L. R. Lewis, "Optimal Path Planning and Control of Tactical Unmanned Aerial Vehicles in Urban Environments," *Proceedings of the AUVSI's Unmanned Systems North America 2007 Conference*, Washington, DC, August 2007.

[12] K. P. Bollino and I. M. Ross, "A Pseudospectral Feedback Method for Real-Time Optimal Guidance of Reentry Vehicles," *Proceedings of the American Control Conference*, 2007, pp. 3861–3867.

[13] K. P. Bollino, L. R. Lewis, P. Sekhavat and I. M. Ross, "Pseudospectral Optimal Control: A Clear Road for Autonomous Intelligent Path Planning," *AIAA Infotech@Aerospace Conference and Exhibit*, Rohnert Park, CA, May 7-10, 2007.

[14] V. G. Boltyanski and A. S. Poznyak, *The Robust Maximum Principle: Theory and Applications*, Birkhäuser, New York, N.Y., 2012.

[15] C. Brif, R. Chakrabarti and H. Rabitz, "Control of Quantum Phenomena: Past, Present and Future," *New Journal of Physics*, Vol. 12, No. 7, 2010, 075008.

[16] A. E. Bryson and Y.-C. Ho, *Applied Optimal Control: Optimization, Estimation and Control*, Revised Printing, Hemisphere Publishing Corporation, New York, N.Y., 1975.

[17] J. Cascio, M. Karpenko, P. Sekhavat, Q. Gong and I. M. Ross, "Smooth Proximity Computation for Collision-Free Optimal Control of Multiple Robotic Manipulators," *Proceedings of the IEEE/RSJ International Conference on Intelligent Robots and Systems*, October 11–15, 2009, St. Louis, MO, pp. 2452–2457.

[18] J. Cascio, M. Karpenko, P. Sekhavat and I. M. Ross, "Optimal Collision-Free Trajectory Planning for Robotic Manipulators: Simulation and Experiments," *Proceedings of the AIAA/AAS Spaceflight Mechanics Meeting*, February 8–12, 2009, Savannah, GA, pp. 1424–1437.

[19] A. Charnes and W. W. Cooper, "Chance-Constrained Programming," *Management Science*, Vol. 6, No. 1, 1959, pp. 73–79.

[20] F. H. Clarke, *Necessary Conditions for Nonsmooth Problems in Optimal Control and the Calculus of Variations*, Ph.D. thesis, University of Washington, 1973.

[21] F. H. Clarke, *Optimization and Nonsmooth Analysis*, SIAM, Philadelphia, PA, 1990.

[22] F. H. Clarke, "The Pontryagin Maximum Principle and a Unified Theory of Dynamic Optimization," *Proceedings of the Steklov Institute of Mathematics*, April 2010, Vol. 268, No. 1, pp. 58–69.

[23] F. H. Clarke, *Functional Analysis, Calculus of Variations and Optimal Control*, Springer-Verlag, London, 2013.

[24] F. H. Clarke and M. R. De Pinho, "Optimal Control Problems with Mixed Constraints," *SIAM Journal of Control Optimization*, Vol. 48, No. 7, 2010, pp. 4500–4524.

[25] P. A. Croley, *Reachable Sets for Multiple Asteroid Sample Return Missions*, Astronautical Engineer Thesis, Naval Postgradudate School, December, 2005.

[26] A. V. Dmitruk, "Maximum Principle for the General Optimal Control Problem with Phase and Regular Mixed Constraints," *Computational Mathematics and Modeling*, October-December 1993, Vol. 4, No. 4, pp 364–377. Translated from Optimal'nost' Upravlyaemykh Dinamicheskikh Sistem, Sbornik Trudov VNIISI, No. 14, pp. 26–42, 1990.

[27] A. V. Dmitruk, "On the Development of Pontryagin's Maximum Principle in the Works of A. Ya. Dubovitskii and A. A. Milyutin," *Control and Cybernetics*, Vol. 38, No. 4A, 2009, pp 923–957.

[28] S. E. Dreyfus, *Dynamic Programming and the Calculus of Variations*, Report R-441-PR, The Rand Corporation, Santa Monica, CA, 1965 (also Academic Press, New York, N.Y., 1965).

[29] F. Fahroo and I. M. Ross, "Pseudospectral Methods for Infinite Horizon Optimal Control Problems," *Journal of Guidance, Control, and Dynamics*, Vol. 31 No. 4, 2008, pp. 927–936.

[30] F. Fahroo and I. M. Ross, "Convergence of the Costates Does Not Imply Convergence of the Control," *Journal of Guidance, Control and Dynamics*, Vol. 31, No. 5, pp. 1492–1497, 2008.

[31] A. Fleming, P. Sekhavat and I. M. Ross, "Constrained, Minimum-Time Maneuvers for CMG Actuated Spacecraft," *Advances in the Astronautical Sciences*, Vol. 135, AAS 09-362, *Proceedings of the 2009 AAS/AIAA Astrodynamics Specialist Conference*, August 9-13, 2009, Pittsburgh, PA.

[32] A. T. Fuller, "Study of an Optimum Nonlinear Control System," *Journal of Electronics and Control*, Vol. 15, No. 1, 1963, pp. 63–71.

[33] R. V. Gamkrelidze, "Discovery of the Maximum Principle," *Journal of Dynamical and Control Systems*, Vol. 5, No. 4, 1999, pp. 437–451.

[34] Q. Gong, W. Kang and I. M. Ross, "A Pseudospectral Method for the Optimal Control of Constrained Feedback Linearizable Systems," *IEEE Transactions on Automatic Control*, Vol. 51, No. 7, July 2006, pp. 1115–1129.

[35] Q. Gong, I. M. Ross and F. Fahroo, "Pseudospectral Optimal Control on Arbitrary Grids," *AAS/AIAA Astrodynamics Specialist Conference*, Pittsburgh, PA. 2009, AAS 09-405.

[36] C. R. Hargraves and S. W. Paris, "Direct Trajectory Optimization Using Nonlinear Programming and Collocation," *Journal of Guidance, Control, and Dynamics*, Vol. 10, No. 4, 1987, pp. 338–342.

[37] A. L. Herman and B. A. Conway, "Direct Optimization Using Collocation Based on High Order Gauss-Lobatto Quadrature Rules," *Journal of Guidance, Control, and Dynamics*, Vol. 19, No. 3, 1996, pp. 592–599.

[38] M. A. Hurni, *An Information-Centric Approach to Autonomous Trajectory Planning Utilizing Optimal Control Techniques*, Ph.D. Dissertation, Naval Postgraduate School, September 2009.

[39] M. A. Hurni, P. Sekhavat, M. Karpenko and I. M. Ross, "Autonomous Multi-Vehicle Formations Using Pseudospectral Optimal Control Framework," *Proceedings of the 2010 IEEE/ASME International Conference on Advanced Intelligent Mechatronics*, July 6-9, 2010, Montréal, QC, pp. 980–986.

[40] M. A. Hurni, P. Sekhavat and I. M. Ross, "An Info-Centric Trajectory Planner for Unmanned Ground Vehicles," *Dynamics of Information Systems: Theory and Applications*, Springer Optimization and its Applications, 2010, pp. 213–232.

[41] Y. Ijiri and G. L. Thompson, "Applications of Mathematical Control Theory to Accounting and Budgeting (The Continous Wheat Trading Model)," *The Accounting Review*, Vol. 45, No. 2, April 1970, pp. 246–258.

[42] S. J. Julier, "The Spherical Simplex Unscented Transformation," *Proceedings of the American Control Conference*, Denver, CO, June 4-6, 2003.

[43] S. J. Julier, J. K. Uhlmann and H. F. Durrant-Whyte, "A New Approach for Filtering Nonlinear Systems," *Proceedings of the American Control Conference*, Seattle, WA, 1995.

[44] R. E. Kalman, "Contributions to the Theory of Optimal Control," *Boletin de la Sociedad Matematica Mexicana*, Vol. 5, 1960, pp. 102–119.

[45] R. E. Kalman, "A New Approach to Linear Filtering and Prediction Problems," *Transcations of the ASME-Journal of Basic Engineering*, Vol. 82 (Series D), 1960, pp. 35–45.

[46] R. E. Kalman, "Toward a Theory of Difficulty of Computation in Optimal Control," *Proceedings of the IBM Scientific Computing Symposium on Control Theory and Applications*, IBM Data Processing Division, White Plains, New York, 1966, pp. 25–43.

[47] M. I. Kamien and N. L. Schwartz, *Dynamic Optimization: The Calculus of Variations and Optimal Control in Economics and Management*, Second Edition, North-Holland, Elsevier, Amsterdam, The Netherlands, 1991.

[48] W. Kang, "Rate of Convergence for the Legendre Pseudospectral Optimal Control of Feedback Linearizable Systems," *Journal of Control Theory and Application*, Vol. 8, No. 4, 2010, pp. 391–405.

[49] W. Kang and N. Bedrossian, "Pseudospectral Optimal Control Theory Makes Debut Flight — Saves NASA $1M in Under 3 hrs," *SIAM News*, Vol. 40, No. 7, September 2007, Page 1.

[50] W. Kang, I. M. Ross and Q. Gong, "Pseudospectral Optimal Control and its Convergence Theorems," *Analysis and Design of Nonlinear Control Systems*, Springer-Verlag, Berlin Heidelberg, 2008, pp. 109–126.

[51] M. Karpenko, S. Bhatt, N. Bedrossian, A. Fleming and I. M. Ross, "First Flight Results on Time-Optimal Spacecraft Slews," *Journal of Guidance Control and Dynamics*, Vol. 35, No. 2, 2012, pp. 367–376.

[52] M. Karpenko, S. Bhatt, N. Bedrossian and I. M. Ross, "Flight Implementation of Shortest-Time Maneuvers for Imaging Satellites," *Journal of Guidance Control and Dynamics*, Vol. 37, No. 4, 2014, pp. 1069–1079.

[53] J. B. Keller, "Optimum Checking Schedules for Systems Subject to Random Failure," *Management Science*, Vol. 21, No. 3, 1974, pp. 256–260.

[54] D. E. Kirk, *Optimal Control Theory: An Introduction*, Dover Publications, Inc., Mineola, N.Y., 2004; originally published by Prentice-Hall, Inc., Englewood Cliffs, N.J., 1970.

[55] A. J. Krener, "The High Order Maximal Principle and Its Application to Singular Extremals," *SIAM Journal of Control and Optimization*, Vol. 15, No. 2, 1977, pp. 256–293.

[56] B. O. Koopman, "The Theory of Search, II: Target Detection," *Operations Research*, Vol. 4, No. 5, 1956, pp. 503–531.

[57] D. F. Lawden, *Optimal Trajectories For Space Navigation*, Butterworths, London, 1963.

[58] L. R. Lewis, *Rapid Motion Planning and Autonomous Obstacle Avoidance for Unmanned Vehicles*, M.S. thesis, Naval Postgraduate School, Monterey, CA, December 2006.

[59] L. R. Lewis and I. M. Ross, "A Pseudospectral Method for Real-Time Motion Planning and Obstacle Avoidance," *Platform Innovations and System Integration for Unmanned Air, Land and Sea Vehicles*, pp. 10-1–10-22, 2007 AVT-SCI Joint Symposium, Meeting Proceedings RTO-MP-AVT-146, Paper 10. Neuilly-sur-Seine, France: RTO.

[60] A. Li, "Space Station: U.S. Life-Cycle Funding Requirements," *U.S. General Accounting Office Report* GAO/T-NSIAD-98-212, 1998.

[61] Jr-S Li and N. Khaneja, "Control of Inhomogeneous Quantum Ensembles," *Physical Review A*, Vol. 73, No. 030302(R), 2006, pp. 030302-1–030303-4.

[62] Jr-S Li, J. Ruths, T-Y Yu, H. Arthanari and G. Wagner, "Optimal Pulse Design in Quantum Control: A Unified Computational Method," *Proceedings of the National Academy of Sciences*, Vol. 108, No. 5, Feb 2011, pp. 1879–1884.

[63] C. E. Linderholm, *Mathematics Made Difficult*, Wolfe Publishing Ltd, London, U.K., 1971.

[64] J. M. Longuski, J. J. Guzmán and J. E. Prussing, *Optimal Control with Aerospace Applications*, Springer, New York, N.Y., 2014.

[65] R. L. McHenry, A. D. Long, B. F. Cockrell, J. R. Thibodeau and T. J. Brand, "Space Shuttle Ascent Guidance, Navigation, and Control," *The Journal of the Astronautical Sciences*, Vol. 27, No. 1, Jan-March 1979, pp. 1–38.

[66] E. J. McShane, "The Calculus of Variations From the Beginning Through Optimal Control Theory," *SIAM Journal of Control and Optimization*, Vol. 27, No. 5, pp. 916–939, September 1989.

[67] J. S. Meditch, "On the Problem of Optimal Thrust Programming for a Lunar Soft Landing," *IEEE Transactions on Automatic Control*, AC-9, No. 4, pp. 477–484, 1964.

[68] C. W. Misner, K. S. Thorne, and J. A. Wheeler, *Gravitation*, W. H. Freeman and Co., San Francisco, CA, 1973.

[69] B. S. Mordukhovich, *Variational Analysis and Generalized Differentiation, I: Basic Theory*, Vol. 330 of Grundlehren der Mathematischen Wissenschaften [Fundamental Principles of Mathematical Sciences] Series, Springer, Berlin, 2005.

[70] B. S. Mordukhovich, *Variational Analysis and Generalized Differentiation, II: Applications*, Vol. 331 of Grundlehren der Mathematischen Wissenschaften [Fundamental Principles of Mathematical Sciences] Series, Springer, Berlin, 2005.

[71] B. S. Mordukhovich and I. Shvartsman, "The Approximate Maximum Principle in Constrained Optimal Control," *SIAM Journal of Control and Optimization*, Vol. 43, No. 3, 2004, pp. 1037–1062.

[72] D. D. Morrison, J. D. Riley and J. F. Zancanaro, "Multiple Shooting Method for Two-Point Boundary Value Problems," *Communications of the ACM*, Vol. 5, No. 12, Dec. 1962, pp. 613–614.

[73] P. J. Nahin, *When Least is Best*, Princeton University Press, Princeton, NJ, 2004.

[74] H. J. Pesch and M. Plail, "The Maximum Principle of Optimal Control: A History of Ingenious Ideas and Missed Opportunities," *Control and Cybernetics*, Vol. 38, No. 4A, pp. 973–995, 2009.

[75] L. S. Pontryagin, V. G. Boltyanskii, R. V. Gamkrelidze and E. F. Mischenko, *The Mathematical Theory of Optimal Processes*, Interscience Publishers, New York, 1962.

[76] R. T. Rockafellar and S. Uryasev, "Optimization of Conditional Value-at-Risk," *The Journal of Risk*, Vol. 2, No. 3, 2000, pp. 21–41.

[77] I. M. Ross, "A Historical Introduction to the Covector Mapping Principle," *Advances in the Astronautical Sciences*, Astrodynamics 2005, Vol. 123, Part II, pp. 1257–1278.

[78] I. M. Ross, "A Roadmap for Optimal Control: The Right Way to Commute," *Annals of the New York Academy of Sciences*, Vol. 1065, Dec 2005, pp. 210-231.

[79] I. M. Ross, "Space Trajectory Optimization and L^1-Optimal Control Problems," Chapter 6 in *Modern Astrodynamics*, edited by P. Gurfil, Elsevier, St. Louis, MO, September 2006, pp. 155–187.

[80] I. M. Ross, "Real-Time Optimal Motion Planning for Unmanned Vehicles," *SIAM Conference on Control and Its Applications*, San Francisco, CA, June 29 – July 1, 2007.

[81] I. M. Ross, *A Beginner's Guide to DIDO: A MATLAB Application Package for Solving Optimal Control Problems*, Elissar Global, Monterey, CA, 2015.

[82] I. M. Ross and C. N. D'Souza, "Hybrid Optimal Control Framework for Mission Planning," *Journal of Guidance, Control and Dynamics*, Vol. 28, No. 4, July-August 2005, pp. 686–697.

[83] I. M. Ross and F. Fahroo, "Issues in the Real-Time Computation of Optimal Control," *Mathematical and Computer Modelling, An International Journal*, Vol. 43, No. 9–10, May 2006, pp. 1172–1188.

[84] I. M. Ross and Q. Gong, *Emerging Principles in Fast Trajectory Optimization*, AIAA Course Notes, Sixth Edition, 2013 (also available from Elissar Global, Monterey, CA).

[85] I. M. Ross, Q. Gong, F. Fahroo and W. Kang, "Practical Stabilization Through Real-Time Optimal Control," *Proceedings of the American Control Conference*, Minneapolis, MN, June 14–16, 2006.

[86] I. M. Ross, Q. Gong and P. Sekhavat, "Low-Thrust, High-Accuracy Trajectory Optimization," *Journal of Guidance Control and Dynamics*, Vol. 30, No. 4, 2007, pp. 921–933.

[87] I. M. Ross and M. Karpenko, "A Review of Pseudospectral Optimal Control: From Theory to Flight," *Annual Reviews in Control*, Vol. 36, No. 2, 2012, pp. 182–197.

[88] I. M. Ross, R. J. Proulx and M. Karpenko, "Unscented Optimal Control For Space Flight," *24th International Symposium on Space Flight Dynamics (ISSFD)*, Laurel, MD, May 5-9, 2014.

[89] I. M. Ross, R. J. Proulx and M. Karpenko, "Unscented Optimal Control for Orbital and Proximity Operations in an Uncertain Environment: A New Zermelo Problem," *AIAA Space and Astronautics Forum and Exposition: AIAA/AAS Astrodynamics Specialist Conference*, San Diego, CA 4-7 August 2014.

[90] I. M. Ross, R. J. Proulx and M. Karpenko, "Unscented Guidance," *Proceedings of the American Control Conference*, Chicago, IL, July 1–3, 2015.

[91] I. M. Ross, R. J. Proulx and M. Karpenko, "Riemann-Stieltjes Optimal Control Problems for Uncertain Dynamcial Systems," *Journal of Guidance, Control and Dynamics*, Vol. 38, 2015.

[92] I. M. Ross, P. Sekhavat, A. Fleming and Q. Gong, "Optimal Feedback Control: Foundations, Examples, and Experimental Results for a New Approach," *Journal of Guidance, Control, and Dynamics*, Vol. 31 No. 2, 2008, pp. 307–321.

[93] S. Russell and P. Norvig, *Artificial Intelligence: A Modern Approach*, Third Edition, Prentice Hall, Upper Saddle River, NJ, 2010.

[94] U. Serres, "On Zermelo-Like Problems: Gauss-Bonnet Inequality and E. Hopf Theorem," *Journal of Dynamical and Control Systems*, January 2009, Vol. 15, No. 1, pp. 99–131.

[95] H. J. Sussmann and J. C. Willems, "The Brachistochrone Problem and Modern Control Theory," *Contemporary Trends in Nonlinear Geometric Control Theory and its Applications* A. Anzaldo-Meneses, B.Bonnard, J.-P. Gauthier, and F. Monroy-Perez Eds; World Scientific Publishers, Singapore, 2000, pp. 113–166.

[96] R. Tapia, "The Isoperimetric Problem Revisited: Extracting a Short Proof of Sufficiency from Euler's 1744 Proof of Necessity," *SIAM Annual Meeting* Minneapolis, Minnesota, July 11, 2012.

[97] W. Thomson (Baron Kevin), "Isoperimetical Problems," *Nature Series: Popular Lectures and Addresses*, Vol. II, Geology and General Physics, Macmillan and Co. and New York, London, 1894, pp. 571–593.

[98] E. C. Tremblay, *Artificial Intelligence Techniques for Space-Based Dynamic Target Selection*, M.S. Thesis, Naval Postgraduate School, December 2014.

[99] R. B. Vinter, *Optimal Control*, Birkhäuser, Boston, MA, 2000.

[100] J. E. Volder, "The Birth of CORDIC," *Journal of VLSI Signal Processing* Vol. 25, 2000, pp. 101–105.

[101] A. R. Washburn, *Search and Detection*, Fourth Edition, INFORMS, Linthicum, Maryland, 2002.

[102] www-history.mcs.st-and.ac.uk/history/HistTopics/Brachistochrone.html.

[103] H. Yan, F. Fahroo and I. M. Ross, "Optimal Feedback Control Laws by Legendre Pseudospectral Approximations," *Proceedings of the American Control Conference*, Vol. 3, pp. 2388–2393, 2001.

[104] H. Yan, I. M. Ross and K. T. Alfriend, "Pseudospectral Feedback Control for Three-Axis Magnetic Attitude Stabilization in Elliptic Orbits," *Journal of Guidance, Control and Dynamics*, Vol. 30, No. 4, pp. 1107–1115, 2007.

[105] L. C. Young, *Lectures on the Calculus of Variations and Optimal Control Theory*, AMS Chelsea Publishing, Providence, RI, First Edition 1969; reprinted by AMS 2000.

[106] P. Zarchan, *Tactical and Strategic Missile Guidance*, Sixth Edition, American Institute of Aeronautics and Astronautics, Reston, VA, 2012.

Index

Made in the USA
Middletown, DE
21 December 2020